TEXTBOOK OF
BASIC MATHEMATICS

(for Forestry, Horticulture, Agriculture Courses and
Related Competitive Examinations)

The Author

S.K. Sheel is Post Graduate in Mathematics as well as in Statistics from HNB Garhwal University, Srinagar - Garhwal.He has more than three decades of experience in research and analysis of statical data emanating from experiments conducted by the scientistsof different biological disciplines of Forest Research Institute, Dehradun. He has also approximately 20 years of teaching experience on Applied Mathematics,Statistical Techniques and Research Methodology to Post Graduate students and Ph.D Scholars of Deemed University - FRI, Dehradun.

At present he is engaged in teaching bio-mathematics and bio-statistics in Dolphin (PG) Institute of Biomedical and Natural Sciences, Dehradun.

TEXTBOOK OF
BASIC MATHEMATICS

(for Forestry, Horticulture, Agriculture Courses and
Related Competitive Examinations)

S.K. SHEEL

Department of Biomathematics & Biostatistics
Dolphin (PG) Institute of Biomedical & Natural Sciences
Manduwala, Dehradun

2015
Daya Publishing House®
A Division of
Astral International Pvt. Ltd.
New Delhi - 110 002

Cataloging in Publication Data--DK
Courtesy: D.K. Agencies (P) Ltd. <docinfo@dkagencies.com>

Sheel, S. K., author.
Textbook of basic mathematics (for degree & honors courses of forestry, horticulture and agriculture) / S.K. Sheel.
     pages   cm
     Includes bibliographical references (pages   ).
     ISBN 978-93-5130-683-2 (International Edition)

     1. Forests and forestry--Mathematics. 2. Horticulture--Mathematics. 3. Agricultural mathematics. I. Title.
DDC 634.90151   23

Published by   :   **Daya Publishing House®**
                 A Division of
                 Astral International Pvt. Ltd.
                 – ISO 9001:2008 Certified Company –
                 4760-61/23, Ansari Road, Darya Ganj
                 New Delhi-110 002
                 Ph. 011-43549197, 23278134
                 E-mail: info@astralint.com
                 Website: www.astralint.com

Laser Typesetting: **Classic Computer Services, Delhi - 110 035**

Printed at   :   **Thomson Press India Limited**

PRINTED IN INDIA

DEDICATED TO

Loving memory
of
My late parents

Preface

Encouraged by the receptions of my recent book on Biomathematics among the students of biotechnology and other allied sciences, the many faculty members and students of forestry and agriculture suggested me to write a course book on Basic Mathematics on similar pattern for them. The present book is indeed the outcome of these suggestions.

I have endeavored to my level best to present a vivid text as required by the syllabi, supplemented with a good number of solved examples for illustration. Practice problems have been included after each topic to assist them in gaining the confidence in mathematics which is now compulsory part of syllabus in theses courses. Seeing the recent trend of university question papers, the objective questions with answers have been given at the end of each chapter. I found this pattern quite successful in my class room teaching of mathematics to students of forestry.

The book encompasses the complete syllabi of Basic Mathematics of forestry, horticulture and agriculture courses of different universities and institutes. I am hopeful that this book would fulfill the demand of the students of the said subjects.

S.K. Sheel

Acknowledgements

I am greatly indebted to Shri Arvind Gupta, the esteemed Chairman of Dolphin (PG) Institute of Biomedical and Natural Sciences, Dehra Dun for providing healthy environment and inspiration for creative work. This book is the outcome of his constant encouragement. His inspiration has been   always a source of energy for me.

I do not find appropriate words for expressing my sense of gratitude to Dr. Arun Kumar, Director, Dolphin (PG) Institute of Biomedical and Natural Sciences, Dehra Dun for all sorts of help and valuable suggestions and ideas in writing this book. Actually it is he who suggested me the topic of the book, since such book is not available for the concerned students.

I wish to express my deepest thanks to Dr. Shailja Pant, Principal, Dolphin (PG) Institute of Biomedical and Natural Sciences, Dehradun for constant support and encouragement in writing the book.

I am thankful to Dr. Sanjay Agarwal, HOD, for encouraging me to write a book on basic mathematics that may be useful to forestry, horticulture and agriculture students.

I wish to thank Mr. Sujith TK for providing me various computer related   assistance.

Lastly, I wish to thank my daughter Aditi Sheel, for preparing the cover page, going through the manuscript and suggesting valuable corrections.

S.K. Sheel

CONTENTS

CHAPTER 3: COMPLEX NUMBERS   83–107

CHAPTER 9: BINOMIAL THEOREM   266 – 279

CHAPTER 10: TRIGONOMETRIC FUNCTIONS   280 – 320

CHAPTER

1

AP and GP Series

1.0 BRIEF HISTORY

There is no specific history as to when arithmetic sequences and series was found or made up but it is known that the Egyptians were the first to develop arithmetic series. There is no specific history as to when sequences were started and studied ,although there was a young math student who created a formula to help solve for the sum of arithmetic sequences. His name was Carl Gauss, he was born in 1777 in a German Empire and at just ten years age he created a formula for summation of arithmetic sequence. His teacher asked him to come to the board to solve the sum of the sequence (also known as a series) $1 + 2 + 3 + ... + 99 + 100$ and he was the only one with the correct answer which was 5050. The formula he used was $M(M+1)/2$. As Gauss grew older he became a very well known mathematician contributing to geometry, number theories, and many more.

Sequences have an ancient history dating back at least as far as rchimedes who used sequences and series in his "Method of Exhaustion" to compute better values of π and areas of geometric figures.

1.1 SEQUENCE

A sequence is a set of natural numbers that are in order. For examples: $\{1, 2, 3, 4, ...\}$, $\{20, 25, 30, 35, ...\}$, $\{1, 2, 4, 8, 16\}$ are the sequences. Symbolically n^{th} element is denoted by the letter a_n.

1.1.1 Finite and Infinite Sequence

If a sequence has finite number of element, then it is called a finite sequence.

Examples

    1. $\{2, 4, 8, 16, ... 256\}$

    2. $\{1, 2, 3, 4, ... 100\}$

If a sequence goes on forever, then it is called an **infinite sequence.**

Examples

    1. $\{2, 4, 8, 16, 32, 64\}$

    2. $\{1, 2, 3, 4, 5, ...\}$

1.1.2 Order of Numbers in a Sequence

1. The order may be **ascending or increasing** i.e. going forward as shown below:

    Example: $\{1, 2, 3, 4, 5, ...\}$, $\{3, 6, 9, 12, ...\}$ etc.

2. The order may be **descending or decreasing. i.e. going backward** as shown below:

    Example: $\{10, 9, 8, 7, 6, ...\}$, $\{243, 81, 27, 9, ...\}$ etc.

1.1.3 Sequence and Set

A sequence is like a set , both use the same notation i.e. the curly bracket {} containing the elements separated by commas.

Sequence: $\{3, 5, 7, 9, 11, ...\}$ The elements are in ascending order.

Set: $\{7, 5, 9, 11, 3, ...\}$ The elements are not in order.

Representation of a sequence

A sequence may be represented in two ways:

1. By giving few terms in a bracket viz. $\{1, 4, 7, 10, ...$ to 15 terms$\}$

2. By writing the nth term. For example in the above sequence, the nth term may be written as $a_n = 3n - 2$. This is also the general term of the sequence.

    Putting 1, 2, 3, 4,.....in place of n terms of the sequence can be obtained.

    Thus we see a sequence has a rule, which helps to find the values of each term.

Solved Examples

1. Find the sequence whose n^{th} term is given by: an $= \{(-1/n)^n\}$

    Solution: Putting $n = 1, 2, 3, ...$ we get,

    $$a_1 = -\{(-1)^1\} = -1$$
    $$a_2 = -\{(-1/2)^2\} = 1/4$$
    $$a_3 = -\{(-1/3)^3\} = -1/27$$
    $$a_4 = -\{(-1/4)^4\} = 1/256$$

    Hence the sequence is $\{a_n\} = \{-1, 1/4, -1/27, 1/256\}$.

2. Find the sequence whose general term is $a_n = 2n + 1$. Find the sequence.

    Solution: Putting $n = 1, 2, 3,$.......we get,

    $$a_1 = \{2 \times 1 + 1\} = 3$$
    $$a_2 = \{2 \times 2 + 1\} = 5$$
    $$a_3 = \{2 \times 3 + 1\} = 7$$
    $$a_4 = \{2 \times 4 + 1\} = 9$$

    Hence the sequence is $\{a_n\} = \{3, 5, 7, 9, ...\}$.

3. Write the first five terms of the sequence whose nth term is $a_n = n(n + 2)$

   Solution: Substituting $n = 1, 2, 3, \ldots\ldots$we get,

   $$a_1 = 1\,(1 + 2) = 3$$
   $$a_2 = 2\,(2 + 2) = 8$$
   $$a_3 = 3\,(3 + 2) = 15$$
   $$a_4 = 4\,(4 + 2) = 24$$
   $$a_5 = 5\,(5 + 2) = 35$$

   Hence the sequence is $\{a_n\} = \{3, 8, 15, 24, 35, \ldots\}$.

4. Find the 17th and 24th terms of the sequence whose n^{th} terms is $a_n = 4n - 3$

   Solution: Putting $n = 17$ we get $a_{17} = 4 \times 17 - 3 = 65$

   Similarly $\qquad\qquad\qquad a_{24} = 4 \times 24 - 3 = 93$

   Hence 17th term is 65 and 24th term is 93.

5. Write the first five terms of the sequence given by $a_1 = 3$, $a_n = 3a_{n-1} + 2, n > 1$

   Solution: Substituting $n = 2, 3, 4, 5$ we get:

   $$a_1 = 3$$
   $$a_2 = 3a_{2-1} + 2 = 3a_1 + 2 = 3 \times 3 + 2 = 8$$
   $$a_3 = 3a_{3-1} + 2 = 3a_2 + 2 = 3 \times 8 + 2 = 26$$
   $$a_4 = 3a_{4-1} + 2 = 3a_3 + 2 = 3 \times 26 + 2 = 90$$
   $$a_5 = 3a_{5-1} + 2 = 3a_4 + 2 = 3 \times 80 + 2 = 242$$

Practice Problems(On Sequence)

1. Find the sequences whose n^{th} terms are:

   (i) $a_n = \dfrac{2n-3}{6}$,

   (ii) $a_n = (-1)_{n-1}5^{n+1}$

2. Find (i) a_7 if $a_n = \dfrac{n^2-1}{2n}$,

   (ii) a_9 if $a_n = (-1)^{n-1}n^3$,

   (iii) a_{20} if $a_n = \dfrac{n(n-2)}{n+3}$

Answers:

1. (i) $\left(\dfrac{-1}{6}, \dfrac{1}{6}, \dfrac{3}{6}, \dfrac{5}{6}, \ldots\right)$,
    (ii) $\{25, -125, 625, -3125\}$

2. (i) $a_7 = \dfrac{24}{7}$,
    (ii) $a_9 = 729$,
    (iii) $a_{20} = \dfrac{360}{23}$

1.2 SERIES

If $\{a_n\}$ is a sequence, then $a_1 + a_2 + a_3 + ... + a_n$ is called a series. In other words, a series is the sum of the terms of a sequence. Sequence is singular as well as a plural word viz. a series and two series.

In a series such as $3 + 7 + 11 + ...+ 100$, the difference between any consecutive terms is same. This is called as **common difference (d)** . The above series represents *an arithmetic series* or *an arithmetic progression*. Series and Progression are synonyms.

Solved Examples

1. Write the following series using summation notation, beginning with $n = 1$ up to five terms. $2 - 4 + 6 - 8 + 10 - 12 + ...$

Solution: We see that each term is twice of the term i.e. 1st term is $2 \times 1 = 2$. In general we can write as 2n.

The terms have alternative signs. The first term is positive. So it can be written as $(-1)^{1+1}$.

The second term is negative. So it can be written as $(-1)^{2+1}$. For third term $(-1)^{3+1}$.

For n^{th} term $(-1)^{n+1}$

Thus the general term $= (-1)^{n+1} \cdot 2n$.

The sum of first five terms $= \displaystyle\sum_{n=1}^{5} (-1)^{n+1} .2n$

2. Write the general term of the following series: $\dfrac{4}{5+2} + \dfrac{4}{6+2} + \dfrac{4}{7+2} + ... + \dfrac{4}{14+2}$

Solution: The number 4 in numerator and 2 in denominator are fixed.

5, 6, 7, ... 14 may be written as $4 + 1, 4 + 2, 4 + 3, ... 4 + 10$. This can be written as $4 + n$ where n varies from 1 to 10.

Hence the general term is $\dfrac{4}{(4+n)+2}$

Sum of the Series $= \displaystyle\sum_{n=1}^{10} \left\{ \dfrac{4}{(4+n)+2} \right\}$

3. **Express** $\displaystyle\sum_{n=2}^{6} 3\left(\dfrac{1}{3}\right)^{n+2}$ **as a series.**

Solution: n varies from 2 to 6 i.e. there are 5 terms.

Hence the series is: $3\left[\left(\dfrac{1}{3}\right)^{2+2}+\left(\dfrac{1}{3}\right)^{3+2}+\left(\dfrac{1}{3}\right)^{4+2}+\left(\dfrac{1}{3}\right)^{5+2}+\left(\dfrac{1}{3}\right)^{6+2}\right]$

or $3\left[\left(\dfrac{1}{3}\right)^{4}+\left(\dfrac{1}{3}\right)^{5}+\left(\dfrac{1}{3}\right)^{6}+\left(\dfrac{1}{3}\right)^{7}+\left(\dfrac{1}{3}\right)^{8}\right]$

4. Find the value of n if $\displaystyle\sum_{k=1}^{n}(84-4k)=0$

Solution: $\displaystyle\sum_{k=1}^{n}(84-4k) = \sum_{k=1}^{n}84 - \sum_{k=1}^{n}4k$

$$= 84 - 4\sum_{k=1}^{n}k = 84 - 4\,(1+2+3-...+n)$$

$$84 - 4.\dfrac{n}{2}[n+1] = 0$$

$$-4.\dfrac{n}{2}[n+1] = -84$$

$$2n\,[n-1] = 84 \Rightarrow n^2 - n - 42 = 0$$

$$(n+7)(n-6) = 0 \Rightarrow n = -7, 6$$

Hence the value of $n = 6$.

Interesting fact about $(1+2+3+...+n) = \dfrac{n}{2}[n+1]$

Mathematician, Karl Fredrich Gauss discovered the proof when he was only 10 years old. His teacher had decided to give his class a problem that would distract them for the entire day by asking them to add all the numbers from 1 to 100. Karl realized how to do this almost instantaneously and shocked the teacher with correct answer (5050).

He wrote $S=\displaystyle\sum_{i=1}^{100}i=\dfrac{n}{2}(n+1)$

Practice Problems

1. Find 19th terms of the sequence whose n^{th} term is $a_n - 4n - 3$.

2. Find the 2nd and 3rd terms of the sequence given by $a_n = 4\left(\dfrac{1}{2}\right)^{n} - 1$.

3. Find the 4th term of the sequence given by $\{a_n\} = \{(1/n)\}^{n-1}$.

4. The nth terms of a sequence is given by $\{a_n\} = \{3n^{2-1}\}$. Which term will be equal to 866?

5. The nth terms of a sequence is given by $a_n = \dfrac{n^2-1}{n+5}$. Which term will be equal to 4.

Answers

1. 73   2. 0 and $-1/2$ 3. 1/64 **4.** 16   **5.** 7th term

1.3 ARITHMETIC PROGRESSION (A.P.)

A finite sequence with elements a, $a + d$, $a + 2d$, ... $a + (n - 1)d$ is known as Arithmetic sequence or Arithmetic Progression , abbreviated as AP. The constant difference between two consecutive terms is known as *common difference* and dented by d.

Example 1. The sequence of real numbers: 1, 3, 5, 9, ... is an AP.

The common difference $(d) = 3 - 1 = 5 - 3 = ... = 2$

Example 2. The sequence of the form: 17, 13, 9, 5, 1 is an AP.

The common difference $d = 13 - 17 = -4$

The first term $a_1 = 17$ and $d = -4$. With the help of first term and common difference an AP can be formed.

The general term of an AP is given by: an $= a_1 + (n-1)d$ where n is the number of terms.

1.3.1   To test whether a sequence is an AP or not

Solved Examples

1. Show that the sequence defined by $a_n - 4n - 3$.

Solution: Put $n = 1, 2, 3$ .....

We get
$$a_1 = 4.1 - 3 = 1$$
$$a_n = 4.2 - 3 = 5$$
$$a_3 = 4.3 = 9 \text{ and so on.}$$

We find $d = a_2 - a_1 = 5 - 1 = 4$

or $d = a_3 - a_2 = 9 - 5 = 4$

As d remains constant, hence the given sequence is an AP

2. Determine whether the sequence $\{a_n\} = \left\{\left(\dfrac{1}{2}\right)2^{n-1}\right\}$

Solution: Put $n = 1, 2, 3,$

The $a_1 = \left\{\left(\dfrac{1}{2}\right)2^{1-1}\right\} = \dfrac{1}{2}$

$$a_2 = \left\{ \left(\frac{1}{2}\right) 2^{2-1} \right\} = \frac{1}{2}.2 = 1$$

$$a_3 = \left\{ \left(\frac{1}{2}\right) 2^{3-1} \right\} = \frac{1}{2}.2^2 = \frac{4}{2} = 2$$

We see $d = a_2 - a_1 = 1/2$

$$d = a_3 - a_2 = 2 - 1 = 1.$$

Since d is not constant, hence the given sequence is not an AP.

Alternative Method

Put $(n + 1)$ in place of n,

$$a_{n+1} = \left(\frac{1}{2}\right) 2^{n+1-1} = \left(\frac{1}{2}\right) 2^n$$

$$d = a_{n+1} - a_n = \frac{1}{2}\left\{2^n - 2^{n-1}\right\} = \frac{1}{2} 2^n \left\{1 - 2^{-1}\right\} = 2^{n-1}\left\{1 - \frac{1}{2}\right\} = 2^{n-2}$$

Since $d = 2^{n-2}$ is not free from n i.e. **d** will change with different value of n, hence the given sequence is not an AP

3. Is the sequence 1, 3, 6, 10, ... is an AP.

Solution: Given $a_1 = 1$, $a_2 = 3$, $a_3 = 6$, $a_4 = 10$

$$d_1 = a_2 - a_1 = 3 - 1 = 2$$
$$d_2 = a_3 - a_2 = 6 - 3 = 3$$
$$d_4 = a_4 - a_3 = 10 - 6 = 4$$
$$\because \qquad d_1 \neq d_2 \neq d_3 \neq ...$$

Hence the given sequence is not an AP

Practice Problems:

Find which of the following sequences are an AP, their nth terms are given as:

(i) $a_n = 114 + (-7)(n - 1)$ (ii) $a_n = 3n + 2$

(iii) $a_n = n^2 + 1$ (iv) $a_n = 3n - 2$

(v) $a_n = \frac{3(n-1)}{2} - \frac{5}{2}$

Answers: Except (iii), all are AP

1.3.2 General term of an AP

The general term of an AP is given by: $a_n = a_1 + (n-1)d$.

Where a_1 is the first term, n is the number of terms in the given AP and d is the common difference. This is also known as the nth term.

Solved Examples

1. Find the tenth term (a_{10}) of the following arithmetic progression: $-1, 3, 7, 11, \ldots$

Solution: Here $a_1 = -1 \quad a_2 = 3 \quad a_3 = 7 \quad n = 10$

$$d = a_2 - a_1 = 3 - (-1) = 4$$

The general term of an AP is given by: $a_n = a_1 + (n-1)\, d$.

$$a_{10} = a_1 + (10-1)\, d$$

$$a_{10} = -1 + (10-1)\, 4 = -1 + 36 = 35$$

2. Given $a_3 = 8$ and $a_6 = 17$. Find the 10th term

Solution: we know that the nth term is given by: $a_n = a_1 - (n-1)\, d$

When $n = 3$ then $a_3 = a_1 + (3-1)d = a_1 + 2d = 8$ ...(1)

When $n = 6$ then $a_6 = a_1 + (6-1)d = a_1 + 5d = 17$ ...(2)

Subtracting (1) from (2) we get $3d = 9 \Rightarrow d = 3$

Substituting $d = 3$ in (1) $\quad a_1 + 2 \times 3 = 8 \Rightarrow a_1 = 8 - 6 = 2$

Hence $a_{10} = a_1 + (10-1)d = 2 + 9 \times 3 = 29$

3. Given $a_1 = 6$ and $d = 3$. Find the 50th term

Solution: $a_1 = 6$, $d = 3$ and $n = 50$

The general term of an AP is given by: $a_n = a_1 + (n-1)d$.

Substituting the given values in general term we get:

$$a_{50} = 6 + (50-1)3 = 6 + 49 \times 3 = 6 + 147 = 153.$$

4. An arithmetic sequence has common difference equal to 5 and its 6th term is equal to 52. Find its 10th term.

Solution: Given $d = 5$ and $a_6 = a_1 + (n-1)d = 52$

Put $a_1 = 5$ and $n = 6$ we $a_1 + (6-1)5 = 52 \Rightarrow a_1 = 52 - 25 = 27$

Now we have to find a_{10}. We have $a_1 = 27$. $d = 5$ and $n = 10$

$$a_{10} = a_1 + (n-1)d$$

$$a_{10} = 27 + (10-1)5 = 27 + 45 = 72$$

5. Find the 30th term of the AP: 10, 7, 4, ...

Solution: First term $a_1 - 10$ and $d = 7 - 10 = -3n - 30$

The nth term of an AP is given by $a_n = a_1 + (n-1)\,d$.

$$a_{30} = 10 + (30-1)(-3) = 10 - 29 \times 3 = 10 - 87 = -77 \,.$$

6. The 5th of an AP is equal to 30 and its 15th term is equal to 70. Find the 25th term.

Solution: Applying the formula of nth term: $a_n = a_1 = (n-1)\,d$.

$$a_5 = a_1 + (5-1)d = 30 \,.$$

or
$$a_1 = 4d = 30 \qquad \qquad \text{...(1)}$$

$$a_{15} = a_1 + (15-1)d = 70 \,.$$

or
$$a_1 + 14d = 70 \qquad \qquad \text{...(2)}$$

Subtracting (1) from (2), we get $10d = 40 \Rightarrow d = 4$

From (1) $a_1 + 4 \times 4 = 30 \Rightarrow a_1 = 30 - 16 = 14$

Now we have to find 25th term i.e. $n = 25$ and $a_1 = 14$, $d = 4$

$$a_{25} = 14 + (25-1) \times 4 = 14 + 96 = 110 \,.$$

The 25th term is 110.

7. Find the number of terms of an AP given as: 3, 7, 11, ..., 407

Solution: The first term $a_1 = 3$ and the last term $a_{25} = 407$ $d = 7 - 3 = 4$

By n^{th} term formula: $a_n = a_1 + (n-1)d$.

$$407 = 3 + (n-1)\,4.$$

$$407 = 3 + 4n - 4.$$

$$407 = 4n - 1 \Rightarrow 4n = 408 \Rightarrow n = \frac{408}{4} = 102 \,.$$

Hence number of terms in the given AP is 102.

8. Which term of the AP: 3, 8, 13, 18, ... is equal to 78

Solution: We have $a_1 = 3$ $\quad d = 8 - 3 = 5$

Let the nth term $= 78$

$$a_n = a_1 + (n-1)\,d = 78.$$

$$3 + (n-1)\,5 = 78$$

$$5n + -2 = 78 \Rightarrow n = \frac{80}{5} = 16$$

Hence 16th term is equal to 78.

9. Find whether -150 is a term of the AP: 11, 8, 5, ... ?

Solution: $a_1 = 11$ and $d = 8 - 11 = -3$

∵
$$a_n = a_1 + (n-1)\, d.$$
$$-150 = 11 + (n-1)(-3).$$
$$-150 = 11 + -3n + 3.$$
$$3n = 150 + 14.$$
$$n = \frac{164}{3} = 54.66.$$

Since n is not an integer, hence -150 can not be a term of the given AP.

10. The 17th term of an AP exceeds its 10th term by 7. Find the common difference (d).

Solution: For 17th term we may write: $a_{17} = a_1 + (17-1)d = a_1 + 16d$

For 10th term we may write: $a_{10} = a_1 + (10-1)d = a_1 + 9d$

∵
$$a_{17} - a_{10} = 7$$
$$\{a_1 + 16d\} - \{a_1 + 9d\} = 7$$
$$7d = 7 \Rightarrow d = 1$$

11. Terms of an AP are: 3, 15, 27, 39, ... Find the term that exceeds its 5th term by 60.

Solution: We have: $a_1 = 3$ and $d = 15 - 3 = 12$

By n^{th} term formula: $a_n = a_1 + (n-1)\, d$
$$a_n = 3 + (n-1)\,12 = 12n - 9.$$
$$a_5 = 3 + (5-1)12 = 51.$$
$$\because a_n - a_5 = 60$$
$$\{12n - 9\} - 51 = 60$$
$$12n - 60 = 60$$
$$n = \frac{120}{12} = 10$$

Hence 10th term exceeds the 5th term by 60.

12. Find the AP whose 3rd term is 16. The 9th term exceeds the 5th term by 20.

Solution: Applying n^{th} term formula: $a_n = a_1 + (n-1)d$.

For 3rd term: $a_3 = a_1 + (3-1)d = a_1 + 2d = 16$ ...(1)

For 9th term: $a_9 = a_1 + (9-1)d = a_1 + 8d$.

For 5th term: $a_5 = a_1 + (5-1)d = a_1 + 4d$.

Given that $a_9 - a_5 = \{a_1 + 8d\} - \{a_1 + 4d\} = 20$

$\therefore \quad 4d = 20 \Rightarrow d = \dfrac{20}{4} = 5$

From (1) $a_1 + 2d = 16$

$$a_1 + 2 \times = 16 \Rightarrow a_1 = 16 - 10 = 6$$

We have: $a_1 = 6$ and $d = 5$

Hence terms of AP are:

$$a_1 = 6$$

$$a_2 = a_1 + d = 6 + 5 = 11$$

$$a_3 = a_1 + 2d = 6 + 10 = 16$$

$$a_4 = a_1 + 3d = 6 + 15 = 21$$

13. The sum of 4th and 8th terms of an AP is 40 and the sum of 6th and 10th term is 44. Find the terms of AP.

Solution: Given that $a_4 + a_8 = 40$

or $\qquad\qquad\qquad \{a_1 + 3d\} + \{a_1 + 7d\} = 40$

$\qquad\qquad\qquad 2a_1 + 10\,d = 40 \Rightarrow a_1 + 5d = 20$ ...(1)

Also $\qquad\qquad\qquad a_6 + a_{10} = 44$

$\qquad\qquad\qquad \{a_1 + 2d\} + \{a_1 + 9d\} = 48$

or $\qquad\qquad\qquad 2a_1 + 14d = 44 \Rightarrow a_1 + 7d = 24$ (2)

Subtracting (1) from (2) we get $2d = 24 - 20 = 4 \Rightarrow d = 2$

From (1) putting d = 5 we get $a_1 + 5 \times 2 = 20 \Rightarrow a_1 = 10$

Hence the required AP is: 10, 12, 14, 16, 18, 20, ...

1.3.3 SUM OF *n* TERMS OF AN AP

The sum of *n* terms of an arithmetic progression is given by:

$$S_n = \frac{n}{2}\left[2a + (n-1)d\right]$$

where *a* is the first term, *d* is common difference and *n* is number of terms.

We can also write : $S_n = \dfrac{n}{2}\left[2a+(n-1)d\right] = \dfrac{n}{2}\left[a+\{a+(n-1)d\}\right]$

$$= \dfrac{n}{2}\left[\text{first term} + \text{last term}\right]$$

$\{a + (n - 1)\,d\}$ is known as n^{th} term or the last term.

Solved Examples

1. Find the sum of 5 terms of an AP, if first term is 5 and common difference is 4.

Solution: Given $a = 5$, $d = 4$ and $n = 5$

The sum of n terms of AP is given by

$$S_n = \dfrac{n}{2}\left[2a+(n-1)d\right]$$

Substituting the given values:

$$S_5 = \dfrac{5}{2}\left[2\times5+(5-1)4\right]$$

$$= \dfrac{5}{2}[10+16] = \dfrac{5}{2}[26] = 65$$

2. Find the value of $1 + 2 + 3 + 4 + 5 + \ldots + 100$

Solution: Given $a = 1$, $d = 1$ and $n = 100$

Applying $S_n = \dfrac{n}{2}\left[2a+(n-1)d\right]$

$$S_{100} = \dfrac{100}{2}\left[2\times1+(100-1)\times1\right] = 500(101) = 5050$$

3. Find the sum of odd numbers between 0 and 60.

Solution: Between 0 and 60 there are 30 odd numbers and 30 even numbers.

The 30 odd numbers are 1, 3, 5, ... 59

Hence $a = 1$ $d = 2$ and $n = 30$

Applying the formula $\quad S_n = \dfrac{n}{2}\left[2a+(n-1)d\right]$

The sum of odd numbers is

$$S_{odd} = \dfrac{30}{2}\left[2\times1+(30-1)\times2\right] = 15[2+29\times2] = 15\times60 = 900$$

4. $x + 2$, $4x - 6$ and $3x - 2$ are first three consecutive terms of an AP. Find the value of x.

Solution: Given $a_1 = x + 2$ $a_2 = 4x - 6$

$$a_3 = 3x - 2$$

In an AP

$$a_2 - a_1 = a_3 - a_2$$
$$(4x - 6) - (x + 2) = (3x - 2) - (4x - 6)$$
$$3x - 8 = -x + 4$$
$$4x = 12 \Rightarrow x = 3$$

5. Find the sum of 20 terms of the series: $1^2 - 2^2 + 3^2 - 4^2 + ...$

Solution: We require $S = 1^2 - 2^2 + 3^2 - 4^2 + 5^2 - ...$ up to 20 terms

We can write $S = 1 - 4 + 9 - 16 + 25 - 36 ...$ up to 20 terms

$$= -3 - 7 - 11 ... \text{ up to 20 terms}$$

We see $a = -3$ $d = (-7) - (-3) = -4$ $n = 20$

$$S_{20} = \frac{n}{2}[2a + (n-1)d] = \frac{20}{2}[2(-3) + (20-1)(-4)]$$

$$= 10[-6 - 76] = -820.$$

6. Find the sum of the AP : 7, 10.5, 14, ……. The last term is 84.

Solution: $a = 7$ $d = 10.5 - 7 = 3.5$ last term $a_n = 84$

Using the formula $a_n = a_1 + (n-1)d$.

$$84 = 7 + (n-1)(3.5) = 7 + 3.5n - 3.5 = 3.5(n+1).$$

$$3.5(n+1) = 84$$

$$n + 1 = \frac{84}{3.5} = \frac{840}{35} = 60$$

$$n = 24$$

We have $a = 7$ $d = 3.5$ $n = 24$

Applying the formula

$$S_n = \frac{n}{2}[2a + (n-1)d]$$

$$S_{24} = \frac{24}{2}[2 \times 7 + (24-1)(3.5)]$$

$$= 12[14 + 23(3.5)]$$

$$= 12[14 + 80.5]$$

$$= 12 \times 94.5 = 774$$

7. Find the value of x. Given that $1 + 6 + 11 + 16 + ... + x = 148$

Solution: From the given series, we have $a = 1$ and $d = 5$ and $S_n = 148$

Sum of an AP is

$$S_n = \frac{n}{2}\left[2a+(n-1)d\right]$$

$$148 = \frac{n}{2}\left[2\times1+(n-1)5\right]$$

$$296 = n\left[2+5n-5\right]$$

$$296 = 5n^2 - 3n$$

$$5n^2 - 3n - 296 = 0$$

$$5n\ -3n - 296 = 0$$

$$5n^2 - 40n + 37n - 37\times8 = 0$$

$$5n(n-8) + 37(n-8) = 0$$

$$(n-8)(5n+37) = 0 \Rightarrow n = 8 \text{ or } \frac{-37}{5}$$

∴ $n = 8$ (since n can not be negative).

Practice Problems:

A. In an AP

1. $a = 5$, $d = 3$ and $a_n = 50$. Find n and S_n.

2. $a = 7$, $d = 3$ and $a_{13} = 35$. Find S_{15}

3. $d = 3$ and $a_{12} = 37$. Find a and S_{12}

4. $a_3 = 15$ and $S_{10} = 125$. Find d and a

5. $d = 5$ and $S_{10} = 75$. Find a and a_{10}

6. If $a_n = 9n - 15$. Find the 10th term

7. Which term of the AP series: 3, 10, 17, ...will be 14 more than the 10th term.

8. If $a = -2$ and $d = -8$, find the expression for nth term

9. Find the general term of the arithmetic sequence: $-2, -8, -14, -20, ...$

10. If $a = 3/2$, $d = -5/2$ and $n = 9$. Find the value of nth term.

Answers: 1. 16,440   2. 420   3. 4,246   4. –1, 17

5. –15, 30   6. 75   7. 10th term   8. $a_n = 6 - 8n$

9. $a_n = 4-6n$ 10. –37/2

B. Find the sum of the AP

1. 3, 8, 13, ……… up to 10 terms. **Ans.** 255

2. 0.5, 1.2, 1.9, ……up to 100 terms. **Ans.** 3515

3. – 47, –43, – 39, …….up to 12 terms. **Ans.** –300

4. 1 + 2 + 3 + …………. + 100. **Ans.** 5050

1.4 GEOMETRIC PROGRESSION (G.P.)

A progression is another way of expressing a sequence. Thus a geometric progression is also known as a geometric sequence. The geometric progression is also abbreviated as GP.

GP has a special property that ratio of two consecutive terms is always constant. This constant ratio is known as common ratio denoted by r. The every succeeding term can be obtained by multiplying the preceding term by common ratio (r).

For example, let a is the first term and the common ratio is r. Then the terms of geometric progression will be as follows: a, ar, ar^2, ar^3, ... ar^{n-1}, ar^n.

Thus the common ration $r = \dfrac{\text{Any term}}{\text{Preceding term}} = \dfrac{ar}{a}$

1.4.1 GENERAL TERM (T_n) OF A GP : $T_n = ar^{n-1}$

This is also known as the nth term.

Solved Examples

1. Find the 7th term of GP: $1, \dfrac{1}{2}, \dfrac{1}{4}, ...$

Solution: First term $a = 1$. $r = \dfrac{1/2}{1} = \dfrac{1}{2}$

$$T_7 = 1\left(\dfrac{1}{2}\right)^{7-1} = \left(\dfrac{1}{2}\right)^6 = \dfrac{1}{64}$$

2. Find 6th term of GP: 36, –12, 4, ...

Solution: First term $a = 36$. $r = \dfrac{-12}{36} = \dfrac{-1}{3}$

$$T_6 = 36\left(\dfrac{-1}{3}\right)^{6-1} = 36\left(\dfrac{-1}{3}\right)^5 = 36 \times \dfrac{-1}{243} = \dfrac{-4}{27}$$

3. Find the common ratio of the GP: $2,\ \sqrt{2},\ 1,\ \dfrac{1}{\sqrt{2}} \ldots\ldots$ Which term will be equal to $\dfrac{1}{4}$?.

Solution: Common ration $r = \dfrac{2^{nd}\text{term}}{1^{st}\text{term}} = \dfrac{2}{\sqrt{2}} = \dfrac{1}{\sqrt{2}}$

$$T_n = ar^{n-1} = \frac{1}{4}$$

$$2.\left(\frac{1}{\sqrt{2}}\right)^{n-1} = \frac{1}{4}$$

$$\left(\frac{1}{\sqrt{2}}\right)^{n-1} = \frac{1}{8} = \left(\frac{1}{\sqrt{2}}\right)^{6}$$

$$\therefore\ n-1 = 6 \Rightarrow n = 7$$

Hence it is the 7th term which is equal to $\dfrac{1}{4}$

4. Which term of GP: $1,\ \sqrt{3},\ 3,....$ will be 81

Solution: First term a = 1, common ration $r = \dfrac{\sqrt{3}}{3} = \dfrac{1}{\sqrt{3}}$

We know that nth term is given by : $T_n = ar^{n-1}$

$$81 = 1.\left(\frac{1}{\sqrt{3}}\right)^{n-1} = 3^{\frac{n-1}{-2}}$$

$$\therefore 3^4 = 3^{\frac{n-1}{2}}$$

or $$\frac{n-1}{2} = 4 \Rightarrow n = 9$$

5. The 3rd and 6th terms of a GP are 12 and 96. Find the first term and common ration.

Solution: Let a and r be the fisrt term and common ratio of a GP

$$T = ar^{3\ 1} = 12 \qquad\qquad ...(1)$$

$$T_6 = ar^{6-1} = 96 \qquad\qquad ...(2)$$

Dividing (2) by (1)
$$\frac{ar^5}{ar^2} = \frac{96}{12} \Rightarrow r^3 = 8 = 2^3$$

$$\therefore r = 2$$

6. The 5th term of a GP is 9 times of the third term. Find the GP if third term is 18.

Solution: Let T_5 and T_3 are the 5th and 3rd terms.

Given that
$$T_5 = 9T_3$$

$$\therefore \frac{T_5}{T_3} = 9$$

$$\frac{ar^{5-1}}{ar^{3-1}} = 9$$

$$\frac{r^4}{r^2} = 9 \Rightarrow r^2 = 9 \Rightarrow r = \pm 3$$

The third term $T_3 = ar^2 = 18$

or
$$a(\pm 3)^2 = 18 \Rightarrow a = 2$$

Hence the GP is 2, ±6, ±18, ±54, ...

7. If a, b, c, d are in GP , show that $a + b, b + c, c + d$ are also in GP.

Solution: Since a, b, c, d are in GP ,

Hence $b = ar$, $c = ar^2$ and $d = ar^3$

$$\frac{b+c}{a+b} = \frac{ar+ar^2}{a+ar} = \frac{ar(1+r)}{a(1+r)} = r$$

$$\frac{c+d}{b+c} = \frac{ar^2+ar^3}{ar+ar^2} = \frac{ar^2(1+r)}{ar(1+r)} = r$$

Thus the common ration of $a + b, b + c$ an $c + d$.

Hence these numbers are in GP

8. a, b, c are in GP. Prove that $1/a, 1/b, 1/c$ are also in GP.

Solution: Given a, b, c are in GP. If r is the common ratio then $b = ar, c = ar^2$

Now $\dfrac{1/b}{1/a} = \dfrac{a}{b} = \dfrac{a}{ar} = \dfrac{1}{r}$

Similarly $\dfrac{1/c}{1/b} = \dfrac{b}{c} = \dfrac{ar}{ar^2} = \dfrac{1}{r}$

Thus $1/a, 1/b, 1/c$ have common ratio as $1/r$.

Hence they are in GP.

9. If a, b, c are in GP, them $\log a$, $\log b$, $\log c$ are also in AP.

Solution: $\log a$, $\log b$, and $\log c$ are in AP, hence their common difference are equal.

So we can write $\log b - \log a = \log c - \log b$

$$\log \frac{b}{a} = \log \frac{c}{b}$$

Taking antilog of both sides: $\dfrac{b}{a} = \dfrac{c}{b} = $ common ratio

Hence a, b, c are in GP.

10. If a, b, c are in GP, then show: $a\left(b^2 + c^2\right) = c\left(a^2 + b^2\right)$

Solution: Let r be the common ratio of given GP, then $b = ar$ and $c = br = ar^2$

$$\left(a^2 + b^2\right) = a^2 + a^2 r^2 = a^2\left(1 + r^2\right)$$

$$\left(b^2 + c^2\right) = a^2 r^2 + a^2 r^4 = a^2 r^2\left(1 + r^2\right)$$

$$\frac{\left(a^2 + b^2\right)}{\left(b^2 + c^2\right)} = \frac{a^2\left(1 + r^2\right)}{a^2 r^2\left(1 + r^2\right)} = \frac{a}{ar^2} = \frac{a}{c}$$

$\therefore$ $c\left(a^2 + b^2\right) = a\left(b^2 + c^2\right)$. Hence Proved.

11. If a, b, c are in GP as well as in AP, then $a = b = c$

Solution: Since a, b, c are in GP , hence $b^2 = ac$

Since a, b, c are in AP , hence $b = (a + c)/2 \rightarrow b^2 = [a + c)^2] /4$

$\therefore$ $(a + c)^2/4 = ac \rightarrow a^2 + c^2 + 2ac = 4ac \rightarrow a^2 + c^2 - 2ac = 0$

$$a^2 + c^2 - 2ac = 0$$

$$(a - c)^2 = 0 \rightarrow a = c$$

$$b^2 = ac = c^2 \rightarrow b = c$$

Hence $a = b = c$.

Practice Problems (On GP)

1. Find the common ratio of the GP, whose first term is 5 and third term is 125.
2. If 2, 4, 8 form a geometric progression then find the 8th term.
3. If first term is 2 and fifth term is 162, then find the third term of the GP.
4. The common ratio of the GP is $\dfrac{1}{2}$ and fourth term is 24. Find the first term.
5. Find the next term of the GP: 1/6, 1/3, 2/3, …..

6. If $-2/7$, x, $-7/2$ are three consecutive terms of a GP, then find the value of x.

7. Find the number of terms of a GP, whose first term is ¾, common ratio is 2 and the last term is 384

8. Determine the value of x, if $(x + 9)$, $(x - 6)$ and (4) are in GP

9. The third term of a GP is 4. Find the product of first five terms.

10. Find the 9th term and general term of the GP: $\dfrac{1}{4}, -\dfrac{1}{2}, 1, -2, ...$

11. If a, b, c are in GP, then $1/a$, $1/b$, $1/c$ are also in GP

Answers

1. 5	**2.** 256	**3.** 18	**4.** 192	**5.** 4/3
6. ± 1.	**7.** n = 10	**8.** 0 or 16	**9.** 4^5	**10.** 64, $(-1)^{n-1} 2^{n-3}$

1.4.2 Sum of n terms of G.P.

Let a be the first term and r be the common ratio of the GP.

The sum of n terms, $\qquad S_n = \dfrac{a\left(r^n - 1\right)}{r - 1} \qquad$ when $r > 1$

or $\qquad\qquad\qquad S_n = \dfrac{a\left(1 - r^n\right)}{1 - r} \qquad$ when $r < 1$

Solved Examples

1. Find the sum of GP: $1 - \dfrac{1}{2} + \dfrac{1}{4} - \dfrac{1}{8} + ...\text{up to 9 terms}$

Solution: The first term $a = 1$, The common ratio $r = \dfrac{-1/2}{1} = -\dfrac{1}{2} < 1$, $n = 9$

When $r < 1$, The sum of GP is given by: $S_n = \dfrac{a\left(1 - r^n\right)}{1 - r}$

$$S_n = \dfrac{1\left(1 - \left(-\dfrac{1}{2}\right)^9\right)}{1 - \left(-\dfrac{1}{2}\right)}$$

$$S_n = \frac{\left(1-\left(-\frac{1}{512}\right)\right)}{1-\left(-\frac{1}{2}\right)}$$

$$S_n = \frac{\left(1+\frac{1}{512}\right)}{1+\frac{1}{2}} = \frac{513/512}{3/2} = \frac{513}{512}\cdot\frac{3}{2} = \frac{171}{256}$$

2. The second term of a GP is $-\frac{1}{2}$ and fifth term is $\frac{1}{16}$. Find the sum of GP up to 8 terms.

Solution: The second term $\qquad T_2 = ar = -\frac{1}{2}$ $\qquad\qquad$ ...(1)

$\qquad$ The fifth term $\qquad\qquad T_5 = ar^4 = \frac{1}{16}$ $\qquad\qquad$ ...(2)

$$\frac{T_5}{T_2} = \frac{ar^4}{ar} = \frac{1/16}{-1/2} = \frac{-2}{16} = \frac{-1}{8}$$

$$r^3 = \left(-\frac{1}{2}\right)^3 \Rightarrow r = -\frac{1}{2}$$

Substituting $r = -\frac{1}{2}$ in (1) we get $a\left(-\frac{1}{2}\right) = -\frac{1}{2} \Rightarrow a = 1$

The sum of GP up to 8 terms is given by:

$$S_8 = \frac{a\left[1-\left(-\frac{1}{2}\right)^8\right]}{1-\left(-\frac{1}{2}\right)}$$

$$S_8 = \frac{1\left[1-\frac{1}{256}\right]}{1+\frac{1}{2}} = \frac{2}{3}\cdot\frac{255}{256} = \frac{85}{128}$$

3. Given: $1+2+2^2+\ldots\ldots\ldots 2^{n-1} = 255$. Find the value of n.

Solution: There are n terms. Sum $S_n = 255$ $r = 2 > 1$

$$S_n = \frac{a(r^n - 1)}{r - 1} = \frac{1(2^n - 1)}{2 - 1} = 255$$

$$(2^n - 1) = 255$$

$$2^n = 256 = 2^8 \Rightarrow n = 8$$

4. Find the sum of the series: $4 + 44 + 444 + \dots$ up to n terms.

Solution: $4 + 44 + 444 + \dots$ up to n terms.

$$= 4(1 + !! + 111 + 1111 + \dots \text{ up to } n \text{ terms.})$$

$$= \frac{4}{9}[9 + 99 + 999 + 9999 + \dots n \text{ } terms]$$

$$= \frac{4}{9}\left[(10 - 1) + (100 - 1) + (1000 - 1) + \dots n \text{ } terms\right]$$

$$= \frac{4}{9}\left[(10 + 10^2 + 10^3 + \dots n \text{ } terms) - (1 + 1 + 1 + \dots n \text{ } terms)\right]$$

$$= \frac{4}{9}\left[\frac{10(10^n - 1)}{10 - 1} - n\right]$$

5. Find the sum of GP: $2 + 4 + 8 + \dots 1024$.

Solution: $1024 = 2^{10}$ *i.e.* there are 10 terms in the given GP

$$a = 2 \quad r = 2 > 1 \quad n = 10$$

Using the formula $S_n = \frac{a(r^n - 1)}{r - 1}$

$$S_{10} = \frac{2(2^{10} - 1)}{2 - 1} = 2(1024 - 1) = 2 \times 1023 = 2046$$

Practice Problems (On Sum of a GP)

1. Find the sum of 7 term of the GP: 3, 6, 12....

2. Determine the number of terms of GP if $a = 3$, n^{th} term $T_n = 96$, $S_n = 189$.

3. Find the sum of GP: 1, 3, 9, 27, ...up to 8 terms.

4. Find the sum of the GP: $0.15 + 0.015 + 0.0015 + \dots$up to 8 terms.

5. Find the sum of GP, if $n = 5$, $r = -2$, $a = 1$

6. Find the sum of S_3, if $T_2 = \frac{-1}{2}$ and $T_3 = \frac{1}{16}$.

Answers:

    1. 381,   **2.** $n = 6$ **3.** 3280   **4.** $\dfrac{1}{6}\left(1 - \dfrac{1}{10^8}\right)$ **5.** 33

    6. 513/144

1.4.3 SUM OF INFINITE (∞) TERMS OF G.P.

If a is the first term and r is the common ratio of GP, then sum of infinite terms are given by:

$$S_\infty = \frac{a}{1-r} \quad if \ \ r < 1$$

Solved Examples

1. Find the sum of GP: $\dfrac{5}{6} + \dfrac{5}{18} + \dfrac{5}{54} + \ldots\ldots up \ to \ \infty$

Solution: First term is a = 5/6. Common ratio $r = \dfrac{5/18}{5/6} = \dfrac{6}{18} = \dfrac{1}{3} < 1$

$$S_\infty = \frac{a}{1-r} = \frac{5/6}{1-5/6} = \frac{5/6}{1/6} = 5$$

2. Find the sum up to infinity of GP : $0.9 + 0.03 + 0.001 + \ldots \ \mu$

Solution: $a = 0.9 = \dfrac{9}{10}$ $r = \dfrac{0.03}{0.9} = \dfrac{3}{90} = \dfrac{1}{30}$

$$\therefore S_\infty = \frac{a}{1-r} = \frac{9/10}{1-1/30} = \frac{9}{10} \cdot \frac{30}{29} = \frac{27}{29}$$

3. Find the sum of GP: $\left(1 + \dfrac{1}{2^2}\right) + \left(\dfrac{1}{2} + \dfrac{1}{2^4}\right) + \left(\dfrac{1}{2^2} + \dfrac{1}{2^6}\right) + \ldots .\infty$

Solution: We can write the given series as:

$$= \left(1 + \frac{1}{2} + \frac{1}{2^2} + \ldots .\infty\right) + \left(\frac{1}{2^2} + \frac{1}{2^4} + \frac{1}{2^6} + \ldots .\infty\right)$$

$$= \left(\frac{1}{1 - 1/2}\right) + \left(\frac{1/2^2}{1 - 1/2^2}\right)$$

$$= \left(\frac{1}{1/2}\right) + \left(\frac{1/4}{3/4}\right)$$

$$= 2 + \frac{1}{3} = \frac{7}{3}$$

4. If $y = x + x^2 + x^3 + \ldots \infty$, then prove $x = \dfrac{y}{1+y}$

Solution: $y = x + x^2 + x^3 + \ldots \infty$

$$x + x^2 + x^3 + \ldots \infty = \frac{x}{1-x}$$

$$y = \frac{x}{1-x}$$

$$\frac{1}{y} = \frac{1-x}{x} = \frac{1}{x} - 1$$

$$\frac{1}{y} + 1 = \frac{1}{x}$$

$$\frac{1+y}{y} = \frac{1}{x} \Rightarrow \frac{y}{1+y} = x \text{ Proved.}$$

5. By using GP for infinity, find the value of 0.33333 ...

Solution: $0.33333\ldots = 0.3 + 0.03 + 0.003 + .0003 + \ldots \infty$

$$= \left[\frac{3}{10} + \frac{3}{100} + \frac{3}{1000} + \frac{3}{10000} + \ldots \infty \right]$$

$$= 3 \left[\frac{1}{10} + \frac{1}{10^2} + \frac{1}{10^3} + \frac{1}{10^4} + \ldots \infty \right]$$

$$= 3 \left[\frac{1/10}{1 - 1/10} \right] = 3 \frac{1}{10} \cdot \frac{10}{9} = \frac{1}{3}$$

6. The sum of an infinite GP is 3 and sum of squares of terms is also 3. Find the first three terms .

Solution: Let a be the first term and the common ratio is r.

Sum of infinite GP is $S_\infty = \dfrac{a}{1-r} = 3 \Rightarrow \dfrac{a^2}{(1-r)^2} = 9$ \qquad ...(1)

The sum of square is : $(a)^2 + (ar)^2 + (ar^3)^2 + (ar^4)^2 + \ldots \infty = 3$

The first term is $= a^2$

Common ratio $= r^2$

$$\therefore S_\infty = \frac{a^2}{1-r^2} = 3 \qquad \text{...(2)}$$

Dividing (1) by (2):

$$\frac{a^2 / (1-r)^2}{a^2 / 1 - r^2} = \frac{9}{3}$$

$$\frac{\left(1 - r^2\right)}{\left(1 - r\right)^2} = 3$$

$$\frac{\left(1 - r\right)\left(1 + r\right)}{\left(1 - r\right)^2} = 3$$

$$\frac{\left(1 + r\right)}{\left(1 - r\right)} = 3 \Rightarrow \left(1 + r\right) = 3\left(1 - r\right) \Rightarrow 4r = 2$$

$$4r = 2 \Rightarrow r = \frac{1}{2}$$

7. The 1st and 2nd term of a GP are x^{-4} and x^n. If x^{52} is the 8th term, then find n.

Solution: Given: $T = a = x$ $\qquad T_2 = ar = x^n$ $\qquad T_8 = ar^7 = x^{52}$

$$\frac{T_2}{T_1} = \frac{ar}{a} = \frac{x^n}{x^{-4}} \Rightarrow r = x^{n+4}$$

Substituting $r = x^{n+4}$ and $a = x^{-4}$ in $T_8 = ar^7 = x^{52}$

We get:

$$x^{-4} \left(x^{n+4} \right)^7 = x^{52}$$

$$\left(x^{n+4} \right)^7 = x^{52+4}$$

$$x^{7(n+4)} = x^{52+4}$$

$$\therefore 7n + 28 = 56 \Rightarrow 7n = 56 - 28 = 28$$

$$\therefore n = \frac{28}{7} = 4$$

8. If $S = a + a^2 + a^3 + \ldots\ldots\ldots to \infty$. Then Show that $a = \dfrac{S}{1+S}$

Solution: $S = a + a^2 + a^3 + \ldots\ldots\ldots to \infty$

$$S = \frac{a}{1-a} \Rightarrow S(1-a) = a$$

$$S - Sa = a \Rightarrow S = a + Sa = a(1 + S)$$

$$S = a(1 + S) \Rightarrow a = \frac{S}{1+S} \qquad \text{Proved.}$$

9. Show that $x^{\frac{1}{2}}.x^{\frac{1}{2}}.x^{\frac{1}{2}}.x^{\frac{1}{2}}........to \infty = x$

Solution: LHS $= x^{\frac{1}{2}}.x^{\frac{1}{2}}.x^{\frac{1}{2}}.x^{\frac{1}{2}}........to \infty$

$$= x^{\frac{1}{2}+\frac{1}{4}+\frac{1}{8}+............to \infty}$$

$$= x^{\frac{1/2}{1-1/2}} = x^{\frac{1/2}{1/2}} = x = RHS$$

Practice Problems (On Sum of infinite terms of GP)

1. Find the sum of GP: $18 + 6 + 2 + ...$

2. Find the sum of infinite GP if $a = 18$ and $r = 1/3$

3. Write the first three terms of the GP whose common ratio is $-\frac{1}{2}$ and sum is 20

4. The first term of a GP is 2, and sum up to infinity is 6. Find the common ratio.

5. Using GP, find the value of $0.55555.....$

Answers:

1. 27   **2.** 27   **3.** 30, −15, 15/2   **4.** 2/3

5. 5/9

Objective Questions: AP & GP

1. A sequence is a set of:
   - (a) Ordered numbers
   - (b) random numbers
   - (c) rational numbers
   - (d) irrational numbers

2. A sequence where difference between two consecutive terms is same is known as:
   - (a) Arithmetic sequence
   - (b) Geometric sequence
   - (c) Harmonic sequence.
   - (d) None of these

3. In AP the difference between two consecutive terms is called:
   - (a) Constant of progression
   - (b) Common difference
   - (c) Common ratio
   - (d) Difference of progression.

4. The general term of a sequence is $T_n = 2n + 3$. Then the second term is:
   - (a) 2
   - (b) 5
   - (c) 7
   - (d) 6

5. The fifth term of a sequence whose general term is $T_n = n(n+3)$ is:
   - (a) 25
   - (b) 30
   - (c) 40
   - (d) 28

6. If $T_n = 2n+3$, then T_{n-1} is:
   - (a) $2n-2$
   - (b) $2n+1$
   - (c) $2n-3$
   - (d) $2n-4$

7. If the general terms is $T_n = 2n^2 + 1$, then 4th term is:
   - (a) 19
   - (b) 27
   - (c) 33
   - (d) 39

8. If $T_n = \dfrac{n}{n+1}$, then $2T_1 + 3T_2$ is equal to:
   - (a) 3
   - (b) 5
   - (c) 4
   - (d) 6

9. The tenth term of an AP 1, 4, 7.......is :
   - (a) 28
   - (b) 31
   - (c) 34
   - (d) 36

10. The common difference of the AP – 6, 0, 6, –9, ...is:
    - (a) 3
    - (b) –5
    - (c) –6
    - (d) 6

11. The common difference of the AP – 3/5, –2/5, –1/5 ....is:
    - (a) 1/5
    - (b) 2/5
    - (c) 3/5
    - (d) 1

12. 3, x, 7 are in AP, then value of x is:
    - (a) 0
    - (b) 4
    - (c) 5
    - (d) 6

13. In an AP, if $d = -4$, $n = 5$, $T_n = 4$, then first term a is :
    - (a) 20
    - (b) 30
    - (c) 35
    - (d) 40

14. In an AP, if $T_{25} - T_{20} = 45$, then common difference d is:
    - (a) 9
    - (b) –9
    - (c) 15
    - (d) 18

15. The sequence 2, 4, 6, 8… represents a :

(a) AP

(b) GP

(c) HP

(d) No. of these

16. The 5th term of the sequence 2, 4, 8, .. is:

(a) 32

(b) 64

(c) 128

(d) 512

17. In a GP $T_5 - T_3 = 15$ and $T_4 - T_2 = 5$, Then the common ratio is

(a) 2

(b) 3

(c) 4

(d) 5

18. The nth term of a sequence is $T_n = 3(2^n)$, represents a:

(a) AP

(b) GP

(c) HP.

(d) None of these

19. The product $6^{\frac{1}{2}}.6^{\frac{1}{4}}.6^{\frac{1}{8}}.........\infty$ is :

(a) 1

(b) 2

(c) 4

(d) 6

20. The sum: $x + x^2 + x^3 + ...\infty = y$, then value of x is:

(a) $\dfrac{1+y}{b}$

(b) $\dfrac{1-y}{b}$

(c) $\dfrac{y}{1+y}$

(d) $\dfrac{y}{1-y}$

21. If $\log_x a$, a, $a^{\frac{x}{2}}$ are in GP, then the value of x is:

(a) $\log_a(\log_b a)$

(b) $-\log_a(\log_b a)$

(c) $\log_a(\log_b a)$

(d) $\log_b(\log_e b)$

22. If x, y, z are in GP, then $(1+\log x), (1+\log y), (1+\log z)$ are in :

(a) AP

(b) GP

(c) HP

(d) None of these

23. Sum of the series: $2 + 6 + 18 + ...+ 4374$. is :

(a) 600

(b) 6000

(c) 6250

(d) 6560

24. If $\log_3 2$, $\log_3\left(2^x - 5\right)$ and $\log_3\left(2^x - \dfrac{7}{2}\right)$, then x is equal to:

(a) 2 (b) 3

(c) 4 (d) 2, 3

25. If $x = 1 + y + y^2 + \ldots\ldots\infty$, then value of y is:

(a) $\dfrac{x}{x-1}$ (b) $\dfrac{x}{x+1}$

(c) $\dfrac{x-1}{x}$ (d) $\dfrac{1-x}{x}$

Answers

1 (a)	2 (a)	3 (b)	4 (c)	5 (c)	6 (b)	7 (c)
8 (a)	9 (a)	10 (d)	11 (a)	12 (b)	13 (a)	14 (a)
15 (a)	16 (a)	17 (b)	18 (b)	19 (d)	20 (c)	21 (a)
22 (b)	23 (d)	24 (d)	25 (c)			

Cartesian Coordinate System - 2D

1.0 BRIEF HISTORY

The invention of Cartesian coordinate system was done by French Mathematician – Rene Descartes' in 17th century. He revolutionized a new concept in mathematics. He provided a systematic link between Euclidean Geometry and algebra. That is why, it is also termed as *Coordinate Geometry.*

Cartesian Coordinate system: In a coordinate system, coordinates of a point are its distances from a set of perpendicular lines called axes- horizontal line is termed as x-axis and the perpendicular line as y-axis. The point of intersection of axes is called the 'origin'.

Cartesian Plane: This consists of the space between two perpendicular lines. Two perpendicular lines create four spaces known as Quadrants. The quadrants are numbered anti-clockwise as Ist , IInd , IIrd and IVth quadrants. Each quadrant represent a Cartesian plane

The term Cartesian is derived from Descartes who first introduced coordinate system. The word Cartesian coordinates was introduce by Leibniz in1677. The phrase Cartesian Coordinate was actually used in 1844.

1.1 COORDINATES AND THEIR SIGN CONVENTIONS

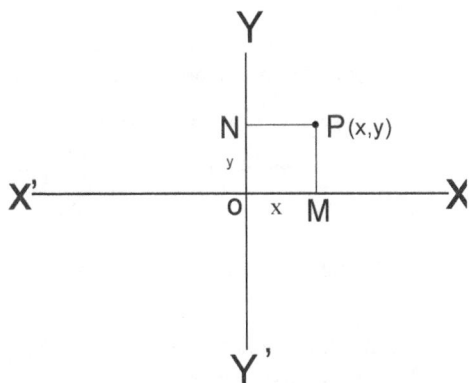

Let XOX' and YOY' be two perpendicular straight lines intersecting at O.

Let P be any point .Draw a perpendicular PM from P on OX and PN on OY. The length OM and ON, usually denoted by x and y, respectively are called *rectangular coordinates* or simply coordinates of point P.

Line OX and OY, with reference to which , the coordinates are measured, are called "the axes of reference".

OX is called the axis of x and OY axis of y.

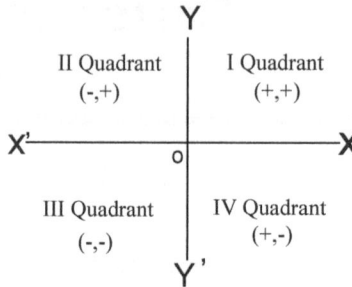

OM or x is called **"abscissa"** or x-coordinate of P.

PM or y is called **"ordinate"** or y-coordinate of P

These are always written in brackets viz. coordinates of point P is (x, y). The point of intersection O is called the **"origin"**.

Sign Convention regarding coordinates

The straight lines XOX' and YOY' the plane into four parts called '**quadrants**'

   (*i*) XOX is the 'first quadrants'

  (*ii*) YOX' is the 'second quadrants'

 (*iii*) $X'OY'$ is the 'third quadrants'

 (*iv*) $Y'OX$ is the ' fourth quadrant'

A point lying in the **1ˢᵗ quadrant**, shall have both x and y as positive i.e. P(x, y)

A point lying in the **2ⁿᵈ quadrant**, shall have x as negative and y as positive i.e. P($-x, y$)

A point lying in the **3ʳᵈ quadrant**, shall have both x and y as negative i.e. P($-x, -y$)

A point lying in the **4ᵗʰ quadrant**, shall have x and y as positive i.e. P(x, y).

Note: Coordinates of the origin are always $(0, 0)$, since it is the point from where we start counting along both axis.

The abscissa means x-coordinates. So abscissa of a point on y-axis is zero. Similarly ordinate means y coordinate. So ordinate of a point on x-axis is zero.

The coordinates of a point on x-axis are $(x, 0)$. The coordinates of a appoint on y-axis are $(0, y)$.

1.2 DISTANCE BETWEEN TWO POINTS

Let P and Q are two points, with coordinates (x_1, y_1) and (x_2, y_2) respectively.

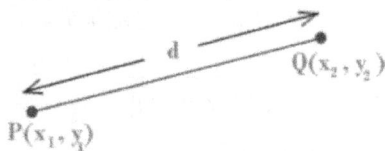

Then the distance between P and Q is given by the following formula:

$$PQ = \sqrt{d_x^2 + d_y^2}$$

where $d_x = (x_2 - x_1)$ and $d_y = (y_2 - y_1)$

∴ or

$$PQ = \sqrt{\left\{(x_2 - x_1)^2 + (y_2 - y_1)^2\right\}}$$

Solved Examples

1. Find the distance between the following points

   (i) P (3,6) , Q (0, 2)

   (ii) P (–2,6) , Q (3, –6),

   (iii) P (a, b) , Q (–b, a)

   (iv) P (a, b) , Q (a + r cos θ, b –r sinθ)

Solutions: (i) P (3,6) , Q (0, 2). Here $x_1 = 3$, $y_1 = 6$ and $x_2 = 0$, $y_2 = 2$

$$\therefore\ PQ = \sqrt{\left\{(x_2 - x_1)^2 + (y_2 - y_1)^2\right\}} = \sqrt{\left\{(0-3)^2 + (2-6)^2\right\}} = \sqrt{\{9+16\}} = \sqrt{25} = 5$$

(ii) P (–2, 6) , Q (3, –6). Here $x_1 = -2$, $y_1 = 6$ and $x_2 = 3$, $y_2 = -6$

$$\therefore\ PQ = \sqrt{\left\{(x_2 - x_1)^2 + (y_2 - y_1)^2\right\}} = \sqrt{\left\{(-2-3)^2 + (6+6)^2\right\}}$$

$$= \sqrt{\{25+144\}} = \sqrt{169} = 13$$

(iii) P (a, b) , Q (–b, a). Here $x_1 = a$, $y_1 = b$ and $x_2 = -b$, $y_2 = a$

∴

$$PQ = \sqrt{\left\{(x_2 - x_1)^2 + (y_2 - y_1)^2\right\}}$$

$$= \sqrt{\left\{(a-(-b))^2 + (b-a)^2\right\}}$$

$$= \sqrt{\left\{(a+b)^2 + (b-a)^2\right\}}$$

$$= \sqrt{\left\{a^2 + b^2 + 2ab + b^2 + a^2 - 2ab\right\}}$$

$$= \sqrt{\left\{2(a^2 + b^2)\right\}}$$

(iv) P (a, b) , Q (a + r cosθ, b –r sinθ).

Here $x_1 = a$, $y_1 = b$ and $x_2 = a + r$ cosθ, $y_2 = b$–rsinθ

∴

$$PQ = \sqrt{\left\{(x_2 - x_1)^2 + (y_2 - y_1)^2\right\}}$$

$$= \sqrt{\left\{(a - a - r\cos)^2 + (b - b + r\sin\theta)^2\right\}}$$

$$= \sqrt{\left\{(-r\cos\theta)^2 + (r\sin\theta)^2\right\}}$$

$$= \sqrt{\left\{r^2 \cos^2\theta) + (r^2 \sin^2\theta)\right\}}$$

$$= \sqrt{\left\{r^2(\cos^2\theta + \sin^2)\right\}} = r$$

2. The ordinate of a point is twice the abscissa. Find the coordinates of the point, if its distance from the point (4, 3) is – 10.

Solution: Let the abscissa (x-coordinate) is k. Hence its ordinate(y-coordinate) is 2k.

Hence points are P (k, 2k) and Q (4, 3).

Here $x_1 = k$ $y_1 = 2k$ and $x_2 = 4$, $y_2 = 3$

and PQ = $\sqrt{10}$

The distance between P and Q is given by:

∴

$$PQ = \sqrt{\left\{(k-4)^2 + (2k-3)^2\right\}} = \sqrt{10}$$

or

$$\left\{(k-4)^2 + (2k-3)^2\right\} = 10$$

or

$$\left\{k^2 + 16 - 8k + 4k^2 + 9 - 12k\right\} = 10$$

or

$$5k^2 - 20k + 15 = 0 - k^2 - 4k + 3 = 0 - k^2 - 3k - k + 3 = 0$$

$$- (k-3)(k-1) = 0 - k = 1 \text{ or } 3$$

Putting k=1 point P (K, 2K) becomes P (1, 2)

Putting k=3 point P (K, 2K) becomes P(3, 6)

Hence coordinates of point P are (1, 2) or (3, 6)

3. Show that the points (3, 8), (–2, 3) and (4, 1) form an isosceles triangle.

Solution: Let RPQ is the triangle where coordinate of vertices are: R (3, 8), P (–2, 3) and Q (4, 1)

The distance between **R** (3, 8), and **P** (–2, 3) is given by:

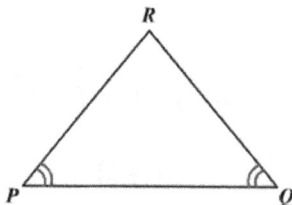

$$RP = \sqrt{\left\{(x_2 - x_1)^2 + (y_2 - y_1)^2\right\}} = \sqrt{\left\{(3+2)^2 + (8-3)^2\right\}} = \sqrt{\left\{(5)^2 + (5)^2\right\}} = \sqrt{\{50\}}$$

Similarly distance between **R** (3, 8), and **Q** (4, 1) is given by"

$$RQ = \sqrt{\left\{(x_2 - x_1)^2 + (y_2 - y_1)^2\right\}} = \sqrt{\left\{(4-3)^2 + (1-8)^2\right\}} = \sqrt{\left\{(1)^2 + (-7)^2\right\}} = \sqrt{\{50\}}$$

$$PQ = \sqrt{\left\{(x_2 - x_1)^2 + (y_2 - y_1)^2\right\}} = \sqrt{\left\{(4+2)^2 + (1-3)^2\right\}} = \sqrt{\left\{(6)^2 + (4)^2\right\}} = \sqrt{\{52\}}$$

Since RQ = RP , hence $\triangle RPQ$ is **an isosceles triangle.**

4. Prove the point (-1, 2) is the centre of the circle which passes through the points (-3, 11), (5, 9) and (6, 8).

Solution: Let points on the circumference be denoted by A (–3, 11), Q (5, 9), R (6, 8).and the centre is O (–1, 2).

$$OP = \sqrt{\left\{(-3+1)^2 + (11-2)^2\right\}} = \sqrt{\left\{(-2)^2 + (9)^2\right\}} = \sqrt{\{4+81\}} = \sqrt{\{85\}}$$

$$OQ = \sqrt{\left\{(5+1)^2 + (9-2)^2\right\}} = \sqrt{\left\{(6)^2 + (7)^2\right\}} = \sqrt{\{36+49\}} = \sqrt{\{85\}}$$

$$OR = \sqrt{\left\{(6+1)^2 + (8-2)^2\right\}} = \sqrt{\left\{(7)^2 + (6)^2\right\}} = \sqrt{\{36+49\}} = \sqrt{\{85\}}$$

Since OP = OQ = OR, hence O is centre of the circle.

5. Vertices of a triangles are (–2, 1), (1, 1) and (1, 2). Prove that it is a right-angle triangle.

Solution: Let A (– 2, 1) , B (1, 1) and C (1, 2). are the vertices of the triangle.

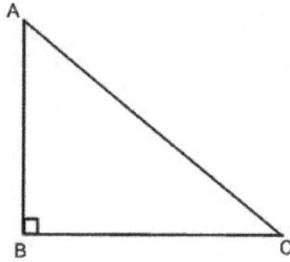

By distance formula $AB = \sqrt{(-2-1)^2 + (1-1)^2} = \sqrt{9+0} = 3$

$$BC = \sqrt{(1-1)^2 + (1-2)^2} = \sqrt{0+1} = 1$$

$$CA = \sqrt{(1+2)^2 + (2-1)^2} = \sqrt{9+1} = \sqrt{10}$$

We find $AB^2 + BC^2 = 9 + 1 = 10 = CA^2$

Hence ABC is a right-angle triangle where $\angle B = 90°$

6. Identify the triangle whose vertices are A $(2, 2)$, B $(-2, -2)$ and $\left(-2\sqrt{3}, 2\sqrt{3}\right)$

Solution: AB $= \sqrt{(-2-2)^2 + (-2-2)^2} = \sqrt{16+16} = \sqrt{16 \times 2} = 4\sqrt{2}$

$$BC = \sqrt{\left(-2\sqrt{3}+2\right)^2 + \left(2\sqrt{3}+2\right)^2} = \sqrt{\left(12+4-8\sqrt{3}\right)+\left(12+4+8\sqrt{3}\right)} = \sqrt{16+16} = 4\sqrt{2}$$

$$CA = \sqrt{\left(-2\sqrt{3}-2\right)^2 + \left(2\sqrt{3}-2\right)^2} = \sqrt{\left(12+4+8\sqrt{3}\right)+\left(12+4-8\sqrt{3}\right)} = \sqrt{16+16} = 4\sqrt{2}$$

Since AB = BC = CA , hence ABC is an equilateral triangle

7. Show that the points A (3, 5) , B(1,1) and C(–2, –5) are collinear.

Solution: AB $= \sqrt{(1-3)^2 + (1-5)^2} = \sqrt{4+16} = \sqrt{20} = \sqrt{4 \times 5} = 2\sqrt{5}$

$$BC = \sqrt{(-2-1)^2 + (-5-1)^2} = \sqrt{9+36} = \sqrt{45} = \sqrt{9 \times 5} = 3\sqrt{5}$$

$$AC = \sqrt{(3+2)^2 + (5+5)^2} = \sqrt{25+100} = \sqrt{125} = \sqrt{25 \times 5} = 5\sqrt{5}$$

Since AB + BC = AC, hence points A, B, C are collinear.

8. Find the circum-centre of the triangle whose vertices are: (2, 1), (5, 2) and (3, 4)

Solution: The circum-center is the point where the perpendicular bisectors of sides of a triangle intersect. The circum-center is the center of the triangle's circumcircle- the circle that passes through all three vertices of the triangle.

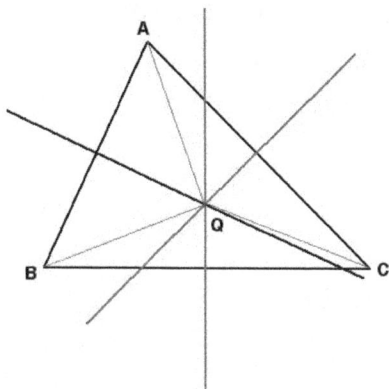

ABC is the triangle and Q is the circum-centre such that QA = QB =QC

Coordinate of vertices are: A (2, 1) , B (5, 2) and C (3, 4). Let coordinates of Q are (x, y)

$$QA = \sqrt{(x-2)^2 + (y-1)^2} \ , \ QB = \sqrt{(x-5)^2 + (y-2)^2} \ , \ QC = \sqrt{(x-3)^2 + (y-4)^2}$$

Since QA = QB $\therefore$ $(x-2)^2 + (y-1)^2 = (x-5)^2 + (y-2)^2 \Rightarrow 3x + y = 12$ ......(1)

Since QB = QC $(x-5)^2 + (y-2)^2 = (x-3)^2 + (y-4)^2 \Rightarrow x - y = 1$ ....(2)

By addition of (1) and (2) $4x = 13 \Rightarrow x = \dfrac{13}{4}$

From (2) $y = x - 1 = \dfrac{13}{4} - 1 = \dfrac{9}{4}$

Hence coordinates of circum-centre are $\left(\dfrac{13}{4}, \dfrac{9}{4}\right)$

Practice Problems (Distance between two points)

1. Find the distance between the points (2, 3) and (0, 6).
2. Find the value of **y** if the distance between the points (−2, −3) and (−5, y) is 5.
3. Find value of y if the point (0, y) is equidistant from (4,−9) and (−5, 6).
4. Find the relationship between x and y if distance between (x, y) and (−2, 4) is equal to 5.

5. Find the hypotenuse of the right triangle whose vertices are A $(-2, 1)$, B $(1, 1)$ and C $(1, 2)$.

6. The point $(4, y)$ is 10 units away from the point $(-2, -1)$. Find the value of y.

7. Find the radius of the circle whose center is $(2, -3)$ and the point $(-1, -2)$ lies on it.

8. Identify the triangle whose vertices are $(4, -3)$, $(3, 0)$ and $(0, 1)$

9. Identify the triangle whose vertices are $(-2, 3)$, $(3, 8)$ and $(4, 1)$

10. Show that the points $(0, 0)$, $(1, 2)$ and $(2, 4)$ are collinear.

Answers:

1. $\sqrt{13}$ 　　　　　　　　 2. $4\sqrt{5}$ 　　　　　　　　 3. $(1, -7)$

4. $y = \dfrac{-93}{4}$ 　　　　　 5. $x^2 + y^2 + 8x - 8y = 5$ 　　 6. $CA = \sqrt{10}$

7. $y = -9$ or 7 　　　　　 8. Isoceles 　　　　　　　 9. Isosceles

1.3 COORDINATES OF THE POINT DIVIDING A LINE IN A GIVEN RATIO

1. The distance between two points(x_1, y_1) and $x_2, y_2)$ can be divided in two ways: **internally** and **externally**.

2. If the point divides the line AB, **internally** in the **ratio m and n**, then the coordinates of dividing point P(x, y) are:

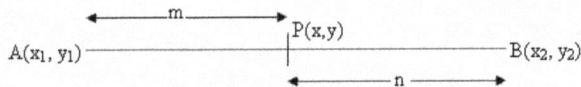

$$x = \frac{mx_2 + nx_1}{m + n} \text{ and } y = \frac{my_2 + ny_1}{m + n}.$$

3. If the point divides the line AB **externally** in the **ratio m and n**, such that **AE = m** and **BE = n**

then the coordinates of dividing point E(x, y) are:

$$x = \frac{mx_2 - nx_1}{m + n} \text{ and } y = \frac{my_2 - ny_1}{m + n}$$

4. Coordinates of **middle point** of the line joining (x_1, y_1) and x_2, y_2 are:

$$x = \frac{x_1 + x_2}{2} \text{ and } y = \frac{y_1 + y_2}{2}$$

Solved Examples

1. Find the coordinates of the point which divides the line joining the points $(3, -5)$ and $(-4, 6)$ internally in the ratio 2:3

Solution: Here $x_1 = 3$ $y_1 = -5$ and $x_2 = -4$, $y_2 = 6$. m = 2 and n = 3

$$\therefore \qquad x = \frac{mx_2 + nx_1}{m+n} = \frac{2(-4) + 3(3)}{2+3} = \frac{-8+9}{5} = \frac{1}{5}$$

$$y = \frac{my_2 + ny_1}{m+n} = \frac{2(6) + 3(-5)}{2+3} = \frac{12-15}{5} = \frac{-3}{5}$$

Hence the coordinates of the dividing point are $\left(\frac{1}{5}, \frac{-3}{5}\right)$

2. Find the coordinates of the point which divides the line joining the points $(4, -5)$ and $(6, 3)$ **internally** and **externally** in the ratio 2:5.

Solution: Here $x_1 = 4$ $y_1 = -5$ and $x_2 = 6$, $y_2 = 3$. m = 2 and n = 5

For internal division:

$$x = \frac{mx_2 + nx_1}{m+n} = \frac{2(6) + 5(4)}{2+5} = \frac{32}{7}$$

$$y = \frac{my_2 + ny_1}{m+n} = \frac{2(3) + 5(-5)}{2+5} = \frac{6-25}{7} = \frac{-19}{7}$$

Hence coordinates of point of internal division are $\left(\frac{32}{7}, \frac{-19}{7}\right)$

For external division:

$$x = \frac{mx_2 - nx_1}{m-n} = \frac{2(6) - 5(4)}{2-5} = \frac{-8}{-3} = \frac{8}{3}$$

$$y = \frac{my_2 - ny_1}{m-n} = \frac{2(3) - 5(-5)}{2-5} = \frac{6+25}{-3} = \frac{-31}{3}$$

Hence coordinates of point of external division are $\left(\frac{8}{3}, \frac{-31}{3}\right)$

3. The coordinates of two points A and B are $(4, -5)$ and $(-3, -2)$ respectively. AB is produced to a point E such that $AE = 3\ BE$. Find the coordinates of E.

Solution: Let coordinates of E are (x, y). E lies beyond AB Point E divides the line AB externally in the ratio 3:1

We are given $x_1 = 4$, $y_1 = -5$, $x_2 = -2$, $y_2 = -2$. m =3, n = 1

For external division:

$$x = \frac{mx_2 - nx_1}{m - n} = \frac{3(-3) - 1(4)}{3 - 1} = \frac{-13}{2}$$

$$y = \frac{my_2 - ny_1}{m - n} = \frac{3(-2) - 1(-5)}{3 - 1} = \frac{-6 + 5}{2} = \frac{-1}{2}$$

Hence coordinates of point E are $\left(\dfrac{-13}{2}, \dfrac{-1}{2}\right)$

4. Find the coordinates of points trisecting a line joining the points $(3, -2)$ and $(-3, -4)$

Solution: Let distance between two points A $(3, -2)$ and B $(-3, -4)$ is trisected by two points, say $P(X_1, Y_1)$ and $Q(X_2, Y_2)$. Thus $AP = PQ = QB$

We see ratio of **AP: PB = 1: 2 i.e. $m_1 : m_2 = 1: 2$**

Hence we have $x_1 = 3$, $y_1 = -2$, $x_2 = -3$, $y_2 = -4$. m = 1, n = 2.

We have to find coordinates of P.

Since division is internal:

$$X_1 = \frac{mx_2 + nx_1}{m + n} = \frac{1(-3) + 2(3)}{1 + 2} = \frac{3}{3} = 1$$

$$Y_1 = \frac{my_2 + ny_1}{m + n} = \frac{1(-4) + 2(-2)}{1 + 2} = \frac{-4 - 4}{3} = \frac{-8}{3}$$

Hence coordinates of point B are $\left(1, \dfrac{-8}{3}\right)$

Again we see that AQ: QB = 2:1

Hence we have $x_1 = 3$, $y_1 = -2$, $x_2 = -3$, $y_2 = -4$. m = 2, n = 1.

We have to find coordinates of Q.

$$X_2 = \frac{mx_2 + nx_1}{m + n} = \frac{2(-3) + 1(3)}{2 + 1} = \frac{-3}{3} = -1$$

$$Y_2 = \frac{my_2 + ny_1}{m + n} = \frac{2(-4) + 1(-2)}{2 + 1} = \frac{-8 - 2}{3} = \frac{-10}{3}$$

Hence coordinates of point B are $\left(-1, \dfrac{-10}{3}\right)$

5. In what ratio x-axis divides the line joining the points $(2, -3)$ and $(5, 6)$?

Solution: Let line be divided in the ratio **k: 1**. Then the coordinates of point dividing the

line can be obtained by using $x = \dfrac{mx_2 + nx_1}{m+n}$ and $y = \dfrac{my_2 + ny_1}{m+n}$

Since the dividing point is on x-axis, so y-coordinate is zero i.e. $y = \dfrac{my_2 + ny_1}{m+n} = 0$

Putting the values of the ratios we get $\dfrac{my_2 + ny_1}{m+n} = 0$

$\Rightarrow$ $\qquad\qquad\qquad\qquad my_2 + ny_1 = 0$

We have : $y_1 = -3$ and $y_2 = 6$ $m = k$ and $n = 1$

Substituting the above values in $my_2 + ny_1 = 0$, we get:

$$k.6 + 1(-3) = 0 \Rightarrow k = \frac{3}{6} = \frac{1}{2}$$

Since ratio is k:1 i.e $\dfrac{1}{2} : 1$ or $1 : 2$.

Hence x-axis divides the line in the ratio **1:2.**

Practice Problems: (Division of a line segment)

1. Find the coordinates of the point which divides the line joining the points $(2, 3)$, $(5, 6)$ internally in the ratio **2:1.**

2. A $(4, 5)$ and B $(7, -1)$ are two given points and the point C divides the line-segment AB externally in the ratio $4 : 3$. Find the co-ordinates of C.

3. Find the ratio in which the line-segment joining the points $(5, -4)$ and $(2, 3)$ is divided by the x-axis.

4. Find the point P which divides the line joining the points A $(9, -1)$ and $(4, 4)$ in the ration $3: 2$.

Answers:

1. $(4, 5)$

2. $\left(\dfrac{4}{3}, \dfrac{-5}{3}\right)$

3. $m = 4, n = 3$

4. $(6, 2)$

1.4 COORDINATES OF CENTROID

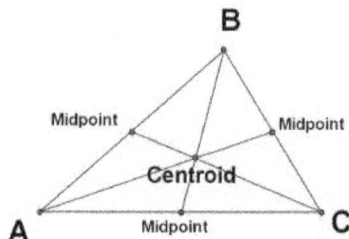

Centroid is the point where the medians of a triangle meet. If $A(x_1, y_1)$, $Bx_2, y_2)$ and $C(x_3, y_3)$ are the coordinates of vertices A, B, C of the triangle, then coordinates of centroid are:

$$\text{Centroid} = \left(\frac{x_1 + x_2 + x_3}{3}, \frac{y_1 + y_2 + y_3}{3} \right)$$

Solved Examples

1. Find the coordinates of centroid of a triangle whose vertices are $(-1, -3)$, $(2, 1)$ and $(8, -4)$.

Solution: $x_1 = -1$, $y_1 = -3$ $x_2 = 2$, $y_2 = 1$ and $x_3 = 8$, $y_3 = -4$

Substitute above values in the formula:

$$\left(\frac{x_1 + x_2 + x_3}{3}, \frac{y_1 + y_2 + y_3}{3} \right)$$

We get: $\left(\dfrac{-1+2+8}{3}, \dfrac{-3+1-4}{3} \right) = \left(\dfrac{9}{3}, \dfrac{-6}{3} \right) = (3, -2)$

Hence coordinates of centroid are (3, –2).

2. Find the centroid of the triangle whose vertices are $(4, -6)$, $(2, -2)$ and $(2, 5)$.

Solution: If (x_1, y_1), (x_2, y_2) and (x_3, y_3) are the coordinates of vertices of the triangle, then

coordinates of centroid are: $\left(\dfrac{x_1 + x_2 + x_3}{3}, \dfrac{y_1 + y_2 + y_3}{3} \right)$

We are given $x_1 = 4$, $y_1 = -6$, $x_2 = 2$, $y_2 = -2$. and $x_3 = 2$, $y_3 = 5$.

Hence coordinates of centroid are:

$$\left(\frac{4+2+2}{3}, \frac{-6-2+5}{3} \right) \text{ i.e. } \frac{8}{3}, \frac{-3}{3}$$

Practice Problems (On Centroid)

1. The angular points of a triangle are (4, – 6), (2, –2) and (2,5). Find coordinates of centroid.

2. The vertices of the triangle are (2, 1), 5, 2) and (3, 4). Find coordinates of centroid.

3. Find the centroid of a triangle whose vertices are (5, 3), (6,1) and (7,8).

Answers:

1. $\left(\dfrac{8}{3}, -1\right)$, 2. $\left(\dfrac{10}{3}, \dfrac{7}{3}\right)$, 3. **(6, 4)**

1.5 Area of a Triangle

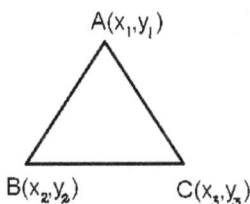

$A(x_1, y_1)$

$B(x_2, y_2)$ $C(x_3, y_3)$

(*i*) Area of a triangle whose vertices are (x_1, y_1), (x_2, y_2) and (x_3, y_3) is

$$\Delta = \frac{1}{2}\left[\left(x_1 y_2 + x_2 y_3 + x_3 y_1\right) - \left(y_1 x_2 + y_2 x_3 + y_3 x_1\right)\right]$$

The above formula may also be expressed as:

$$\Delta = \frac{1}{2}\left[x_1\left(y_2 - y_3\right) + x_2\left(y_3 - y_1\right) + x_3\left(y_1 - y_2\right)\right]$$

(*ii*) Three points will be collinear i.e. in straight line if area of Δ formed by them is zero.

Solved Examples

1. Find the area of the triangle whose vertices are:

(*i*) (4, 6), (0, 4), (6, 2)   (*ii*) (a, b), (b, c), (c, a)

Solution: (*i*) We are given $x_1 = 4$, $y_1 = 6$, $x_2 = 0$, $y_2 = 4$. $x_3 = 6$, $y_3 = 2$.

$$\text{Hence } \Delta = \frac{1}{2}\left[\left(x_1 y_2 + x_2 y_3 + x_3 y_1\right) - \left(y_1 x_2 + y_2 x_3 + y_3 x_1\right)\right]$$

$$= \frac{1}{2}\left[\left(4.4 + 0.2 + 6.6\right) - \left(6.0 + 4.6 + 2.4\right)\right]$$

$$= \frac{1}{2}\left[\left(16 + 0 + 36\right) - \left(0 + 24 + 8\right)\right]$$

$$= \frac{1}{2}\left[(52) - (32)\right] = \frac{20}{2} = 10 \text{ sq units.}$$

(*ii*) We are given $x_1 = a$, $y_1 = b$, $x_2 = b$, $y_2 = c$. $x_3 = c$, $y_3 = a$.

Hence $\Delta = \dfrac{1}{2}\left[(x_1 y_2 + x_2 y_3 + x_3 y_1) - (y_1 x_2 + y_2 x_3 + y_3 x_1)\right]$

$\qquad = \dfrac{1}{2}\left[(ab + bc + ca) - (b.b + c.c + a.a)\right]$

$\qquad = \dfrac{1}{2}\left[(ab + bc + ca) - (b^2 + c^2 + a^2)\right]$ sq units.

2. Prove that the points (2, –2), (–3, 8) and (–1, 4) are collinear.

Solution: We are given $x_1 = 2$, $y_1 = -2$, $x_2 = -3$, $y_2 = 8$. $x_3 = -1$, $y_3 = 4$.

Hence from $\Delta = \dfrac{1}{2}\left[x_1(y_2 - y_3) + x_2(y_3 - y_1) + x_3(y_1 - y_2)\right]$ we get

$\qquad \Delta = \dfrac{1}{2}\left[2(8 - 4) + (-3)(4 - (-2)) + (-1)((-2) - 8)\right]$

$\qquad = \dfrac{1}{2}\left[2 \times 4 - 3 \times 6 + 10\right] = \dfrac{1}{2}\left[8 - 18 + 10\right] = 0$

Hence the given points are collinear.

3. Find the value of k if the points (k, 4), (2, –2) and (–3, 8) are collinear.

Solution: (k, 4), (2, -2) and (-3, 8) will be collinear if the area of triangle formed by them is zero.

We are given $x_1 = k$, $y_1 = 4$, $x_2 = 2$, $y_2 = -2$, $x_3 = -3$, $y_3 = 8$.

Putting these values in

$\qquad \Delta = \dfrac{1}{2}\left[(x_1 y_2 + x_2 y_3 + x_3 y_1) - (y_1 x_2 + y_2 x_3 + y_3 x_1)\right] = 0$

$\qquad = \dfrac{1}{2}\left[\{k(-2) + 2 \times 8 + (-3) \times 4\} - \{4 \times 2 + (-2)(-3) + 8 \times k\}\right] = 0$

$\qquad = \dfrac{1}{2}\left[\{-2k + 16 - 12\} - \{8 + 6 + 8k\}\right] = 0$

$\Rightarrow \qquad\qquad (-2k + 4) - (14 + 8k) = 0$

$\qquad\qquad\qquad -10k - 10 = 0$

$\Rightarrow \qquad\qquad\qquad k = -1$

Hence value of $k = -1$.

4. Find the area of the triangle whose angular points are (t, t–2), (t +2, t +2) and (t +3, t).

Solution: We are given $x_1 = t$, $y_1 = t-2$, $x_2 = (t+2)$, $y_2 = (t+2)$, $x_3 = (t+3)$, $y_3 = t$

$$\Delta = \frac{1}{2}\left[x_1(y_2 - y_3) + x_2(y_3 - y_1) + x_3(y_1 - y_2)\right]$$

$$= \frac{1}{2}\left[t(t+2-t) + (t+2)(t-t+2) + (t+3)(t-2-t-2)\right]$$

$$= \frac{1}{2}\left[2t + 2(t+2) - 4(t+3)\right]$$

$$= \frac{1}{2}\left[2t + 2t + 4 - 4t - 12\right] = \left|\frac{-8}{2}\right| = 4$$

Practice Problems (On Area of triangle)

1. Three vertices of a triangle are (1, 2), (7, 5), and (7, 9). Find the area of the triangle.
2. Find the area of the triangle whose vertices are (0, 9), (0, –4) and (5, –4)
3. If the points (4, 2), (–1, k), (1, 6) are collinear. Find the value of k.
4. What is co-linearity? Are the points A (4, 11) B (6, 8) and C (14, 14) collinear?

Answers:

    1. 12 sq. units   2. 32.5 sq. units   3. k = 8.67

    4. No

1.6 AREA OF QUADRILATERAL

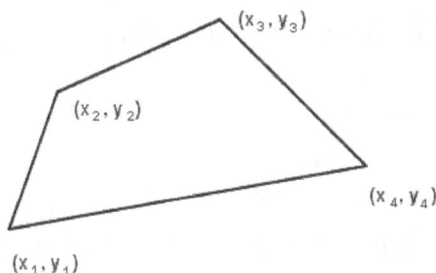

Let the vertices of quadrilateral are: $A(x_1, y_1)$, $B(x_2, y_2)$, $C(x_3, y_3)$ and $D(x_4, y_4)$.
Its area is given by:

$$\frac{1}{2}\left[(x_1y_2 + x_2y_3 + x_3y_4 + x_4y_1) - (y_1x_2 + y_2x_3 + y_3x_4 + y_4x_1)\right]$$

Solved Examples

1. Find the area of quadrilateral whose vertices are: A (−4, −2), B (−3, −5), C (3, −2), and D(2, 3).

Solution: Given:

$$x_1 = -4 \quad x_2 = -3 \quad x_3 = 3 \quad x_4 = 2$$

$$y_1 = -2 \quad y_2 = -5 \quad y_3 = -2 \quad y_4 = 3$$

∴ Area of quadrilateral $= \dfrac{1}{2} \left[(x_1 y_2 + x_2 y_3 + x_3 y_4 + x_4 y_1) - (y_1 x_2 + y_2 x_3 + y_3 x_4 + y_4 x_1) \right]$

$$= \frac{1}{2} \left[\{(-4 \times -5) + (-3 \times -2) + (3 \times 3) + (2 \times -2)\} - \{(-2 \times -3) + (-5 \times 3) + (-2 \times 2) + (3 \times -4)\} \right]$$

$$= \frac{1}{2} \left[\{(20) + (6) + (9) + (-4)\} - \{(6) + (-15) + (-4) + (-12)\} \right]$$

$$= \frac{1}{2} \left[\{31\} - \{-25\} \right] = \frac{1}{2} [56] = 28 \text{ sq.units}$$

2. Find the area of quadrilateral whose vertices are: A (2, 2), B (−1, 4), C (−3, 1), and D(1, −1)

Solution: Given:

$$x_1 = 2 \quad x_2 = -1 \quad x_3 = -3 \quad x_4 = 1$$

$$y_2 = 4 \qquad\qquad y_4 = -1$$

∴ Area of quadrilateral $= \dfrac{1}{2} \left[(x_1 y_2 + x_2 y_3 + x_3 y_4 + x_4 y_1) - (y_1 x_2 + y_2 x_3 + y_3 x_4 + y_4 x_1) \right]$

$$= \frac{1}{2} \left[\{(2 \times 4) + (-1 \times 1) + (-3 \times -1) + (1 \times 2)\} - \{(2 \times -1) + (4 \times -3) + (1 \times 1) + (-1 \times 2)\} \right]$$

$$= \frac{1}{2} \left[\{(8) + (-1) + (3) + (2)\} - \{(-2) + (-12) + (1) + (-2)\} \right]$$

$$= \frac{1}{2} \left[\{12\} - \{-15\} \right] = \frac{27}{2} = 13.5$$

Practice Problems (On Area of Quadrilateral)

1. The vertices of a quadrilateral are: A (2, 1), B (-7, 1), C (9, 5), and D(4, 5). Determine the area of the quadrilateral.

2. The vertices of a quadrilateral are: A (0, 0), B (-6, 8), C (9, 5), and D(2, 11). Determine the area of the quadrilateral.

Answers:

1. $\Delta = 21$ sq. units   **2.** $\Delta = 65$ sq units

1.7   EQUATION OF A STRAIGHT LINE: (PASSING THROUGH TWO POINTS)

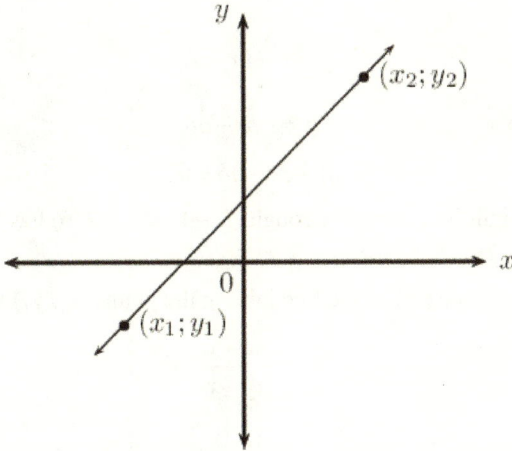

(*a*) Equation of a line joining the points (x_1, y_1) and (x_2, y_2) is:

$$(y - y_1) = \frac{y_2 - y_1}{x_2 - x_1}(x - x_1)$$

(*b*) Gradient or slope (m) of the line joining the points (x_1, y_1) and (x_2, y_2) is

$$m = \frac{y_2 - y_1}{x_2 - x_1}$$

(*c*) Gradient or slope (m) of a line is obtained by $m = \dfrac{-\text{Coefficient of } x}{\text{Coefficient of } s}$, if the line is written in the form of $ay + bx + c = 0$

Solved Examples

1. Obtain the equation of straight line between: **(i)** (2, –3) and (–4, 9), **(ii)** (a, 0) and (0, b)

Solution: (i) line between (2, –3) and (–4, 9) . Here $x_1 = 2$, $y_1 = -3$, $x_2 = -4$, $y_2 = 9$,

Substituting the given values in $(y - y_1) = \dfrac{y_2 - y_1}{x_2 - x_1}(x - x_1)$, we get:

$$(y - (-3)) = \frac{9 - (-3)}{(-4) - 2}(x - 2)$$

$$\Rightarrow \qquad (y + 3) = \frac{12}{-6}(x - 2)$$

$$\Rightarrow \qquad y + 3 = -2(x - 2)$$

$$\Rightarrow \qquad y + 2x = 1$$

(*ii*) Equation of line joining (a,0) and (0, b)

$$(y-0) = \frac{b-0}{0-a}(x-a)$$

$\Rightarrow$ $\qquad\qquad\qquad y = \frac{b}{-a}(x-a)$

or $\qquad\qquad\qquad ay = -bx + ab$

$\Rightarrow$ $\qquad\qquad\qquad ay + bx - ab = 0$.

2. Show that the straight line passing through (3, –4) and (–2, 6) has the same gradient as that of the line y + 2x + 6 = 0.

Solution: Gradient or slope (m) of the line joining the points (x_1, y_1) and (x_2, y_2) is

$$m = \frac{y_2 - y_1}{x_2 - x_1}$$

We have $x_1 = 3$ $y_1 = -4$, $x_2 = -2$, $y_2 = 6$.

Hence $m = \dfrac{6-(-4)}{(-2)-3} = \dfrac{6+4}{-2-3} = \dfrac{10}{-5} = -2$

Given line is y + 2x + 6 = 0. This can be written as $y = -2x - 3$

Comparing with $y = mx + c$, we see $m = -2$.

Hence gradient of both lines are same. The lines having same gradient are parallel.

3. Show that the points (a + 2b, –2a + b), (a – b, a + b) and (a, b) are collinear.

Solution: Let the points are A (a + 2b, –2a + b), B (a – b, a + b) and C (a, b).

Here $x_1 = a + 2b$, $y_1 = -2a + b$, $x_2 = a - b$, $y_2 = a + b$

Hence the equation of straight line AB is :

$$(y-(-2a+b)) = \frac{(a+b)-(-2a+b)}{(a-b)-(a+2b)}(x-(a+2b))$$

or $\qquad\qquad (y+2a-b) = \dfrac{(a+b+2a-b)}{(a-b-a-2b)}(x-a-2b))$

or $\qquad\qquad (y+2a-b) = \dfrac{(3a)}{(-3b)}(x-a-2b)$

Substituting the coordinates of point C (a, b) i.e. x = a, y = b in the above equation of AB

We get $\qquad\qquad (b+2a-b) = \dfrac{(3a)}{(-3b)}(a-a-2b)$

$$(2a) = \frac{(a)}{(-b)}(-2b)$$

Since left hand side = right hand side, hence coordinates of point C satisfy the equation of AB. Hence point C lies on the line AB.

Practice Problems: (Equation of a straight line passing through two points).

1. Find the equation of the line that passes through the points (–3, 4) and (1, 7).

2. Find the equation of a line with a slope equal to 3 and passes through the point (1, 2).

3. Find the equation of the straight line whose slope is $m = 4$ and passes through the point (–1, –6).

4. Find the slope of the line $3x + 2y = 12$.

5. Find the slope of the line passing through points (-2,3) and (4,-5).

Answers:

1. $y = \left(\frac{3}{4}\right)x + \left(\frac{25}{4}\right)$,

2. $y = 3x - 1$,

3. $y = 4x - 2$

4. Slope = –3/2

5. Slope = –4/3

1.8   POINT OF INTERSECTION OF TWO LINES

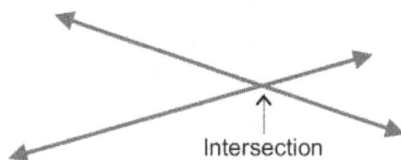

Intersection

Point of intersection of two straight lines is obtained by solving the equations together exactly like the solution of linear simultaneous equations.

Solved Examples

1. Find the point of intersection of two straight lines whose equations are:

(*i*) $x + 3y = 5$ and $x - 2y + 5 = 0$ (*ii*) $\frac{x}{a} + \frac{y}{b} = 1$ and $\frac{x}{a} - \frac{y}{b} = 1$

Solution: (*i*) Two lines are: $x + 3y = 5$ ...(a)

$x - 2y = -5$ ...(b)

Subtracting (b) from (a) we get 5y =10 $\Rightarrow$ $y = 2$.

Substituting $y = 2$ in equation (a) we get $x = -1$

Hence coordinates of point of intersection are $(-1, 2)$.

(ii) Two lines are:
$$\frac{x}{a} + \frac{y}{b} = 1 \qquad \qquad \text{...(a)}$$

and
$$\frac{x}{b} - \frac{y}{a} = 1 \qquad \qquad \text{...(b)}$$

Multiplying the equation.– (a) by **b** and equation (b) by **a** we get:
$$\frac{bx}{a} + \frac{by}{b} = b \qquad \qquad \text{...(a)}$$

and
$$\frac{ax}{b} - \frac{ay}{a} = a \qquad \qquad \text{...(b)}$$

Adding (a) and (b) $\dfrac{bx}{a} + \dfrac{ax}{b} = a + b$

$\Rightarrow$
$$x\left(\frac{b}{a} + \frac{a}{b}\right) = (a+b)$$

$\Rightarrow$
$$x = \frac{ab(a+B)}{a^2 + b^2}$$

Now multiplying equation.(a) by **a** and equation (b) by **b** we get:
$$\frac{ax}{a} + \frac{ay}{b} = a \qquad \qquad \text{...(a)}$$

and
$$\frac{bx}{b} - \frac{by}{a} = b \qquad \qquad \text{...(b)}$$

Subtracting (b) from (a) we get
$$\frac{by}{a} + \frac{ay}{b} = a - b$$

$\Rightarrow$
$$y\left(\frac{b}{a} + \frac{a}{b}\right) = (a-b)$$

$\Rightarrow$
$$y = \frac{ab(a-b)}{a^2 + b^2}$$

Hence coordinates of point of intersection are $\left(\dfrac{ab(a+B)}{a^2 + b^2}, \dfrac{ab(a-b)}{a^2 + b^2}\right)$.

2. Find the value of **m** if the following lines are concurrent:
$$y = x+1 , \quad y = 2x+2 \quad \text{and} \quad y = mx+3$$

Solution: Lines are said to be concurrent if they pass through a point. In such case, the coordinates of point of intersection must satisfy the third equation.

Equating first two equations we get $x+1=2x+2$

$\Rightarrow$ $\qquad\qquad\qquad\qquad x=-1$

Now putting $x=-1$ in $y=x+1$ we get y $=0$

Thus coordinates of point of intersection are $(-1, 0)$

Substitute these coordinates in third equation $y=mx+3$,we get $0=m(-1)+3$ or $m=3$

Hence the value of slope (m) is 3.

3. Find the point of intersection of circle: $x^2+y^2=5$ and the straight line $2x+y=3$

Solution: $2x+y=3$. It can be written as $y=3-2x$

Substituting y in the Circle $x^2+y^2=5$

We get $\qquad\qquad\qquad\qquad x^2+(3-2x)^2=5$

or $\qquad\qquad\qquad\qquad x^2+(9+4x^2-12x)=5$

or $\qquad\qquad\qquad\qquad 5x^2-12x+9=5$

or $\qquad\qquad\qquad\qquad 5x^2-12x+4=0$

$5x^2-10x-2x+4=0$ by factorization.

or $\qquad\qquad\qquad\qquad (5x^2-10x)-(2x-4)=0$

$\qquad\qquad\qquad\qquad 5x(x-2)-2(x-2)=0$

$\qquad\qquad\qquad\qquad (x-2)(5x-2)=0 \Rightarrow x=2$ or $\dfrac{5}{2}$

Put x $=2$ in $2x+y=3$, we get $4+y=3 \Rightarrow y=-1$

So one pair of coordinates is (2, –1).

Now put x $=\dfrac{5}{2}$ in $2x+y=3$ we get in $2.\dfrac{5}{2}+y=3 \Rightarrow y=3-5=-2$

Hence another pair of coordinates is $(\dfrac{5}{2}, -2)$

The line will intersect the circle at two points (2, -1) or $(\dfrac{5}{2}, -2)$

Practice Problems (Point of intersection).

1. Find the point of intersection of the two lines $y = 3x+2$, and $y = 2x-1$.

2. Find the point of intersection of the straight lines: $y = -x+1$ and $y = x^2 - 1$.

3. Find the point of intersection of two lines: $y = 3x - 6$ and $2y = 3 - 5x$

Answers

    1. (–3,–7),   **2.** (–2, 3) , (1,0),   **3.** (15, 41)

1.9 ANGLE BETWEEN TWO STRAIGHT LINES

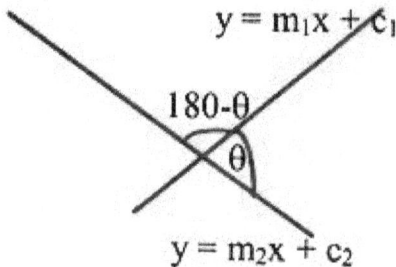

(i) The angle between the two lines $y = m_1 x + c_1$ and $y = m_2 x + c_2$ is given by:

$$\tan \theta = \frac{m_1 - m_2}{1 + m_1 m_2}$$

(ii) If $m_1 = m_2$, then the lines are **parallel** to one another.

(ii) If $m_1 . m_2 = -1$, then the lines are **perpendicular** to each other.

Solved Examples

1. Find the equation of a straight line passing through the point (1, 1) and parallel to the line $3x - 4y = 7$

Solution: The gradient of equation $3x - 4y = 7$ is

$$m = \frac{-\text{ Coefficient of } x}{\text{Coefficient of } y} = \frac{-(3)}{(-4)} = \frac{3}{4}$$

A line passing through (x_1, y_1) and having gradient m is given by $y - y_1 = m(x - x_1)$

We have $x_1 = 1$, $y_1 = 1$ and $m = \frac{3}{4}$.

Then equation of line is $(y-1) = \frac{3}{4}(x-1)$ or $4y - 3x + 1 = 0$.

2. Find the angle between the straight lines: $3x - y = 6$ and $2y + 5x = 3$.

Solution: let m_1 and m_2 are the gradient of the given lines.

$$\therefore \qquad m_1 = \frac{-\text{coeff.of } x}{\text{coeff.of } y} = \frac{-3}{-1} = 3 \,.$$

Similarly

$$m_2 = \frac{-\text{coeff. of } x}{\text{coeff. of } y} = \frac{-5}{2}$$

Angle between two straight lines is given by

$$\tan \theta = \frac{m_1 - m_2}{1 + m_1 m_2} = \frac{3 - \left(\dfrac{-5}{2}\right)}{1 + (3)\left(\dfrac{-5}{2}\right)}$$

$$= \frac{\dfrac{11}{2}}{\dfrac{-13}{2}} = \frac{-11}{13}$$

Practice Problems: (Angle between two straight lines)

1. Given the lines $3x + y - 1 = 0$ and $2x + my - 8 = 0$, calculate the value of m so that the two lines form an angle of 45°.

2. Find the equation of a straight line which passes through the point $(4, -5)$ and is perpendicular to the straight line $3x + 4y + 5 = 0$.

3. Find the angle between two lines whose slopes are

$$m_1 = \frac{1}{2} \,, \ m_2 = 2 \,.$$

Answers:

    1. m = −1   **2.** 4x - 3y = 31,   **3.** $\tan \theta = \dfrac{3}{4}$

1.10   EQUATION OF STRAIGHT LINE (Cutting Intercepts on both axes)

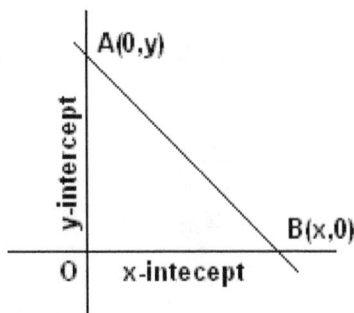

Let AB the line that cuts x-axis at B. Hence OB = x-intercept

Let AB the line that cuts y-axis at A. Hence OA = y-intercept

(i) If x- intercept = **a** and y-intercept = **b**, then equation of AB is given by:

$$\frac{x}{a} + \frac{y}{b} = 1$$

(ii) To determine the intercepts, convert the given equation in the form of $\frac{x}{a} + \frac{y}{b} = 1$, then **a** and **b** are the intercepts.

Solved Examples

1. Find the equation of the straight line which cut off following intercepts from the axes of x and y respectively: (*i*) 3, 4; (*ii*) 3, –4; (*iii*) k, 2k;

Solution:

(*i*) Intercepts: a = 3 and b = 4. Hence the equation is $\frac{x}{3} + \frac{y}{4} = 1 \Rightarrow 4x + 3y = 12$

(*ii*) Intercepts: a = 3 and b = –4. Hence the equation is $\frac{x}{3} - \frac{y}{4} = 1 \Rightarrow 4x - 3y = 12$

(*iii*) Intercepts: a = k and b = 2k. Hence the equation is $\frac{x}{k} + \frac{y}{2k} = 1 \Rightarrow 2x + y = 2k$

2. Find the equation of the straight line which passes through (3, 4) and cuts intercept on the y-axes twice as long as the intercept on x-axes.

Solution: Let x-axes intercept a = k and y-axes intercept b = 2k.

Hence equation is $\frac{x}{k} + \frac{y}{2k} = 1 \Rightarrow 2x + y = 2k$

Since, the line passes through (3, 4). From the above equation 2(3) + (4) = 2k $\Rightarrow$ k = 5

Hence the required equation is $2x + y = 10$.

3. Find the equation of the straight line passing through (3, 4) cutting intercepts whose sum is 14.

Solution: Let a and b are the intercepts; a + b = 14 ∴ a = 14 – b

The equation of the line is

$$\frac{x}{a} + \frac{y}{b} = 1$$

$\Rightarrow$ $\qquad\qquad\qquad bx + ay = ab$

$\Rightarrow$ $\qquad\qquad\qquad 4b + 3a = ab$

Now substitute a = 14 − b in the above equation. We get $4b + 3(14 - b) = (14 - b)b$

Simplifying the new equation:

$$4b + (42 - 3b) = (14b - b^2)$$

or
$$b^2 - 13b + 42 = 0$$

or
$$b^2 - 6b - 7b + 42 = 0$$

$$(b-6)(b-7) = 0 \Rightarrow b = 6 \text{ or } 7$$

∵ a = 14 − b ∴ **a = 14 − 6 = 14 or a = 14 − 7 = 7.**

If a = 6 or 7 then b = 8 or 7. These will give two equations

The required equations are $\dfrac{x}{6} + \dfrac{y}{8} = 1$ or $\dfrac{x}{7} + \dfrac{y}{7} = 1 \Rightarrow x + y = 7$.

Practice Problems (Equation with x and y-intercepts)

1. Find the equation of a line with x-intercept of 5 and y-intercept of −6.

2. Find the x and y intercepts of the line with the given equation: 2x − 6y =12

3. Give an equation for the line whose x-intercept is 5 and whose y-intercept is −4

4. Find the x- and y-intercepts of the line that passes through the given points: (−6, 2) and (−3, 4).

5. Find the x and y intercepts of the following straight lines

  (*i*)   3x + 8y = 24

  (*ii*)   8x + 3y = 24

  (*iii*)   16x + 3y = 48,

  (*iv*)   3x + y = 6

Answers:

    1. $6x - 5y = 30$,   **2.** 6 and 2   **3.** 4x − 5y =20 **4.** (−3, −4)

    5. (*i*) (8,3) , (*ii*) (3, 8) (*iii*) (3, 16) (*iv*) (2, 6).

1.11   EQUATION OF A STRAIGHT LINE: (when Intercept and slope are given)

  (*i*)   Equation of a line which cuts off an intercept c from y-axis and is inclined at an angle θ with x-axis is given by $y = mx + c \ (m = \tan\theta)$.

  (*ii*)   A line passing through the origin (0, 0) cuts no intercept from y-axis i.e. c = 0, hence the equation is represented as $y = mx$

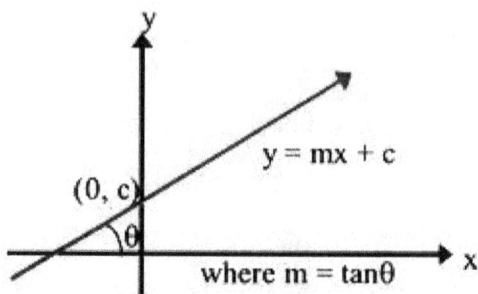

(*iii*) To find out the intercept c and the angle of inclination θ of a given equation, first convert the equation in the form $y = mx + c$ (the coefficient of y is unity)

Solved examples

1. Find the equation of a straight line which cuts off an intercept of 5 units from y-axis and makes with x-axis an angle of 30° .

Solution: Given that c = 5 and $m = \tan 30 = \dfrac{1}{\sqrt{3}}$

Hence required equation $y = \dfrac{1}{\sqrt{3}} x + 5 \Rightarrow \sqrt{3} y = x + 5\sqrt{3}$

2. Find the length of the intercept on y-axis and the angle with x-axis of the line $y = x - \dfrac{3}{2}$

Solution: Equation of the line is $y = x - \dfrac{3}{2}$

Comparing the given equation with $y = mx + c$, we find m = 1 and $c = -\dfrac{3}{2}$

$m = \tan\theta = 1 \Rightarrow \tan\theta = \tan 45$ or $\theta = 45^0$

Hence the given line cuts off an intercept of c = $\dfrac{3}{2}$ from the negative side of y-axis, and its inclination on x-axis is 45^0.

3. Find the equation of the straight line which cuts off an intercept of 3 from the y-axis and makes an angle of -60^0 with x-axis.

Solution: Here c = 3 and $m = \tan(-60) = -\tan 60 = -\sqrt{3}$

Now using $y = mx + c$, the required equation of the straight line is $y = -\sqrt{3}x + 3$.

4. Find the inclination and intercept from y-axis cut off by the following equations:

    (*i*) $y - x\sqrt{3} + 4 = 0$ (*ii*) $x - y = 3$

Solution; (*i*) The given equation is $y - x\sqrt{3} + 4 = 0$. This can be written as $y = x\sqrt{3} - 4$

Comparing it with $y = mx + c$, we get $m = \sqrt{3}$ and $c = -4$

$$m = \sqrt{3} = \tan 60 \Rightarrow \theta = 60^0$$

    Hence the given line is inclined at 60° on x-axis and cuts off an intercept of 4 units from negative side of y-axis.

    (*ii*)   The given equation is $x - y = 3$. This can be written $y = x - 3$

         Comparing it with $y = mx + c$, we get $m = 1$ and $c = -3$

$$m = 1 = \tan 45 \Rightarrow \theta = 45$$

Practice Problems: (Straight Line in slope-intercept form)

1. Find the equation of the straight line that has slope $m = 4$ and passes through the point (–1, –6).

2. Find the slope and y-intercept for the equation $3y = -9x + 15$.

3. Find the equation of the line whose slope is 4 and crosses the y-axis at (0,2).

4. Find the intercept of the line which passes through the point (-2, 3) and makes an with the angle of 45^0 with positive direction of x axis. (Hint: $m = \tan\theta = \tan 45 = 1$)

5. Find the equation of a line whose (i) gradient = -1, y-intercept = 3 and (ii) slope = – 2/7, y-intercept = -3.

Answers:

    1. $y = 4x - 2$ 2. m = -3, C = 5,   3. $y = 4x + 2$

    4. C = 5   5. (i) x + y -3 = 0, (ii) 2x + 7y +21 = 0

1.12   EQUATION OF STRAIGHT LINE (Parallel or Perpendicular to a Given Line)

Parallel Lines   Perpendicular Lines

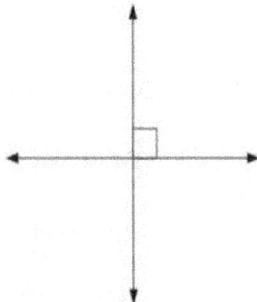

Solved Examples

1. Find the equation of a straight line parallel to $3x-4y=7$ and passes through the point $(1, 1)$

Solution: The given equation can be expressed as $-4y=-3x+7$ or $y=\dfrac{3}{4}x-7$

Comparing it with $y=mx+c$

Slope of the given line is $m=\dfrac{3}{4}$

The line passing through $(1, 1)$ is parallel to given line, hence its slope is also $m=\dfrac{3}{4}$

Therefore the equation of line from $y-y_1=m(x-x_1)$ is:

$$y-1=\dfrac{3}{4}(x-1) \Rightarrow 4y-3x-1=0$$

2. Find the equation of a straight line perpendicular to the line $7x+8y=5$ and passing through the point $(-6, 10)$.

Solution: The slope of the given line is

$$m_1=\dfrac{-\text{coeff. of } x}{\text{coeff. of } y}=\dfrac{-7}{8}$$

Let the slope the line perpendicular to given line be m_2

If two lines are perpendicular, then

$$m_1.m_2=-1 \Rightarrow m_2=-1/m_1$$

Since $m_1=\dfrac{-7}{8}$. Hence $\Rightarrow m_2=8/7$

The equation of the straight line:

$$y-(-6)=\dfrac{8}{7}(x-10) \Rightarrow 7(y+6)=8(x-10) \Rightarrow 7y-8x=122$$

3. Let A $(6, 4)$ and B $(2, 12)$ be two points. Find the slope of a line perpendicular to AB.

Solution: Let m_1 be the slope of AB. $\therefore m_1=\dfrac{y_2-y_1}{x_2-x_1}=\dfrac{12-4}{2-6}=\dfrac{8}{-4}=-2$

If m_2 be the slope of perpendicular lime then we know that

$$m_1m_2=-1 \Rightarrow m_2=\dfrac{-1}{m_1}=\dfrac{-1}{-2}=\dfrac{1}{2}$$

4. If 2 is the slope of line passing through (2, 5) and (x, 3), then find the value of x.

Solution: Let m_1 be the slope the line passing through (2, 5) and (x, 3)

Then
$$m_1 = \frac{y_2 - y_1}{x_2 - x_1} = \frac{3-5}{x-2} = \frac{-2}{x-2}$$

or
$$= \frac{-2}{x-2} \Rightarrow x-2 = -1 \text{ or } x = 1$$

5. If the line through (3, y) and (2, 7) is parallel to the line through (-1, 4) and (0, 6). Find the value of y.

Solution: Let m_1 and m_2 be the slopes of the lines.

Then
$$m_1 = \frac{y_2 - y_1}{x_2 - x_1} = \frac{7-y}{2-3} = \frac{(7-y)}{-1} = (y-7)$$

$$m_2 = \frac{y_2 - y_1}{x_2 - x_1} = \frac{6-4}{0-(-1)} = \frac{2}{+1} = 2$$

Since the lines are parallel, therefore

$$m_1 = m_2 \Rightarrow (y-7) = 2 \Rightarrow y = 9$$

Practice Problems (Line Parallel or Perpendicular to a line)

1. Is the straight line $y = 3x + 2$ is parallel to the straight line $2y + 3x = 3$? Explain.

2. Two lines are perpendicular. If the slope of first line is $\frac{3}{4}$, and the slope of second is $\frac{8}{k+6}$. Find the value of k. (Hint.: $m_1.m_2 = -1$)

3. Find the slope of a line parallel to a line whose slope is. $\frac{-2}{3}$ (Hint.: $m_1.m_2 = -1$)

4. Find the slope of a line parallel to the line whose equation is $3y + 2x = 6$.

5. Find the equation of a line perpendicular to the line $3y + 2x = 6$ and passes through (3,5).

Answers:

1. No,   **2.** x = -12,   **3.** $\frac{3}{2}$,

4. $\frac{-2}{3}$ **5.** $(y-5) = \frac{3}{2}(x-3)$

1.13 LENGTH OF PERPENDICULAR

Length of perpendicular from the point (x', y') on the line $ax + by + \lambda = 0$ is given by:

$$p = \frac{|ax' + by' + \lambda|}{\sqrt{a^2 + b^2}}$$

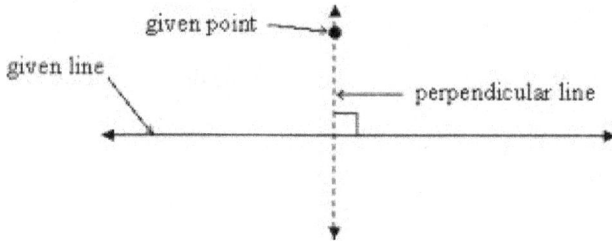

Solved Examples

1. Find the length of perpendicular from the point $(-2, 1)$ on the line $3x - 4y - 1 = 0$.

Solution: $x' = -2$, $y' = 1$, $a = 3$, $b = -4$, $\lambda = -1$

$$\therefore p = \frac{|ax' + by' + \lambda|}{\sqrt{a^2 + b^2}} = \frac{|3(-2) + (-4)(1) + (-1)|}{\sqrt{3^2 + 4^2}} = \frac{|-6 - 4 - 1|}{\sqrt{25}} = \frac{11}{5}$$

2. Find the length of the perpendicular drawn from $(-4, 1)$ on $3(x - 4) = 4(y + 3)$

Solution: The given equation of the line is:

$$3(x - 4) = 4(y + 3)$$
$$3x - 12 = 4y + 12$$

or $\qquad\qquad\qquad\qquad 3x - 4y - 24 = 0$

$x' = -4$, $y' = 1$, $a = 3$, $b = -4$, $\lambda = -24$

$$\therefore p = \frac{|ax' + by' + \lambda|}{\sqrt{a^2 + b^2}} = \frac{|3(-4) + (-4)(1) + (-24)|}{\sqrt{3^2 + 4^2}} = \frac{|-12 - 4 - 24|}{\sqrt{5}} = \frac{40}{8} = 8$$

3. If p is the perpendicular from the origin to the line $\dfrac{x}{a} + \dfrac{y}{b} = 1$. Then prove that:

$$\frac{1}{p^2} = \frac{1}{a^2} + \frac{1}{b^2}$$

Solution: $x' = 0$, $y' = 0$, $a = \dfrac{1}{a}$, $b = \dfrac{1}{b}$, $\lambda = -1$

$$\therefore p = \frac{|ax' + by' + \lambda|}{\sqrt{a^2 + b^2}} = \frac{\left|\left(\frac{1}{a}\right)(0) + \left(\frac{1}{b}\right)(0) + (-1)\right|}{\sqrt{\left(\frac{1}{a}\right)^2 + \left(\frac{1}{b}\right)^2}} = \frac{|-1|}{\sqrt{\left(\frac{1}{a}\right)^2 + \left(\frac{1}{b}\right)^2}}$$

$$\therefore p = \frac{1}{\sqrt{\left(\frac{1}{a}\right)^2 + \left(\frac{1}{b}\right)^2}}$$

Squaring both sides $p^2 = \dfrac{1}{\left(\dfrac{1}{a^2} + \dfrac{1}{b^2}\right)}$

$$\therefore \left(\frac{1}{a^2} + \frac{1}{b^2}\right) = \frac{1}{p^2} \quad \text{Hence Proved.}$$

4. Find the perpendicular distance between two parallel lines $3x - 2y = 1$ and $6x + 9 = 4y$.

Solution: The equations in required forms are

$$3x - 2y - 1 = 0 \qquad\qquad\qquad ...(1)$$

$$6x - 4y + 9 = 0 \qquad\qquad\qquad ...(2)$$

Put x = 1 in (1) we get

$$3 - 2y - 1 = 0 \Rightarrow 2y = 2 \Rightarrow y = 1$$

Hence point **(1, 1) lies on** $3x - 2y - 1 = 0$

Now we have to calculate the length of perpendicular from (1, 1) to the line $6x - 4y + 9 = 0$.

$x' = 1$, $y' = 1$, $a = 6$, $b = -4$, $\lambda = 9$

$$\therefore p = \frac{|ax' + by' + \lambda|}{\sqrt{a^2 + b^2}} = \frac{|6 \times 1 - 4 \times 1 + 9|}{\sqrt{6^2 + 4^2}} = \frac{11}{\sqrt{52}} = \frac{11}{2\sqrt{13}}$$

Practice Problems: (On Length of perpendicular from a point on a line).

1. Find the value of c if the perpendicular from (3, 1) to a line 4x + 3y + c = 0 is equal to 5

2. Find the equation of the perpendicular to the line 3x −4y +8 = 0 and passing through (3,4)

3. Find the perpendicular distance if the point (4, 2) from the line 2x − 3y + 8 = 0

4. Find the length of perpendicular from the point (5, -3) to the line joining points (-2,-6) and (8,4)

5. Find the equation of the line parallel to the line to $5x + 2y - 3 = 0$ from the point $(5, 2)$

Answers:

1. 10   2. $4x + 3y - 24 = 0$,   3. $\dfrac{10}{\sqrt{3}}$ units   4. $2\sqrt{2}$ units

5. $5x + 2y - 29$.

1.14   EQUATION OF A CIRCLE

The equation of circle with centre at (h, k) and radius r is given by:

$$(x-h)^2 + (y-k)^2 = r^2$$

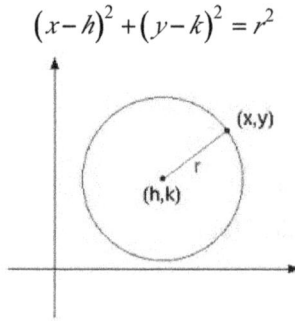

Solved Examples

1. Find the equation of a circle whose centre is (3, -4) and radius is 5.

Solution: The equation of the circle is:

$$(x-h)^2 + (y-k)^2 = r^2$$

Given: Radius r = 5 h = 3 and k = – 4.

Substituting these values in the above equation:

$$(x-h)^2 + (y-k)^2 = r^2$$

We get $(x-3)^2 + (y-(-4))^2 = 5^2$

or $\left(x^2 - 6x + 9\right) + \left(y^2 + 8y + 16\right) = 25$

or $x^2 + y^2 - 6x + 8x = 0$

Hence the required equation of circle is $x^2 + y^2 - 6x + 8x = 0$.

2. Given Radius a = 5 h = 0 and k = 3. Find the equation of the circle.

Solution: The equation of circle is:

$$(x-0)^2 + (y-3)^2 = 5^2$$

or $$\left(x^2\right)+\left(y^2-6y+9\right)=25$$

or $$x^2+y^2-6y=16$$

Hence the required equation of circle is $x^2+y^2-6y=16$

3. Find the equation of a circle whose radius is (a − b) and centre is at (a, b)

Solution: Given Radius = (a − b) Coordinates of centre are: h = a and k = b

The equation of circle is:

$$(x-a)^2+(y-b)^2=(a-b)^2$$

or $$\left(x^2-2ax+a^2\right)+\left(y^2-2by+b^2\right)=(a^2-2ab+b^2)$$

or $$x^2+y^2-2ax-2by+2ab=0$$

Hence the required equation of circle is $x^2+y^2-2ax-2by+2ab=0$

4. Find the center and radius of the circle: $(x-2)^2+(y+3)^2=16$.

Solution: Comparing the given equation with the standard form $(x-h)^2+(y-k)^2=r^2$

We find $r^2=16$.

Hence radius $$r=\sqrt{16}=4$$
$$-h=-2 \text{ or } h=2 \text{ and } -k=+3 \text{ or } k=-3.$$

Hence coordinates of centre are (2, −3) and r =4.

5. Convert $x^2+y^2-6x+4y+9=0$ into center-radius form.

Solution: $$x^2-6x+9-9+y^2+4y+4-4+9=0$$
$$(x-3)^2+(y+2)^2=4$$

This is the require centre-radius form

Comparing with $(x-h)^2+(y-k)^2=r^2$

Centre is (3, -2) and radius is 2.

6. Write the equation of a circle whose diameter has endpoints (4, −1) and (−6, 7).

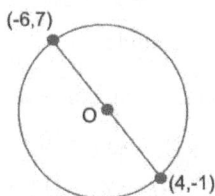

Coordinates of centre O are:

$$\frac{-6+4}{2} = -1 \text{ and } \frac{7-1}{2} = 3$$

Length of diameter is $D = \sqrt{(-6-4)^2 + (7+1)^2} = \sqrt{100+64} = 2\sqrt{41}$

Hence radius $r = \dfrac{2\sqrt{41}}{2} = \sqrt{41}$.

Equation of the circle: $(x+1)^2 + (y-3)^2 = \left(\sqrt{41}\right)^2 = 41$

1.15 EQUATION OF CIRCLES (when endpoints of diameter are given):

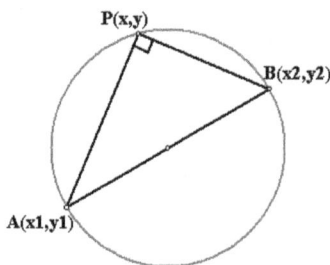

If (x_1, y_1) and (x_2, y_2) are the coordinates of end points of a diameter AB of a circle, then its equation is $(x - x_1)(x - x_2) + (y - y_1)(y - y_2) = 0$

Solved Examples

1. Find the equation of a circle with end points of its diameter as: $(-5, 5)$ and $(5, -5)$.

Solution: If (x_1, y_1) and (x_2, y_2) are the end points of the diameter, then equation of the circle is: $(x - x_1)(x - x_2) + (y - y_1)(y - y_2) = 0$

$$x_1 = -5, y_1 = 5 \; x_2 = 5, y_2 = -5$$

Substituting the given coordinates in the formula:

$$(x - x_1)(x - x_2) + (y - y_1)(y - y_2) = 0$$

We get: $(x + 5)(x - 5) + (y - 5)(y + 5) = 0$

$$x^2 - 25 + y^2 - 25 = 0$$

$$x^2 + y^2 = 50$$

This is the required equation of the circle.

2. Find the equation of a circle with end points of its diameter as: $(2, -3)$ and $(-8, 10)$.

Solution: Here $x_1 = 2$, $x_2 = -8$ and $y_1 = -3$, $y_2 = 10$

Equation of circle is: $(x - x_1)(x - x_2) + (y - y_1)(y - y_2) = 0$

$$(x-2)(x+8)+(y+3)(y-10)=0$$

$$\Rightarrow \qquad x^2+y^2+6x-7y=46$$

2.16 RELATIVE POSITION OF A POINT WITH REGARD TO THE CIRCLE

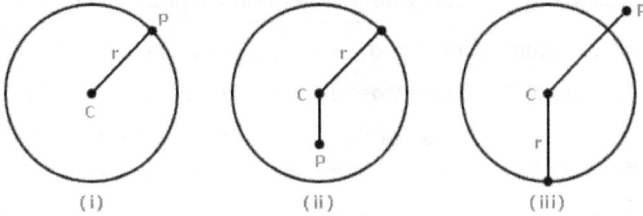

(i)　　　　　　(ii)　　　　　　(iii)

Let $P(x_1,y_1)$ be the point and $x^2+y^2+2gx+2fy+c=0$ be the equation of the circle.

Substitute the value of coordinates (x_1,y_1) in the given equation:

(*i*) If $x_1^2+y_1^2+2gx_1+2fy_1+c=0 \Rightarrow$ the point is exactly on the given circle.

(*ii*) If $x_1^2+y_1^2+2gx_1+2fy_1+c<0 \Rightarrow$ the point is inside the given circle.

(*iii*) If $x_1^2+y_1^2+2gx_1+2fy_1+c>0 \Rightarrow$ the point is outside the given circle.

Solved Examples

1. Determine, whether the point (3, 4) lies outside, inside or on the circle:

$$x^2+y^2+2x+2y-7=0$$

Solution: Substituting the values (3, 4) in $x^2+y^2+2x+2y-7$

$$3^2+4^2+2.3+2.4-7=9+16+6+8-7=32>0$$

The point (3,4) is outside the circle.

2. Discuss the position of the point (–2, 3) with respect to the circle $x^2+y^2=16$

Solution: Substituting the values (3, 4) in x^2+y^2-16

$$x^2+y^2-16=2^2+3^2-16=4+9-16=-3<0 .$$

Hence the given point lies inside the circle.

3. Find whether the point (2, 3) lies inside, outside or on the circle: $x^2+y^2-4x+2y=11$

Solution: Substituting the coordinates (2, 3) in equation $x^2+y^2-4x+2y=11$

We get $\quad x^2+y^2-4x+2y-11=2^2+3^2-4.2+2.3-11=19-19=0$

∴ The point lies on the circle

Practice Problems: (Equation of Circle)

1. Find the equation of a circle with centre (4, –2) and radius 3.

2. Find the equation of a circle whose center is at the point (–2, 3) and its diameter has a length of 10. (hint: h = –2, k = 3 and r = 5)

3. Find an equation of the circle whose center is at the point (–4, 6) and passes through the point (1 , 2).(hint: radius $r = \sqrt{(-4-1)^2 + (6-2)^2} = \sqrt{25+16} = \sqrt{41} \Rightarrow r = 41$)

4. Find an equation of the circle whose diameter has endpoints at (–5 , 2) and (3 , 6).

5. Find the points of intersection of the circle with equation $(x-2)^2 + (y-6)^2 = 40$ and the line with equation y = 3x (hint: put y =3x in the equation of the circle).

6. Find the centre and radius of the circle whose equation is $x^2 + y^2 = 16$

7. Find the centre and radius of the circle whose equation is $(x-2)^2 + (y+1)^2 = 16$

8. Discuss the position of the points A (1,1) and B (5,1) with respect to the circle:

$$x^2 + y^2 + 2x + 2y - 7 = 0$$

Answers:

1. $x^2 + y^2 - 8x + 4y + 11 = 0$,
2. $(x+2)^2 + (y-3)^2 = 25$

3. $(x+4)^2 + (y-6)^2 = 41$,
4. $x^2 + y^2 + 2x - 8y - 3 = 0$

5. $x = 5, \dfrac{4}{5}$ and $y = 15, \dfrac{12}{5}$
6. Center (0, 0) and radius = 4

7. Center (2, –1), radius = 4
8. A-Inside, B-outside.

1.17 PARABOLA

A parabola is the locus (path) of appoint P which moves in such a way that its distance from a fixed point S (called focus) and its perpendicular distance from a straight line called directrix bears a constant ratio.

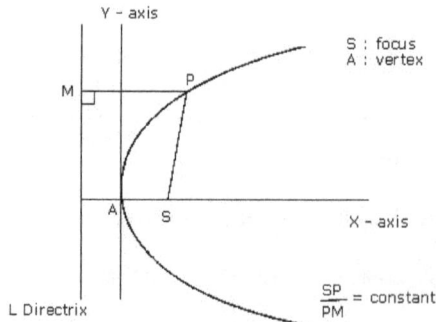

1.18   SOME TERMS OF PARABOLA

Axis of Parabola: The straight line passing through the focus and perpendicular to a line (known as directrix) is called the axis of parabola. It is also known as *Axis of symmetry*.

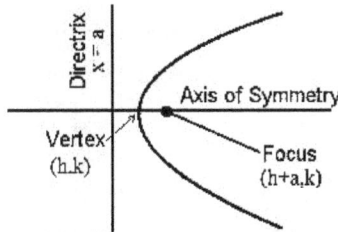

Focus: The fixed point on the axis of parabola is called the focus.

Vertex: The point of intersection of parabola and its axis is called the vertex. Its distances from focus and directrix are equal.

Directrix: The straight line perpendicular to the axis of parabola such that its distance from the vertex is equal to the distance of vertex from the focus.

Distance between vertex and focus = Distance between vertex and directrix

Eccentricity: The constant ratio between the distance of the point on parabola from the focus and its perpendicular distance from the directrix.is known as the *eccentricity.*

Letus rectum: The chord of a parabola passing through the focus and perpendicular to the axis of parabola is called the letus rectum.

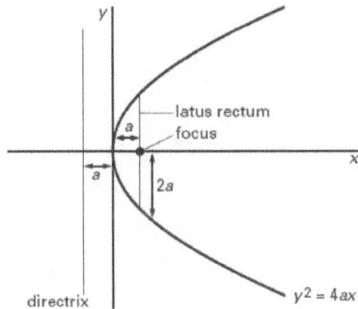

Let (x, y) be the coordinates of the moving point, say P in any position . The locus of the point P is then described by the equation - $y^2 = 4ax$ where **a** is the distance of vertex from the focus.

This is equation of parabola.

1.19 HORIZONTAL PARABOLA

If the Axis of symmetry is along *x*- **axis**, then the equation of parabola is $y^2 = 4ax$.

This is known as horizontal parabola

If $a < 0$ i.e. negative then the parabola opens to left.

If $a > 0$ i.e. positive then the parabola opens to right

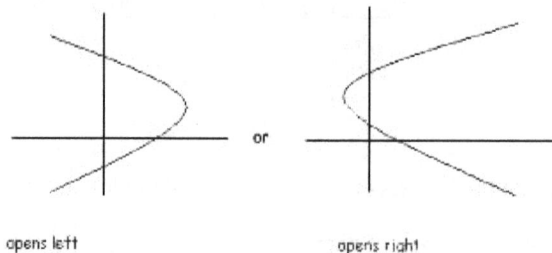

or

opens left   opens right

1.20 VERTICAL PARABOLA

If the Axis of symmetry is along *y*- **axis**, then the equation of parabola is $x^2 = 4ay$

This is known as vertical parabola

If $a < 0$ i.e. negative then the parabola opens downward.

If $a > 0$ i.e. positive then the parabola opens upward.

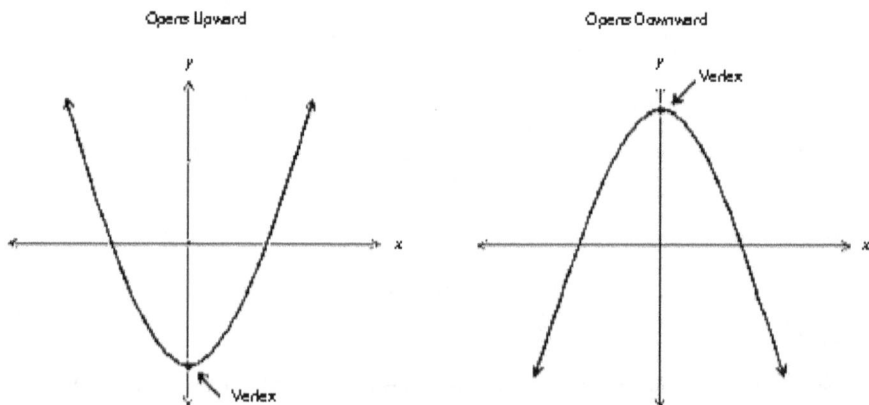

Opens Upward   Opens Downward

Solved Examples:

1. Express the following equation of parabola in standard form: $6y^2 - 12x = 0$. Also find the focus and directrix.

Solution: $6y^2 - 12x = 0$

The above equation can be written as:

$$6y^2 = 12x \text{ or } y^2 = 2x$$

Comparing with the standard form $y^2 = 4ax$

We get $4a = 2 \Rightarrow a = \dfrac{1}{2}$

Focus of parabola is given by : (a,0). Hence focus of given parabola is $\left(\dfrac{1}{2}, 0\right)$

Equation of directrix is given by: $X + a = 0$.

Hence directrix of given parabola is $X + \dfrac{1}{2} = 0$

The Length of letus rectum of parabola is given by:

$$4a = 4.\dfrac{1}{2} = 2$$

2. Find the focus and the letus rectum of the parabola: $y^2 = 8x$

Solution: Comparing with $y^2 = 4ax$

We get $\qquad\qquad\qquad 4a = 8 \Rightarrow a = 2$

Focus is given by (a, 0) i.e. (2, 0)

Letus rectum = 4a = 4 × 2 = 8

3. Find the equation of parabola whose focus is at (0, a) and directrix is $y + a = 0$.

Solution: Let (x', y') be any point on the parabola.

Its distance from the focus $= \sqrt{(x'-0)^2 + (y'-a)^2} = \sqrt{x'^2 + (y'-a)^2}$

Its distance from the directrix $= \dfrac{y'+a}{\sqrt{1+0}} = y' + a$

Since distances of P from focus and the directrix are equal:

$$\sqrt{x'^2 + (y'-a)^2} = y' + a$$

Squaring both sides we get:

$$x'^2 + (y'-a)^2 = (y'+a)^2$$

$$x'^2 + y'^2 - 2ay' + a^2 = y'^2 + 2ay' + a^2$$

$$x'^2 = 4ay'$$

Hence equation of the parabola is $x^2 = 4ay$. The parabola is on y-axis.

* The perpendicular distance of a point (x', y') from the line $ax + by + c - 0$ is given by:

$$p = \frac{ax' + by' + c}{\sqrt{a^2 + b^2}}$$

4. Determine the point of intersection of straight line $x + y = 5$ and the parabola $y^2 = 16x$

Solution: $x + y = 5 \Rightarrow x = 5 - y$

Substituting the value of x in the equation of parabola:

$$y^2 = 16(5 - y)$$
$$y^2 + 16y - 80 = 0$$
$$y^2 + 20y - 4y - 80 = 0$$
$$y(y + 20) - 4(y + 20) = 0$$
$$(y - 4)(y + 20) = 0$$
$$\Leftrightarrow y = 4 \text{ or } -20$$

The equation of parabola: $x = \dfrac{y^2}{16}$

When y = 4 then $x = \dfrac{4^2}{16} = 1$

When y = −20 then $x = \dfrac{(-20)^2}{16} = \dfrac{400}{16} = 25$

Hence points of intersections are (1, 4) and (25, −20).

5. Find the equation of horizontal parabola, which passes through the point (3, 4) and vertex (0, 0).

Solution: The equation of parabola vertex is at (0, 0) is $(y - 0)^2 = 4a(x - 0)$

Since it passes through the point (3, 4) hence we get:

$$(4 - 0)^2 = 4a(3 - 0) \Rightarrow a = \frac{16}{12} = \frac{4}{3}$$

The required equation of parabola is

$$y^2 = 4 \cdot \frac{4}{3} x = \left(\frac{16}{3}\right) x$$

6. State whether the parabola $y = -5x^2$ is vertical or horizontal?

Solution : The given parabola can be written as $x^2 = -\dfrac{1}{5} y = 4\left(\dfrac{-1}{20}\right) y$

This indicates that axis of symmetry is the y-axis. Hence the parabola is vertical. Since $a < 0$ hence its opening is downward.

7. Find whether the parabola: $7x^2 = 8y$ opens up, down, left or right?

Solution: The given parabola can be written as $x^2 = \dfrac{8}{7}y$.

This shows that axis of symmetry is y-axis

Comparing with standard form of parabola $x^2 = 4ay$

We find $4a = \dfrac{8}{7} \Rightarrow a = \dfrac{2}{7}$

This shows that a > 0

Thus the parabola opens up.

8. The focal distance of a point (x, y) on the parabola: $y^2 = 8x$ is 4. Find the values of x and y.

Solution : Comparing the given parabola with the standard form: $y^2 = 4ax$

We get $4a = 8 \Rightarrow a = 2$

We know that the focal distance of a point (x, y) on the parabola $y^2 = 4ax$ is given by $x + a$

Hence focal distance is $x + 2 = 4 \therefore x = 2$

Putting the value of $x = 2$ in $y^2 = 8x$

we get $y^2 = 8 \times 2 = 16 \Rightarrow y = \pm 4$

Hence coordinates of the point are ($2, \pm 4$).

9. Find the vertex of the parabola: $y^2 - 2y + 6x + 13 = 0$

Solution: The given parabola can be written as:

$$y^2 - 2y = -6x - 13$$

$$\left(y^2 - 2y + 1\right) - 1 = -6x - 13$$

$$(y-1)^2 = -6x - 13 + 1 = -6(x+2)$$

If vertex is (h, k), then the equation of parabola is given by:

$$(y-k)^2 = 4a(x-h)$$

Comparing the above equation with the given equation, $(y-1)^2 = -6(x+2)$

We get $k = 1$ and $h = -2$.

Hence coordinates of vertex are (–2, 1).

Practice Problems: (On parabola)

1. Find the coordinates of focus of the parabolas;

   (i) $y^2 = 8x$ **(ii)** $x^2 = 6y$ **(iii)** $y^2 = -12x$ **(iv)** $x^2 = -16y$

2. Determine the coordinates of focus and vertex of the parabola: $y^2 - 8y - x + 19 = 0$

3. At what point on the parabola: $x^2 = 9y$ its abscissa is three times of ordinate.

4. Find the vertex, focus and the *letus rectum* of the parabola: $y^2 = 8x$

5. Find whether the parabola: $7x^2 = 8y$ opens up, down, left or right.

6. Find the focus and directrix of parabola: $x^2 = -24y$

7. Find the equation of parabola whose focus is (3, 0) and directrix is x = -3

8. Find the equation of the horizontal parabola which passes through the point (3, 4) and its vertex is (0, 0).

9. Show that the parabola: $9y^2 = x$ is horizontal.

Answers:

1. (i) (2, 0) , **(ii)** (0,3/2) , **(iii)** (-3,0) , **(iv)** (0-4)

2. F (1/4, 0) , V (3, 4)   **3.** (3, 1)   **4.** V(0, 0), F((2, 0) , LR = 8

5. Opens up.   **6.** Focus: (0, -6) , Directrix: y = 6   **7.** $y^2 = 12x$

8. $3y^2 = 16x$

1.21   ELLIPSE

An ellipse is the locus (path) of a point P which moves in such manner that its distance from fixed point S (called the focus) bears to its distance from a fixed straight line NZ (called directrix) a constant ratio.

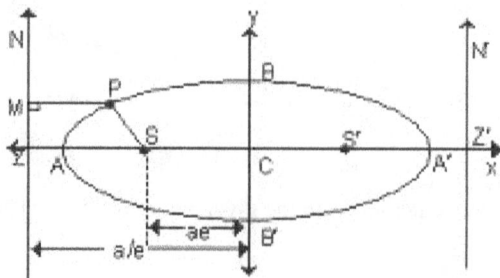

1.22 SOME IMPORTANT TERMS

1. **Major and Minor Axes**: The axes are denoted by a and b. They are along the x-axis and y-axis respectively. If **a > b**, then **a** is the major axis and the ellipse is horizontal.

   If **b > a** , then **b** is the major axis along the y-axis and the ellipse is vertical.

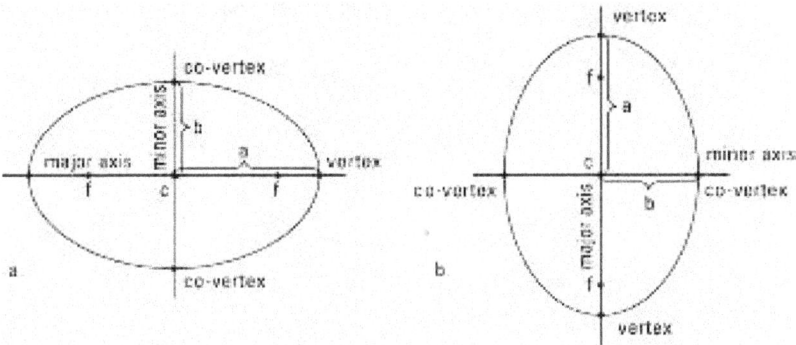

2. **Vertex and Co-vertex:** The end points of major axis are known as vertices and the end points of minor axis are known as co-vertices. In case of horizontal ellipse major axis is along the axis. Then coordinates of vertices are (a, 0) and (-a, 0)

   In case of horizontal ellipse major axis is along the axis. Then coordinates of vertices are (a, 0) and (-a, 0)

   In case of vertical ellipse major axis is along the y-axis. Then coordinates of vertices are (0,b) and (0, -b)

3. **Foci (Plural of focus):** There are two foci. Foci are always on major axis. In case of horizontal ellipse foci will be on x-axis. The coordinates are (ae, 0) and (-ae,0). In case of vertical axis foci will be on y-axis. The coordinates are (0, be) and (0, -be).

   Distance between foci is **2ae.**

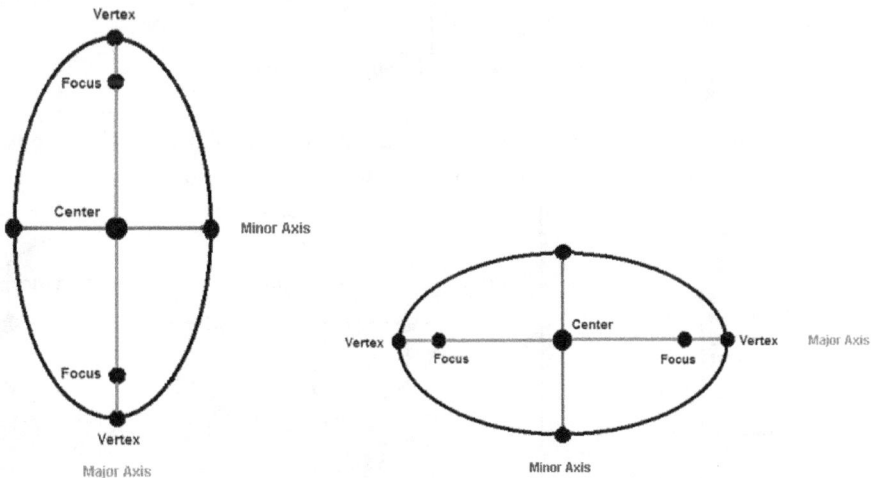

4. **Focal Distance:** The distances of a point from the foci are known as focal distances. The sum of focal distances of a point on the ellipse is equal to the length of major axis i.e. 2a or 2b

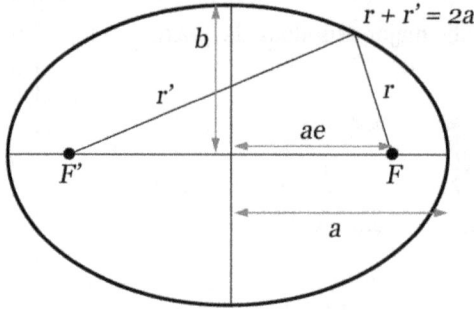

$a = semimajor\ axis$
$b = semiminor\ axis$
$e = eccentricity$
$F\ and\ F' = focal\ points$

5. **Directrix:** There are two directrices of an ellipse. The distances of directrices from the vertices are $X = \pm\ ae$ in case of horizontal ellipse and $Y = \pm\ be$ in case of vertical ellipse.

$$P_1 F_1 = e P_1 D_1$$

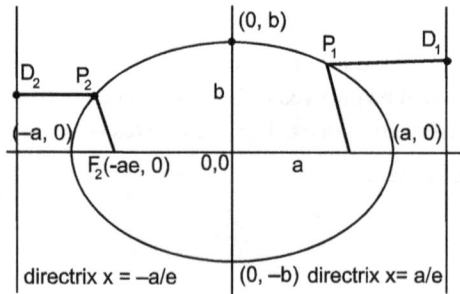

6. **Letus rectum**: A chord passing through a focus and perpendicular to the major axis is called the letus- rectum. Its length is given by $2b^2/a$.

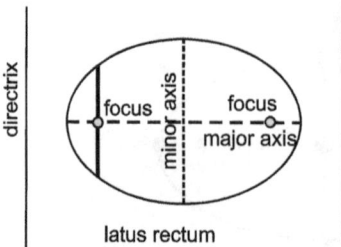

Fig. 1

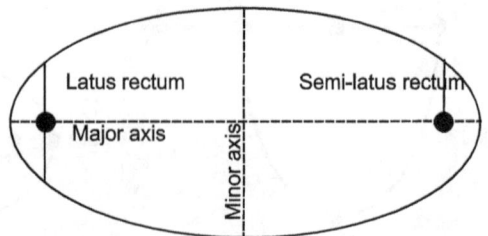

Fig. 2

7. Eccentricity: It is denoted by e . It is given by

$$e = \sqrt{1 - \left(\frac{\text{semi minor axis}}{\text{semi major axis}}\right)^2}$$

$$e = \sqrt{1 - \frac{b^2}{a^2}} \quad a > b \qquad \qquad \text{(horizontal)}$$

$$e = \sqrt{1 - \frac{a^2}{b^2}} \quad b > a \qquad \qquad \text{(vertical)}$$

Equation of ellipse: $\dfrac{x^2}{a^2} + \dfrac{y^2}{b^2} = 1$

Where a = semi major axis and b = semi minor axis .

(Semi minor axis)^2 = (Semi major axis)^2 $(1 - e^2)$

or $b^2 = a^2 \left(1 - e^2\right)$ e is called the eccentricity which is < 1

Points to remember:

The foci are given by (ae, 0) and (-ae, 0) if major axis is along x-axis.

The foci are given by (0, be) and (0, -be) if major axis is along y-axis.

The equation of directrix are $X \pm \dfrac{a}{e} = 0$ if major axis is along x-axis.

The equation of directrix are $Y \pm \dfrac{b}{e} = 0$ if major axis is along y-axis.

The length of letus rectum $= \dfrac{2b^2}{a}$ if a > b i.e. horizontal ellipse

The length of letus rectum $= \dfrac{2a^2}{b}$ if b > a i.e. vertical ellipse

The length of major axis = 2a if a > b i.e. horizontal ellipse

The length of major axis = 2b if b > a i.e. vertical ellipse

Solved Examples

1. Find the equation of the ellipse whose major axis is 3 and minor axis is 8/5

Solution: $\qquad\qquad\qquad\qquad$ 2a = 3 $\qquad\qquad$ ∴ $\quad$ a = 3/2

$\qquad\qquad\qquad\qquad\qquad\quad$ 2b = 8/3 $\qquad\qquad$ ∴ $\quad$ b = 4/3

$\qquad\qquad\qquad\qquad\qquad\quad$ a > b

Substituting the values of a and b in the equation of the ellipse

$$\frac{x^2}{a^2} + \frac{y^2}{b^2} = 1$$

We get the required equation of ellipse:

$$\frac{x^2}{(3/2)^2} + \frac{y^2}{(4/3)^2} = 1$$

$$\frac{4x^2}{9} + \frac{9y^2}{16} = 1$$

2. Find the eccentricity of the ellipse:

$$\frac{x^2}{9} + \frac{y^2}{25} = 1$$

Solution: Comparing the given equation with standard equation $\frac{x^2}{a^2} + \frac{y^2}{b^2} = 1$

We find $a^2 = 9$ and $b^2 = 25$ $b > a$

The eccentricity $e = \sqrt{1 - \frac{a^2}{b^2}}$ $\because$ $b >$ (it is a vertical ellipse)

$$e = \sqrt{1 - \frac{9}{25}} = \sqrt{\frac{16}{25}} = \frac{4}{5}$$

3. The equation of ellipse is $4x^2 + 6y^2 - 9 = 0$. Find the length of the axes and eccentricity.

Solution: Writing the equation in standard for; $\frac{x^2}{9/4} + \frac{y^2}{9/6} = 1$

We see that $a^2 = 9/4 \Rightarrow a = 3/2$ and $b^2 = 9/6 = 3/2 \Rightarrow b = \sqrt{3/2}$

a > b (it is a horizontal ellipse)

The length of major axis $= 2a = 2 \times \frac{3}{2} = 3$

The length of major axis $= 2b = 2 \times \sqrt{\frac{3}{2}} = \sqrt{2}\sqrt{3} = \sqrt{6}$

The eccentricity $e = \sqrt{1 - \frac{b^2}{a^2}}$ $a > b$

$$e = \sqrt{1 - \frac{9/6}{9/4}} = \sqrt{1 - \frac{4}{6}} = \sqrt{\frac{2}{6}} = \sqrt{\frac{1}{3}}$$

4. The letus rectum is half of the major axis of the ellipse. Find the eccentricity.

Solution: We know that major axis = 2a and letus rectum = $2b^2/a$

It is given that letus rectum is half of its major axis.

$$\frac{2b^2}{a} = \frac{2a}{2} \Rightarrow b^2 = \frac{a^2}{2} \Rightarrow \frac{b^2}{a^2} = \frac{1}{2}$$

The eccentricity $\qquad\qquad e = \sqrt{1-\frac{b^2}{a^2}}\qquad\qquad$ (it is a horizontal ellipse, a > b)

$$= \sqrt{1-\frac{1}{2}} = \sqrt{\frac{1}{2}}$$

5. Show that if eccentricity is zero, then ellipse becomes a circle.

Solution: we know that $b^2 = a^2\left(1-e^2\right)$

Since e = 0 $\qquad\qquad\qquad b^2 = a^2\left(1-0\right)$

$$b^2 = a^2$$

The equation of the ellipse $\frac{x^2}{a^2}+\frac{y^2}{b^2}=1$ becomes $\frac{x^2}{a^2}+\frac{y^2}{a^2}=1 \Rightarrow x^2+y^2=a^2$

$x^2+y^2=a^2$ is the equation of a circle.

6. Find the equation of ellipse with centre (0, 0) and height is 10 units and width 20 units.

Solution: The equation of ellipse with centre (0, 0) is:

$$\frac{(x-0)^2}{a^2}+\frac{(y-0)^2}{b^2}=1$$

Its height i.e. 2b = 10 b = 5

Its width i.e. 2a = 20 a = 10

Now substituting the values of a and b in the standard form of ellipse, we get

$$\frac{x^2}{(10)^2}+\frac{y^2}{(5)^2}=1$$

$$\frac{x^2}{100}+\frac{y^2}{25}=1 \Rightarrow x^2+4y^2=100$$

7. The equation of the ellipse is $4x^2+9y^2=36$. Find its foci, eccentricity and letus rectum.

Solution: Dividing the given equation by 36 we get:

$$\frac{4x^2}{36}+\frac{9y^2}{36}=\frac{36}{36}$$

or
$$\frac{x^2}{9}+\frac{y^2}{4}=1$$

Comparing with the standard equation of ellipse:

$$a^2=9 \Rightarrow a=3 \text{ and } b^2=4 \Rightarrow b=2$$

Eccentricity $e=\sqrt{1-\frac{b^2}{a^2}}=\sqrt{1-\frac{4}{9}}=\sqrt{\frac{5}{9}}=\frac{\sqrt{5}}{3}$

Foci are given by (ae, 0) and (-ae, 0) i.e. $\left(3.\frac{\sqrt{5}}{3},0\right)$ and $\left(-3.\frac{\sqrt{5}}{3},0\right)$

Hence the foci are: $\left(\sqrt{5},0\right)$ and $\left(-\sqrt{5},0\right)$

Length of letus rectum $=\dfrac{2b^2}{a}=\dfrac{2.4}{3}=\dfrac{8}{3}$

8. Determine the equation of ellipse whose center is at (0,0) and passes through (2,1) and its minor axis is **8**.

Solution: The equation of ellipse with center (0,0) is:

$$\frac{x^2}{a^2}+\frac{y^2}{b^2}=1$$

Since it passes through the point (2, 1) hence we get:

$$\frac{2^2}{a^2}+\frac{1^2}{b^2}=1$$

Also given the minor axis; 2b = 8 ∴ b = 4 ⟹ b^2 = 16

Put b = 4 in the equation of ellipse, we get: $\dfrac{2^2}{a^2}+\dfrac{1^2}{4^2}=1$

$$\frac{2^2}{a^2}=1-\frac{1}{16}=\frac{15}{16}$$

$$\frac{a^2}{2^2}=\frac{15}{16}$$

$$\therefore a^2=\frac{15}{16}\times 4=\frac{15}{4}$$

Hence the required equation of ellipse is: $\dfrac{4x^2}{15}+\dfrac{y^2}{16}=1$

9. Determine the equation of ellipse whose center is at (0, 0) and it passes through (0, 4) and has eccentricity 3/5.

Solution: The equation of ellipse with center (0,0) is : $\dfrac{x^2}{a^2}+\dfrac{y^2}{b^2}=1$

Since it passes through the point (0, 4) hence we get: $\dfrac{0^2}{a^2}+\dfrac{4^2}{b^2}=1\Rightarrow b=4$

∴ $b=4\Rightarrow b^2=16$

As we know that
$$b^2=a^2\left(1-e^2\right)$$

$$4^2=a^2\left(1-\left(\frac{3}{5}\right)^2\right)=a^2\left(1-\frac{9}{25}\right)$$

$$4^2=a^2\left(\frac{16}{25}\right)$$

$$a^2=25$$

Hence the required equation of ellipse is : $\dfrac{x^2}{25}+\dfrac{y^2}{16}=1$

10. The letus rectum is half of the major axis of an ellipse. Find eccentricity (e).

Solution: Let the equation of ellipse is $\dfrac{x^2}{a^2}+\dfrac{y^2}{b^2}=1$

We know that:

Major axis = 2a

Letus rectum = $\dfrac{2b^2}{a}$

It is given that letus rectum is half of its major axis.

$$2\frac{b^2}{a}=\frac{1}{2}.2a$$

$$2b^2=a^2$$

The eccentricity $e=\sqrt{1-\dfrac{b^2}{a^2}}=\sqrt{1-\dfrac{b^2}{2b^2}}=\sqrt{1-\dfrac{b^2}{2b^2}}=\sqrt{\dfrac{1}{2}}$

11. Calculate the co-ordinates of foci, vertices and eccentricity of the ellipse:

$$\frac{x^2}{16}+\frac{y^2}{12}=$$

Solution: Comparing the given equation with standard equation of ellipse:

$$\frac{x^2}{a^2}+\frac{y^2}{b^2}=1$$

We get $$a^2 = 16 \Rightarrow a = 4$$
$$b^2 = 12 \Rightarrow b = 2\sqrt{3}$$

$a > b$ i.e. it is a horizontal ellipse.

a is semi major axis

Eccentricity $e = \sqrt{\left(1 - \dfrac{(\text{minor axis})^2}{(\text{major axis})^2}\right)} = \sqrt{1 - \left(\dfrac{2\sqrt{3}}{4}\right)^2} = \sqrt{1 - \left(\dfrac{12}{16}\right)} = \sqrt{\dfrac{4}{16}} = \sqrt{\dfrac{1}{4}} = \dfrac{1}{2}$

Coordinates of foci are: $(-ae, 0)$ and $(ae, 0)$ i.e. $\left(-4.\dfrac{1}{2}, 0\right)$ and $\left(4.\dfrac{1}{2}, 0\right)$

or $(-2, 0)$ and $(2, 0)$

Coordinates of Vertices are:: $(-a, 0)$ and $(a, 0)$ i.e. $(-4, 0)$ and $(4, 0)$

Coordinates of Co-vertices are: $(0, -b)$ and $(0, b)$ i.e. $\left(0, -2\sqrt{3}\right)$ and $\left(0, 2\sqrt{3}\right)$

Equation of directrices are : $X = \pm a / e = \pm \dfrac{4}{\frac{1}{2}} = \pm 8$

Length of letus rectum: $\dfrac{2b^2}{a} = 2.\dfrac{12}{4} = 6$

12. Find the equation of ellipse if its Center is (0, 0), Vertices (0, ±5) Co-vertices (±4, 0).

Solution: The vertices are along y-axis since x-coordinate is 0 and y-coordinate is ±5 **i.e. b = 5**

The Co-vertices are along x-axis since y-coordinate is 0 and x-coordinate is ±4 **i.e. a = 4**

It shows that the major axis is along y-axis. Hence it is a vertical ellipse.

Since its center is at (0,0) and axes are 4,5 hence the equation of ellipse is :

$$\dfrac{(x-0)^2}{4^2} + \dfrac{(y-0)^2}{5^2} = 1.$$

or $$\dfrac{(x)^2}{16} + \dfrac{(y)^2}{25} = 1.$$

Practice Problems (On Ellipse)

1. If major axis is 12 and minor axis is 8. Find the equation of the ellipse.

2. Find the vertices of the ellipse: $2x^2 + 3y^2 - 6 = 0$. (Hint: (a,0) and (-a,0))

3. Find semi major and semi minor axes of the ellipse $\dfrac{x^2}{25} + \dfrac{y^2}{16} = 1$

4. Find whether the major axis of the ellipse: $\dfrac{x^2}{9}+\dfrac{y^2}{4}=1$, is horizontal or vertical.

5. The equation of ellipse is : $9x^2+4y^2=36$.

   Find (i) coordinates of foci, (ii) Length of major and minor axis.

6. The equation of ellipse is $64x^2+9y^2=576$

   (i) Write the equation in Standard form.

   (ii) Determine: foci, vertices and co-vertices.

7. Find the eccentricity of the ellipse: $\dfrac{x^2}{9}+\dfrac{y^2}{25}=1$

8. Find the length of the axes and eccentricity of the ellipse: $4x^2+6y^2-9=0$

9. Given center (0, 0), vertices (0, 5) and (0, -5), Co-vertices (4, 0) and (-4, 0).

   Find the equation of ellipse.

10. Determine the equation of ellipse from the following information:

    Center (0, 0) , foci (2, 0) and co-vertex (3, 0)

11. Find the equation of an ellipse whose center is at origin, the major axis is 10 along x-axis and minor axis is 6 along y – axis.

12. The equation of ellipse is: $\dfrac{(x-1)^2}{9}+\dfrac{(y-4)^2}{16}=1$.

    Find: Center, lengths of axes, Vertices and Foci.

Answers:

1. $\dfrac{x^2}{36}+\dfrac{y^2}{16}=1$,　　　　2. $\left(\pm\sqrt{3},0\right)$,　　　　3. 5, 4

4. Horizontal.　　　　5. $(\pm\,2\sqrt{5}/3,\,0)$

6. (i) $\dfrac{x^2}{9}+\dfrac{y^2}{64}=1$ (ellipse is vertical)

   (ii) Vertices (0, ±8), Co-vertices (±3, 0), Foci (0, ±8e)

7. e = 4/5　　　　8. 2a = 3 , 2b = √6, e = 1/√3　9. $\dfrac{x^2}{16}+\dfrac{y^2}{25}=1$,

10. $\dfrac{x^2}{9}+\dfrac{y^2}{5}=1$,　　　　11. $\dfrac{x^2}{25}+\dfrac{y^2}{9}=1$.

12. Center (1, 4), major axis =16, minor axis = 9,

    Vertices (0, ± 4) and foci (0, ± √7)

Objective Questions (On Cartesian System)

1. The point (–3, 3) lies in the:
   (a) I Quadrant
   (b) II Quadrant
   (c) III Quadrant
   (d) IV Quadrant

2. The co-ordinates of a point on x – axis, 3 units away from origin is expressed as:
   (a) (3, 0)
   (b) (0, 3)
   (c) (–3, 0)
   (d) (–3, 3)

3. The point whose both coordinates are negative shall lie in:
   (a) I Quadrant
   (b) II Quadrant
   (c) III Quadrant
   (d) IV Quadrant

4. The point whose abscissa is –3 and ordinate in 5 , lies in:
   (a) I Quadrant
   (b) II Quadrant
   (c) III Quadrant
   (d) IV Quadrant

5. The distance between the origin and the point (3, 4) is:
   (a) 2 units
   (b) 3 Units
   (c) 4 Units
   (d) 5 Units

6. The mid-point of the line joining the points (4, -2) and (-8, 6) is:
   (a) (6, 4)
   (b) (–6, 4)
   (c) (2, 2)
   (d) (–2, 2)

7. The distance between (4, –3) and the originis:
   (a) 2 units
   (b) 3 Units
   (c) 4 Units
   (d) 5 Units

8. Value of K if the distance between the points (0, 4) and (k, 1) is 5
   (a) 2 units
   (b) 3 Units
   (c) 4 Units
   (d) 5 Units

9. The slope of the straight line joining the points (4, 6) and (–1, –2) is:
   (a) 4/3
   (b) 3/4
   (c) 8/5
   (d) 5/8

10. The slope of the line $2x + 3y + 5 = 0$ is:
    (a) 2/3
    (b) – 2/3
    (c) 3/2
    (d) – 4/3

11. The slope of the line parallel to X – axis is:
    (a) 1/2
    (b) 1
    (c) 2
    (d) 0

12. The slope of the line perpendicular to the line $2y + 6x = 24$ is:

   (a) –3
   (b) –1/3
   (c) 1/3
   (d) 1/6

13. The angle between the lines $y - \sqrt{3}x = 5$ and $\sqrt{3}y + x = 6$ is:

   (a) 30°
   (b) 45°
   (c) 60°
   (d) 90°

14. The angle between the lines whose slopes are 3 and $\dfrac{1}{2}$

   (a) 30°
   (b) 45°
   (c) 60°
   (d) 90°

15. The intercepts cut by the line $2x - 3y + 6 = 0$ from coordinate axes are:

   (a) 3, 2
   (b) –3, 2
   (c) –3, –2
   (d) None of these.

16. The equation of the line whose intercepts on the axes are 4 and 3 is:

   (a) $3x - 4y = 10$
   (b) $3x + 4y = 12$
   (c) $4x - 3y = 12$
   (d) $4x + 3y = 12$

17. The line $3x - 7y = 21$ cuts the $y - axis$ at the point:

   (a) $(0, -3)$
   (b) $(3, -1)$
   (c) $(1, -3)$
   (d) $(-3, 0)$

18. The length of perpendicular from the point (-2,3) to the line $x - y = 5$ is:

   (a) 5 units
   (b) $5\sqrt{2}$ units
   (c) 4 units
   (d) 5 units

19. The centre of the circle $(x-1)^2 + (y+3)^2 = 25$ is:

   (a) $(-1, 3)$
   (b) $(3, -1)$
   (c) $(1, -3)$
   (d) $(-3, 1)$

20. The end points of the diameter of a circle are $(-3, -2)$ and $(7, 8)$. The radius is:

   (a) 5 units
   (b) $5\sqrt{2}$ units
   (c) 4 units
   (d) 5 units

21. The point of intersection of the circle $x^2 + y^2 = 25$ and the straight line $x = 4$ are:

   (a) $(3,4),(-3,-4)$
   (b) $(3,4),(-3,4)$
   (c) $(3,-4),(-3,-4)$
   (d) $(-3,-4),(3,-4)$

22. The vertices of triangle are (1, 4) ,(5, 2) and(0, –3). The centroid is:

(a) $(0,3)$

(b) $(2,1)$

(c) $(1,-2)$

(d) $(-2,1)$

23. Identify the standard equation of parabola:

(a) $y^2 = 4x^2$

(b) $y^2 = 4ax$

(c) $4x^2 = y^2$

(d) $x = 4y + c$

24. The point of intersection of the parabola $y^2 = 4x$ and the straight line $x = 4$ are:

(a) $(2,3),(4,4)$

(b) $(3,4),(4,4)$

(c) $(4,4),(4,-4)$

(d) $(4,5),(4,-4)$

25. Does the point (-1, 7) inside, on or outside the parabola $y^2 = 12x$

(a) Inside

(b) Outside

(c) On the parabola

(d) None of these

26. The eccentricity of the ellipse $4x^2 + 9y^2 = 1$ is:

(a) $e = \dfrac{\sqrt{5}}{3}$

(b) $e = 5\sqrt{3}$

(c) $e = \dfrac{\sqrt{5}}{3}$

(d) 5

Answers

1. b	2. a	3. c	4. b	5. d	6. d	7. d
8. c	9. d	10. b	11. d	12. c	13. d	14. d
15. b	16. b	17. b	18. b	19. b	20. d	21. b
22. b	23. b	24. b	25. b	26. a		

3

Complex Numbers

1.0 BRIEF HISTORY

Greek mathematician Heron of Alexandria is said to have conceived these numbers, but Rafael Bombelli, an Italian mathematician, was first to set down the rules for addition, subtraction, multiplication and division of complex numbers in 1572. The concept appeared in the work of Gerolams Cardano. At that time, such numbers were poorly understood and regarded as fictitious or useless. Even mathematicians like 'Rene Descartes' were slow to adopt the use of imaginary numbers.

The use of imaginary numbers was not widely accepted until the work of Leonard Euler (1707-83) and Carl Fredric Gauss (1777-1855). The geometric significance of complex numbers was first found by Casper Wessel (1745 – 1818)

1.1 IMAGINARY NUMBER — i(iota)

If $x^2 + 1 = 0 \Rightarrow x^2 = -1 \Rightarrow x = \sqrt{-1} = i$(iota) . i denotes the imaginary number.

Thus $$i = \sqrt{-1} \Rightarrow i^2 = -1$$

Then $$i^3 = i^2 i = -i$$

$$(i^2)^2 = (-1)^2 \Rightarrow i^4 = 1.$$

Similarly $i^5 = i^4 i = i$

This indicates that if power of iota (i) is multiple of 4 then it will reduce to 1.

In order to find out i^n where n > 4, we divide n by 4 and obtain remainder r. Let m be the quotient and r be the remainder when n is divided by 4.

Hence $n = 4m + r$, $0 < r < 4$

Thus we can write $i^n = i^{4m+r} = i^{4m} i^r = i^r$

Solved Examples

1. Evaluate the following: **(i)** i^{135} **(ii)** i^{17} **(iii)** $(-i)^{19}$ **(iv)** $(-i)^{4m+3}$

Solutions:

(i) $i^{139}s = i^{4 \times 34 + 3} = i^3 = i^2 i = -i \quad \because i^2 = -1$

(ii) $i^{17} = i^{4 \times 4 + 1} = i$

(iii) $(-i)^{19} = (-i)^{4 \times 4 + 3} = 1.i^3 = i^2 i = -i$

(iv) $(-i)^{4m+3} = (-i)^{4m}(-i)^3 = (-i)^3 = (-i)^2(-i) = +i^2(-i) = -(-i) = i$

2. Evaluate $\left[i^{21} + \left(\dfrac{1}{i}\right)^{25} \right]^2$

Solution: $\left[i^{4 \times 5 + 1} + (-i)^{4 \times 6 + 1} \right]^2 = \left[i + (-i) \right]^2 = 0$

3. Find the value of $\dfrac{i^{494} + i^{390} + i^{388} + i^{684}}{i^{382} + i^{180} + i^{278} + i^{574}}$.

Solutions: Divide the power of I and obtain the remainder as shown below:

$$= \frac{i^{4 \times 123 + 2} + i^{4 \times 97 + 2} + i^{4 \times 97} + i^{4 \times 171}}{i^{4 \times 95 + 2} + i^{4 \times 95} + i^{4 \times 69 + 2} + i^{4 \times 141 + 2}}$$

$$= \frac{i^2 + i^2 + i^0 + i^0}{i^2 + i^0 + i^2 + i^2}$$

$$= \frac{-1 - 1 + 1 + 1}{-1 + 1 - 1 - 1} = 0$$

4. Find the value of $i^n + i^{n+1} + i^{n+2} + i^{n+3}$

Solution: $i^n + i^{n+1} + i^{n+2} + i^{n+3} = i^n \left[1 + i + i^2 + i^3 \right]$

$$= i^n \left[1 + i - 1 - i \right] = i^n \times 0 = 0$$

$\because i^2 = 0$, $i^3 = i^2.i = -i$

5. Show that $i^{n+100} + i^{n+50} + i^{n+48} + i^{n+46} = 0$

Solution: $i^{n+100} + i^{n+50} + i^{n+48} + i^{n+46} = i^n \left[i^{100} + i^{50} + i^{48} + i^{46} \right]$

$$= i^n \left[i^{4 \times 25} + i^{4 \times 12 + 2} + i^{4 \times 12} + i^{4 \times 11 + 2} \right]$$

$$= i^n \left[1 + 1.i^2 + 1 + 1.i^2 \right] = i^n \left[1 - 1 + 1 + -1 \right] = i^n \times 0 = 0$$

6. Evaluate; $\left(4+i^3\right)-\left(6-i^5\right)+\left(9+7i^2\right)$

Solution: $\left(4+i^3\right)-\left(6-i^5\right)+\left(9+7i^2\right)=\left(4-i\right)-\left(6-i\right)+\left(9-7\right)=4-i-6+i+2=0$

1.2 IMAGINARY QUANTITIES

The square root of a negative real number is called an imaginary number. For example $\sqrt{-3}$, $\sqrt{-4}$... are imaginary quantities.

Solved examples:

1. Express the following in terms of i.

(*i*) $\sqrt{-121}$ (*ii*) $\sqrt{\dfrac{-64}{144}}$ (*iii*) $4\sqrt{-121}$

(*iv*) $\sqrt{-0.0016}$ (*v*) $-13\sqrt{-4}-15\sqrt{-49}$ (*vi*) $\sqrt{-225}$

(*vii*) $4\sqrt{-121}$ (*viii*) $\sqrt{-0.0016}$

Solutions:

(*i*) $\sqrt{-121}=\sqrt{-11^2}=\sqrt{-1\times11^2}=\sqrt{-1}\times\sqrt{11^2}=i.11$

(*ii*) $\sqrt{\dfrac{-64}{144}}=\sqrt{-1}\times\sqrt{\left(\dfrac{8}{12}\right)^2}=\dfrac{8}{12}i=\dfrac{2i}{3}$

(*iv*) $\sqrt{-0.0016}=\sqrt{\dfrac{-16}{10000}}=\dfrac{4}{10}i=\dfrac{2i}{5}$

(*v*) $-13\sqrt{-4}-15\sqrt{-49}=-13\times2i-15\times7i=\left(-26-105\right)i=-131i$

(*vi*) $\sqrt{-225}=\sqrt{-15^2}=i15$

(*vii*) $4\sqrt{-1\times11^2}=4\times11i=44i$

(*viii*) $\sqrt{-0.0016}=\sqrt{-1\times\left(0.04\right)^2}=\sqrt{i_x^2\left(0.04\right)^2}=0.04i$

1.3 COMPLEX NUMBER

A **complex number** is a number that can be expressed in the form $a+bi$, where a and b are real numbers and i is the imaginary unit, which satisfies the equation $i^2=-1$. In this expression, a is the *real part* and b is the *imaginary part* of the complex number.

1.3.1 Standard Form of Complex Numbers:

If a, b are two real numbers, then number of the form $a + ib$ is known as the standard form of complex number. For example $7 + 2i$, $-3 + 2i$ etc. represents standard form of complex number.

1.3.2 Real and Imaginary Parts of Complex Numbers:

Let $z = a + ib$ is a complex number, then a is called *real part of z* and denoted by $\text{Re}(z)$ and b is called the *imaginary part* and denoted by $\text{Im}(z)$.

If $\text{Re}(z) = 0$, then z is purely an imaginary number.

If $\text{Im}(z) = 0$, then z is purely a real number.

1.4 OPERATIONS ON COMPLEX NUMBERS

1.4.1 Addition of Complex Numbers:

Let $z_1 = a_1 + ib_1$ and $z_2 = a_2 + ib_2$ be two complex numbers. Then their difference will be

$$z_1 - z_2 = (a_1 + ib_1) - (a_2 + ib_2) = (a_1 - a_2) + i(b_1 - b_2)$$

1.4.2 Subtraction of Complex Numbers:

Let $z_1 = a_1 + ib_1$ and $z_2 = a_2 + ib_2$ be two complex numbers. Then their sum will be

$$z_1 + z_2 = (a_1 + a_2) + i(b_1 + b_2)$$

1.4.3 Multiplication of Complex Numbers:

Let $z_1 = a_1 + ib_1$ and $z_2 = a_2 + ib_2$ be two complex numbers. Then their product will be

$$z_1 z_2 = (a_1 + ib)(a_2 + ib_2) = a_1 a_2 + i(a_1 b_2 + b_1 a_2) - b_1 b_2$$

or
$$z_1 z_2 = (a_1 + ib)(a_2 + ib_2) = a_1 a_2 + i(a_1 b_2 + b_1 a_2) - b_1 b_2$$

1.4.4 Division of Complex Numbers:

Let $z_1 = a_1 + ib_1$ and $z_2 = a_2 + ib_2$ be two complex numbers. Their division will be

$$\frac{z_1}{z_2} = \frac{a_1 + ib_1}{a_2 + ib_2} \times \frac{a_2 - ib_2}{a_2 - ib_2}$$

Rationalising by the conjugate of denominator

$$= \frac{(a_1 + ib_1)(a_2 - ib_2)}{(a_2^2 + b_2^2)} = \frac{(a_1 a_2 + b_1 b_2)}{(a_2^2 + b_2^2)} + i\frac{(a_2 b_1 - a_1 b_2)}{(a_2^2 + b_2^2)}$$

Solved Examples:

1. Express the following complex numbers in standard form $(a + ib)$

   (i) $3+\sqrt{-16}$ **(ii)** $-5i+3i^2$ **(iii)** $\left(3+i^{11}\right)+\left(9i^7+6i^5\right)$

Solutions:

  (i) $3+\sqrt{-16}=3+\sqrt{-4^2}=3+i4$

  (ii) $-5i+3i^2=-5i+3(-1)=-3-i5$

  (iii) $\left(3+i^{11}\right)+\left(9i^7+6i^5\right)=\left(3+i^{4\times2+3}\right)+\left(9i^{4+3}+6i^{4+1}\right)$

$$=\left(3+i^3\right)+\left(9i^3+6i^1\right)=\left(3+10i^3+6i\right)=3-10i+6i=3-4i$$

2. Perform the operation and then write in standard form.

  (i) $(2+3i)-(6-4i)$ **(ii)** $\left(-3+\sqrt{-18}\right)+\left(7-\sqrt{-8}\right)$

  (iii) $(2+3i)(1-5i)$ **(iv)** $7i(7-3i)$

  (v) $(3+2i)^2+(4-3i)$

Solutions:

  (i) $(2+3i)-(6-4i)=2+3i-6+4i=-4+7i$

  (ii) $\left(-3+\sqrt{-18}\right)+\left(7-\sqrt{-8}\right)=-3+\sqrt{-9\times2}+7-\sqrt{-4\times2}$

$$=-3+3\sqrt{2}i+7-2\sqrt{2}i$$

$$=4+i\sqrt{2}$$

  (iii) $(2+3i)(1-5i)=2+3i-10i+15=17-7i$

  (iv) $7i(7-3i)=49i-21i^2=49i-21(-1)$

$$=49i+21=21+49i$$

  (v) $(3+2i)^2+(4-3i)=\left(9+12i+4i^2\right)+(4-3i)$

$$=9-4+12i+4-3i=9+9i$$

3. Write the following complex numbers in standard form $(a+ib)$

  (i) $\dfrac{4}{2+3i}$ **(ii)** $\dfrac{1}{3-4i}$ **(iii)** $\dfrac{1}{-2+\sqrt{-3}}$

  (iv) $\dfrac{3+4i}{3-2i}$ **(v)** $\dfrac{5+4i}{4+5i}$ **(vi)** $\dfrac{(1+i)}{3-i}$

(vii) $\dfrac{1+i}{1-i}$ **(viii)** $\dfrac{(3-2i)(2+3i)}{(1+2i)(2-i)}$ **(ix)** $(1-2i)^{-3}$

(x) $\dfrac{3-\sqrt{-36}}{1-\sqrt{-25}}$

Solutions: (i) $\dfrac{4}{2+3i}$

Multiplying the above expression by $\dfrac{2-3i}{2-3i}$

$$\frac{4}{2+3i}\times\frac{2-3i}{2-3i}=\frac{4(2-3i)}{(2)^2-(3i)^2}=\frac{8-12i}{4-9i^2}=\frac{8-12i}{4+9}=\left(\frac{8}{13}\right)-\left(\frac{12}{13}\right)i$$

(ii) $\dfrac{1}{3-4i}$.

Multiplying by $\dfrac{3+4i}{3+4i}$, we get

$$\frac{1}{(3-4i)}\times\frac{3+4i}{3+4i}=\frac{3+4i}{(3-4i)(3-4i)}=\frac{3+4i}{(3)^2-(4i)^2}.$$

$$=\frac{3+4i}{(9)-(16i^2)}=\frac{3+4i}{(9)-(-16)}\qquad\because i^2=-1$$

$$=\frac{3+4i}{(9)+(16)}=\left(\frac{3}{25}\right)+\left(\frac{4}{25}\right)i$$

(iii) $\dfrac{1}{-2+\sqrt{-3}}=\dfrac{1}{-2+\sqrt{3}i}$

Multiplying by $\dfrac{-2-\sqrt{3}i}{-2-\sqrt{3}i}$, we get

$$\frac{1}{-2+\sqrt{3}i}\times\frac{-2-\sqrt{3}i}{-2-\sqrt{3}i}=\frac{-2-\sqrt{3}i}{(-2)^2-(\sqrt{3}i)^2}=\frac{-2-3i}{4-(-3)}=\left(\frac{-2}{7}\right)+\left(\frac{-\sqrt{3}}{7}\right)i.$$

(iv) $\dfrac{3+4i}{3-2i}$

Multiplying the above expression by $\dfrac{3+2i}{3+2i}$

$$\frac{3+4i}{3-2i} \times \frac{3+2i}{3+2i} = \frac{(3+4i)(3+2i)}{(3)^2 -(2i)^2} = \frac{9+12i+6i+8i^2}{(9)-(-4)}$$

$$= \frac{9+12i+6i+8i^2}{(9)-(-4)} = \frac{9+18i-8}{13} = \left(\frac{1}{13}\right) + \left(\frac{18}{13}\right)i$$

(v) $\dfrac{5+4i}{4+5i} = \dfrac{5+4i}{4+5i} \times \dfrac{4\ 5}{4\ 5} = \dfrac{(5+4i)}{(4+5i)} \times \dfrac{(4-5i)}{(4-5i)}$

$$= \frac{20+16i-25i+20}{(4)^2 -(5i)^2} = \frac{40-9i}{16+25}$$

$$= \frac{40-9i}{41} = \left(\frac{40}{41}\right) - \left(\frac{9}{41}\right)i$$

(vi) $\dfrac{(1+i)(3+i)}{(3-i)(3+i)} = \dfrac{3+3i+i+i^2}{(3)^2 -(i)^2} = \dfrac{3+4i-1}{9-(-1)}$

$$= \frac{2+4i}{10} = \left(\frac{2}{10}\right) + \left(\frac{4}{10}\right)i$$

(vii) $\dfrac{(1+i)}{(1-i)} \times \dfrac{(1+i)}{(1+i)} = \dfrac{(1+i)^2}{(1-i^2)} = \dfrac{1+2i+i^2}{1+1} = \dfrac{1+2i-1}{2} = i$

(viii) $\dfrac{(3-2i)(2+3i)}{(1+2i)(2-i)} = \dfrac{6-4i+9i-6i^2}{2+4i-i-2i^2}$

$$= \frac{6+5i-6(-1)}{2+3i-2(-1)} = \frac{12+5i}{4+3i}$$

Now we have to express $\dfrac{12+5i}{4+3i}$ in standard form.

Multiplying $\dfrac{12+5i}{4+3i}$ by $\dfrac{4-3i}{4-3i}$

$$\frac{12+5i}{4+3i} \times \frac{4-3i}{4-3i} = \frac{48+20i-36i-15i^2}{(4)^2 -(3i)^2} = \frac{48-16i+15}{16+9} \qquad \because i^2 = -1$$

$$= \frac{63-16i}{25} = \left(\frac{63}{25}\right) - \left(\frac{16}{25}\right)i$$

(ix) $(1-2i)^{-3} = \dfrac{1}{(1-2i)^3}$

$\because (a-b)^3 = a^3 - b^3 - 3ab(a-b)$

$\therefore (1-2i)^3 = 1 - (2i)^3 - 3(1)(2i)(1-2i)$

$\qquad = 1 - 2(-i) - 6i(1-2i)$

$\qquad = 1 + 2i - 6i + 12i^2 \qquad\qquad \because i^3 = i^2 . i = -i$

$\qquad = 1 - 4i - 12 = -11 - 4i$

$\therefore \dfrac{1}{(1-2i)^3} = \dfrac{1}{-11-4i}$

Now multiplying the above expression by $\dfrac{-11+4i}{-11+4i}$

$$\frac{1}{-11-4i} \times \frac{-11+4i}{-11+4i} = \frac{-11+4i}{(-11)^2 - (4i)^2} = \frac{-11+4i}{121+16} = \left(\frac{-11}{137}\right) + \left(\frac{4}{137}\right)i$$

(x) $\dfrac{3-\sqrt{-36}}{1-\sqrt{-25}} = \dfrac{3-6i}{1-5i}$

Multiplying it by $\dfrac{1+5i}{1+5i}$

$$\therefore \frac{3-6i}{1-5i} \times \frac{1+5i}{1+5i} = \frac{(3-6i)(1+5i)}{(1)^2 - (5i)^2} = \frac{3-6i+15i-30i^2}{1+25}$$

$$= \frac{3+9i-30(-1)}{26} = \left(\frac{33}{26}\right) + \left(\frac{9}{26}\right)i$$

4. Prove that $\left(\dfrac{2+3i}{3+4i}\right)\left(\dfrac{2-3i}{3-4i}\right)$ is purely a real number.

Solution: $\left(\dfrac{2+3i}{3+4i}\right)\left(\dfrac{2-3i}{3-4i}\right) = \dfrac{(2)^2 - (3i)^2}{(3)^2 - (4i)^2} = \dfrac{4+9}{9+16} = \dfrac{13}{25}$

$\dfrac{13}{25}$ is purely a real number.

5. Prove that $\left(\dfrac{1+i}{1-i}\right)^{4n+1}$ is an imaginary number.

Solution: $\left(\dfrac{1+i}{1-i}\right)^{4n+1} = \left(\dfrac{1+i}{1-i} \times \dfrac{1+i}{1+i}\right)^{4n+1} = \left(\dfrac{1+2i+i^2}{1-i^2}\right)^{4n+1}$

$$= \left(\dfrac{1+2i-1}{1-(-1)}\right)^{4n+1}$$

$$= \left(\dfrac{2i}{2}\right)^{4n+1} = i^{4n}.i = i \quad \because i^{4n} = 1$$

∴ Given number is an imaging number.

Practice Problems

Express the following in the Standard Form

1. *(i)* $\sqrt{4}\left[\cos\dfrac{\pi}{2} + i\sin\dfrac{\pi}{2}\right]$,   *(ii)* $\sqrt{-36} + 3i^4$,   *(iii)* $\dfrac{5-3i}{(3-2i)(1+i)}$,

(iv) $\dfrac{(1+i)}{(1-i)}$ *(v)* $\dfrac{1}{3+5i}$,   *(vi)* $\dfrac{4+5i}{5+4i}$

(vii) $(\cos\theta - i\sin\theta)^2$

2. Write the following in Standard Form $(a + bi)$:

(i) $(3+2i)+(2+4i)$ *(ii)* $(4+3i)-(2+5i)$ *(iii)* $(4+2i)+(4-2i)$

(iv) $(3+2i)(4-3i)$ *(v)* $(3+4i)^2$ *(vi)* $(1+2i)(1-i)(2+i)$

(vii) $\dfrac{3+5i}{2+6i}$ *(viii)* $\dfrac{6-2i}{5+6i}$ *(ix)* $\dfrac{3-2i}{3+4i}$

(x) $\dfrac{5+2i}{7+4i}$

Answers:

1. *(i)* $\sqrt{4}i$,   *(ii)* $(3+6i)$,   *(iii)* $\left(\dfrac{11}{13}\right) - \left(\dfrac{10}{13}\right)i$

(iv) $(0+i)$,   *(v)* $\left(\dfrac{3}{34}\right) - \left(\dfrac{5}{34}\right)i$ *(vi)* $\left(\dfrac{40}{41}\right) - \left(\dfrac{16}{41}\right)i$

(vii) $(1 - i\sin 2\theta)$

Answers:

2. (i) $(5+6i)$ **(ii)** $(2-2i)$ **(iii)** 8

(iv) $(18-i)$ **(v)** $(7+5i)$ **(vi)** (9/10)–(1/5)i

(vii) (10/13)I   **(viii)** (5/13)–(6/13)i   **(ix)** (43/65)–6/65)i

1.5 CONJUGATE OF A COMPLEX NUMBER

A Complex Number z is represented as $z = a + ib$. Its conjugate is denoted by $\bar{z}$ and represented as $\bar{z} = a - ib$

The real parts of a complex number and its conjugate remain the same but imaginary part becomes negative i.e. i is replaced by $-i$

The product of a complex number and its conjugate is purely real number.

$$z.\bar{z} = (a+bi)(a-bi) = (a)^2 - (bi)^2 = a^2 + b^2$$

Complex Number (z)	Conjugate ($\bar{z}$)
7 – 2i	7 – 2i
4i + 3	–4i + 3
–5i	+5i
12	12

It may be noticed that the sign of imaginary number changes while the real part remains the same.

Solved Examples

1. Find the conjugate of following complex numbers:

  (i) $8 + 5i$ **(ii)** $-7 - 6i$ **(iii)** $-4 + 5i$

Solutions: (i) Let $z = 8 + 5i$ ∴ $\bar{z} = 8 - 5i$

  (ii) Let $z = -7 - 6i$ ∴ $\bar{z} = -7 + 6i$

  (iii) Let $z = -4 + 5i$ ∴ $\bar{z} = -4 - 5i$

2. Given $z = -i$ and $z_2 = -3 + 4i$. Find $\left(\dfrac{z_1 z_2}{\bar{z}_1} \right)$

Solution: $z_1 z_2 = (1-i)(-3+4i) = -3 + 3i + 4i - 4i^2 = -3 + 7i + 4 = 1 + 7i$

  If $z_1 = 1 - i$, then $\bar{z}_1 = 1 + i$

$$\therefore \left(\frac{z_1 z_2}{\bar{z}_1} \right) = \frac{1 + 7i}{1 + i}$$

We can write $\left(\dfrac{z_1 z_2}{\overline{z}_1}\right) = \dfrac{1+7i}{1+i} \times \dfrac{1-i}{1-i} = \dfrac{1+7i-1-7i^2}{1-i^2} = \dfrac{7i+7}{2} = \left(\dfrac{7}{2}\right) + \left(\dfrac{7}{2}\right)i$

3. If $z = 12 + 5i$, then $z.\overline{z} = |z|^2$

Solution: $z = 12 + 5i \quad \overline{z} = 12 - 5i$

$\therefore z.\overline{z} = (12 + 5i)(12 - 5i) = 12^2 - (5i)^2 = 144 + 25 = 169$

$$|z| = \sqrt{12^2 + 5^2} = \sqrt{169}$$

$$|z|^2 = \left(\sqrt{169}\right)^2 = 169$$

$\therefore z.\overline{z} = |z|^2$ Hence Proved.

4. If $z_1 = 6 + 7i$ and $z_2 = 1 - 3i$, then show $\overline{z}_1 + \overline{z}_2 = \overline{(z_1 + z_2)}$

Solution: $\overline{z}_1 + \overline{z}_2 = (6 - 7i) + (1 + 3i) = 7 - 4i$

$$\overline{z_1 + z_2} = 7 - 4i$$

Hence $\overline{z}_1 + \overline{z}_2 = \overline{(z_1 + z_2)}$

1.6 MODULUS OR ABSOLUTE VALUE OF A COMPLEX NUMBER

The modulus of a complex number $z = a + ib$ is denoted by $|z|$ and defined as $|z| = \sqrt{a^2 + b^2}$

Solved Examples:

1. $z = 4 + i3$. Find the value of $z.\overline{z}$, $|z|$ and $|\overline{z}|$

Solution: $\qquad\qquad z.\overline{z} = (4 + 3i)(4 - 3i) = 4^2 + 3^2 = 25$

Modulus of z : $|z| = \sqrt{4^2 + 3^2} = \sqrt{25} = 5$

$$|\overline{z}| = \sqrt{4^2 + 3^2} = \sqrt{25} = 5$$

It may be noticed that $|z| = |\overline{z}|$

2. Find the modulus of following complex numbers: (*i*) $7 - i^3$ and (*ii*) $6 + 3i^5$

Solution: (*i*) $7 - i^3 = 7 - (i^2 . i) = 7 - (-i) = 7 + i$

$\therefore |7 + i| = \sqrt{7^2 + 1^2} = \sqrt{49 + 1} = 5\sqrt{2}$

(ii) $6+3i^5 = 6+3.i^4.i = 6+3i \quad \because i^4 = 1$

$$|6+3i| = \sqrt{6^2+3^2} = \sqrt{36+9} = 3\sqrt{5}$$

3. Find the conjugate of $z = \dfrac{1}{3+4i}$ .Also find $|z|$

Solution: First express it in standard form i.e. multiply the numerator and denominator by the conjugate of $3+4i$ .The conjugate of $3+4i$ is $3-4i$.

$$z = \frac{1}{3+4i} \times \frac{3-4i}{3-4i} = \frac{3-4i}{9+16} = \left(\frac{3}{25}\right) - \left(\frac{4}{25}\right)i$$

Hence conjugate of z is $\overline{z} = \left(\dfrac{3}{25}\right) + \left(\dfrac{4}{25}\right)i$

$$|z| = \sqrt{(a)^2 + (b)^2} = \sqrt{\left(\frac{3}{25}\right)^2 + \left(-\frac{4}{25}\right)^2} = \sqrt{\frac{9+16}{25^2}} = \sqrt{\frac{25}{25^2}} = \frac{1}{5}$$

4. If $z_1 = -3+4i$ and $z_2 = 12-5i$. then show that

(i) $|z_1.z_2| = |z_1|.|z_2|$ and **(ii)** $\left|\dfrac{z_1}{z_2}\right| = \dfrac{|z_1|}{|z_2|}$

Solutions: (i) $z_1 = -3+4i$.

$$\therefore |z_1| = \sqrt{(-3)^2 + (4)^2} = \sqrt{9+16} = \sqrt{25} = 5$$

$$z_2 = 12-5i$$

$$\therefore |z_2| = \sqrt{(12)^2 + (-5)^2} = \sqrt{144+25} = \sqrt{169} = 13$$

$$z_1.z_2 = (-3+4i)(12-5i) = -36+15i+48i-20i^2$$

$$= -36+63i+20 \quad \therefore i^2 = -1$$

$$= -16+36i$$

$$|z_1.z_2| = |-16+63i| = \sqrt{(-16)^2 + (63)^2} = \sqrt{256+3969} = \sqrt{4225} = 65$$

$$|z_1|.|z_2| = 5.13 = 65$$

Hence $|z_1.z_2| = |z_1|.|z_2|$

(ii) $\dfrac{z_1}{z_2} = \dfrac{-3+4i}{12-5i} \times \dfrac{12+5i}{12+5i} = \dfrac{-36+63i-20i^2}{(12)^2-(5i)^2}$

$$= \dfrac{-16+63i}{144+25} = \dfrac{-16+63i}{169} = \left(\dfrac{-16}{169}\right) + \left(\dfrac{63}{169}\right)i$$

$$\left|\dfrac{z_1}{z_2}\right| = \sqrt{\left(\dfrac{-16}{169}\right)^2 + \left(\dfrac{63}{169}\right)^2} = \sqrt{\dfrac{256+3969}{(169)^2}} = \sqrt{\dfrac{4225}{169^2}} = \dfrac{65}{169} = \dfrac{5}{13}$$

$$\left|\dfrac{z_1}{z_2}\right| = \dfrac{5}{13}$$

Hence $\left|\dfrac{z_1}{z_2}\right| = \dfrac{|z_1|}{|z_2|}$

Practice Problems: (On Conjugate and Modulus)

1. Calculate $\bar{z}, \bar{\bar{z}}, |\bar{z}|, z.\bar{z}$ if **(i)** $z = 4+3i$ **(ii)** $z = 3-15i$

2. If $z_1 = 6+7i$ and $z_1 = 1-3i$ then show-

   (i) $\overline{z_1-z_2} = \bar{z_1} - \bar{z_2}$, **(ii)** $\overline{\bar{z_1}.\bar{z_2}} = \overline{z_1.z_2}$, **(iii)** $\overline{\left(\dfrac{z_1}{z_2}\right)} = \dfrac{\bar{z_1}}{\bar{z_2}}$

3. If $z_1 = 3+2i$ and $z_2 = 2-4i$, then show $\overline{z_1.z_2} = |z_1.z_2|$

4. Evaluate : **(i)** $\left|8+6i^3\right|$, **(ii)** $\left|4+3i^5\right|$

Answers 1:

   (i) $\bar{z} = 4-3i$, $\bar{\bar{z}} = 4+3i$, $|\bar{z}| = 5$, $z.\bar{z} = 25$

   (ii) $\bar{z} = 3+15i$, $\bar{\bar{z}} = 3-15i$, $|\bar{z}| = \sqrt{234}$, $z.\bar{z} = -216$

Answers 4:

   (i) 10   **(ii)** 5

1.7 EQUALITY OF COMPLEX NUMBERS

Let $z_1 = a_1 + ib_1$ and $z_2 = a_2 + ib_2$ be two complex numbers.

They will be equal if $a_1 = a_2$ and $b_1 = b_2$

Solved Examples

1. If $(3x-7)+2iy = -5y+(5+x)i$. Find the values of x and y.

Solution: Equating the real part:

$$(3x-7) = -5y \Rightarrow 3x+5y = 7 \qquad \text{...(i)}$$

Equating the imaginary part:

$$2y = 5+x \Rightarrow -x+2y = 5 \qquad \text{...(ii)}$$

Multiplying (ii) by 3 and adding to (i)

$$3x+5y = 7$$
$$\underline{-3x+6y = 15}$$

$\therefore$ $\qquad\qquad 11y = 22 \Rightarrow y = 2$

Put $y = 2$ in (*i*) we get

$$3x + 5x^2 = 7 \quad \Rightarrow \quad x = -1$$

Hence $x = -1, \; y = 2$

2. Given: $(1-i)x+(1+i)y = 1-3i$. Find the values of x and y.

Solution: The given equation can be written as:

$$(x+y)-(x-y)i = 1-3i$$

Equating the real parts:

$$(x+y) = 1 \qquad \text{...(i)}$$

Equating the imaginary parts:

$$(x-y) = 3 \qquad \text{...(ii)}$$

Adding (i) and (ii) we get $2x = 4 \Rightarrow x = 2$

Putting $x = 2$ in (i) we get y = -1

Hence $x = 2, \; y = -1$

Practice Problems: (Equality of Complex numbers:)

1. Find m and n that make following equation true.

(*i*) $\quad 18+7i = 3m+2ni$ $\qquad$ (*ii*) $3m+ni = 5m+1+2i$ $\qquad$ (*iii*) $3+5i+x-yi = 6-2i$

(*iv*) $\quad x+yi = (1-i)(2+8i)$

Answers:

(*i*) $\;$ m = 6, n = 7/2, $\qquad$ (*ii*) m = $-(1/2)$ $\qquad$ (*iii*) x = 3, y = 7

(*iv*) $\;$ x = 10, y = 6

1.8 ARGUMENT OR AMPLITUDE OF A COMPLEX NUMBER

Let $z = a + ib$, then argument of z or arg (z) is given by

$$\tan^{-1}\left(\frac{b}{a}\right) \text{ or } \tan^{-1}\left(\frac{\text{Im } g(z)}{\text{Re}(z)}\right)$$

where Img(z) indicates imaginary part of z and Re(z) the real par of z. It is the angular value in radians. This angle is subtended by the complex number z at the centre with x – axis. If tangle is θ then:

$$\tan\theta = \left|\frac{b}{a}\right| \text{ or } \theta = \tan^{-1}\left|\frac{b}{a}\right|$$

Solved Examples

Example 1: Find the modulus and argument of the complex numbers

(*i*) $z = 4 + 3i$,   (*ii*) $z = 3 - 2i$ (*iii*) $z = (-2 + i)(1 + 2i)$ (*iv*) $z = \dfrac{1 + 3i}{3 - 2i}$

(*v*) Find the arguments of $z = \sqrt{\ } + i$, $z_2 = -1 - i\sqrt{3}$, $z_1.z_2$ and $\left(\dfrac{z_1}{z_2}\right)$

Solution: (*i*) $z = 4 + 3i$ Here $a = 4$ and $b = 3$

$$\therefore |z| = |4 + 3i| = \sqrt{(4)^2 + (3)^2} = \sqrt{16 + 9} = \sqrt{25} = 5$$

$$\because \text{arg}(z) = \tan^{-1}\left(\frac{b}{a}\right)$$

$$\text{arg}(z) = \tan^{-1}\left|\frac{3}{4}\right|$$

Solution: (*ii*) $z = 3 - 2i$ $|z| = \sqrt{(3)^2 + (-2)^2} = \sqrt{9 + 4} = \sqrt{13}$

$$\text{arg}(z) = \tan^{-1}\left|\frac{2}{3}\right|$$

Solution: (*iii*) $z = (-2 + i)(1 + 2i) = -4 - 3i$

$$\therefore |z| = \sqrt{(-4)^2 + (-3)^2} = \sqrt{16 + 9} = \sqrt{25} = 5$$

$$\text{arg}(z) = \tan^{-1}\left|\frac{3}{4}\right|$$

Solution: (iv) $z = \dfrac{1+3i}{3-2i}$.

Multiplying the numerator and denominator by conjugate of denominator i.e (3+2i)

$$z = \frac{1+3i}{3-2i} \times \frac{3+2i}{3+2i} = \frac{3+2i+9i+6i^2}{(3)^2 - (2i)^2}.$$

$$= \frac{3+11i-6}{9-(-4)} = \frac{-3+11i}{13} = \left(\frac{-3}{13}\right) + \left(\frac{11}{13}\right)i.$$

$$|z| = \sqrt{\left(\frac{-3}{13}\right)^2 + \left(\frac{11}{13}\right)^2} = \sqrt{\frac{9+121}{13^2}} = \sqrt{\frac{130}{13^2}} = \sqrt{\frac{10}{13}}$$

$$\arg(z) = \tan^{-1}\left|\frac{11/13}{3/13}\right| = \tan^{-1}\left|\frac{11}{3}\right|$$

(v) Find the arguments of $z_1 = \sqrt{3}+i$, $z_2 = -1-i\sqrt{3}$, $z_1.z_2$ and $\left(\dfrac{z_1}{z_2}\right)$

$$\arg(z_1) = \tan^{-1}\left|\frac{\operatorname{Im} g(z_1)}{\operatorname{Re}(z_1)}\right| = \tan^{-1}\left(\frac{1}{\sqrt{3}}\right)$$

$$\arg(z_2) = \tan^{-1}\left|\frac{\operatorname{Im} g(z_2)}{\operatorname{Re}(z_2)}\right| = \tan^{-1}\left|\frac{-1}{-\sqrt{3}}\right| = \tan^{-1}\left|\frac{1}{\sqrt{3}}\right|$$

$$z_1.z_2 = \left(\sqrt{3}+i\right)\left(-1-i\sqrt{3}\right) = -\sqrt{3}-3i-i-i^2\sqrt{3} = 0-4i$$

$$\arg(z_1.z_2) = \tan^{-1}\left|\frac{-4i}{0}\right| = \tan^{-1}(\infty)$$

$$\frac{z_1}{z_2} = \frac{\left(\sqrt{3}+i\right)}{-\left(1+i\sqrt{3}\right)} \times \frac{\left(1-i\sqrt{3}\right)}{\left(1-i\sqrt{3}\right)} = \frac{\sqrt{3}-i3+i-i^2\sqrt{3}}{-\left[(1)^2 - \left(i\sqrt{3}\right)^2\right]}$$

$$= \frac{\sqrt{3}-4i-(-1)\sqrt{3}}{-(1+3)}$$

$$= \frac{2\sqrt{3}-4i}{-4} = \frac{2\left(\sqrt{3}-2i\right)}{-4}$$

$$= \frac{-\sqrt{3}}{2} + \frac{1}{2}i$$

$$\arg(z_1.z_2) = \tan^{-1}\left|\frac{1/2}{-\sqrt{3}/2}\right| = \tan^{-1}\left|\frac{1}{-\sqrt{3}}\right|$$

(*vi*) Find the modulus and argument of complex number $(1+i)$ and express it in the polar Form.

Solution: Let $z = 1+i$. Comparing with standard form $z = a+ib$

We find $a = 1$ and $b = 1$.

Hence modulus $r = |z| = \sqrt{a^2 + b^2} = \sqrt{1+1} = \sqrt{2}$

The argument is: $\tan\theta = \dfrac{b}{a} = \dfrac{1}{1} = 1 = \tan 45 \Rightarrow \theta = 45^0$

The polar form is given by: $z = |z|(\cos\theta + i\sin\theta) = \sqrt{2}\left[\cos\dfrac{\pi}{4} + i\sin\dfrac{\pi}{4}\right]$

Practice Problems (Argument)

1. Find the modulus and argument of the following complex numbers:

(i) $z = 3+4i$, (ii) $z = -5+12i$, (iii) $z = 5-12i$, (iv) $z = -3-4i$,

(v) $z = 4+3i$ (vi) $z = 3-2i$, (vii) $z = \dfrac{3-2i}{4i+(1+i)^2}$,

Answers: 1

(i) $5, \tan^{-1}\left|\dfrac{4}{3}\right|$, (ii) $13, \tan^{-1}\left|\dfrac{12}{5}\right|$, (iii) $13, \tan^{-1}\left|\dfrac{12}{5}\right|$, (iv) $5, \tan^{-1}\left|\dfrac{3}{4}\right|$

(v) $5, \tan\left|\dfrac{-}{-}\right|$, (vi) $\sqrt{13}, \tan^{-1}\left|\dfrac{2}{3}\right|$, (vii) $\dfrac{\sqrt{13}}{6}, \tan^{-1}\left|\dfrac{3}{2}\right|$

1.9   POLAR FORM OF COMPLEX NUMBER

The polar form of a complex number $z = a+bi$ is represented as $z = r(\cos\theta + i\sin\theta)$ where

$$r = |z| = \sqrt{a^2 + b^2} \text{ and } a = r\cos\theta \text{ and } b = r\sin\theta.$$

$\tan\theta = \dfrac{b}{a} \Rightarrow \theta = \tan^{-1}\theta$ (Note that here θ is measured in radians.)

Solved Examples

1. Convert the complex number z = 4 + 3 *i* into polar form

Solution:

The polar form of a complex number $z = 4 + 3i$ as $z = r(\cos\theta + i\sin\theta)$.

The first task is to find the absolute value of r.

$$r = |z| = \sqrt{a^2 + b^2} = \sqrt{4^2 + 3^2} = \sqrt{25} = 5$$

Now we have to find the argument θ.

Since $a > 0$, use the formula

$$\theta = \tan^{-1}\left(\frac{b}{a}\right).$$

$$\therefore \theta = \tan^{-1}\left(\frac{3}{4}\right)$$

Therefore, the polar form of $z = 4 + 3i$ is

$$z = 5(\cos\theta + i\sin\theta)$$

where $\theta = \tan^{-1}\left(\frac{3}{4}\right)$

2. Find the polar form and represent graphically the complex number $z = 7 - 5i$

Solution: The first task is to find the absolute values of r *and* θ

$$r = |z| = \sqrt{a^2 + b^2} = \sqrt{12^2 + 5^2} = \sqrt{144 + 25} = \sqrt{169} = 13$$

$$\theta = \tan^{-1}\left(\frac{b}{a}\right) = \tan^{-1}\left(\frac{5}{12}\right)$$

Therefore, the polar form of $z = 12 + 5i$ is

$$z = 13 (\cos\theta + i\sin\theta)$$

where $\theta = \tan^{-1}\left(\frac{5}{12}\right)$

1.10 DE MOIVRE'S THEOREM

De Moivre's Theorem is a simple formula is used for calculating the power of a complex number. It states that for any integer n the following trigonometric relation ship is true:

$$(\cos x + i\sin x)^n = \cos(nx) + i\sin(nx)$$

The De Moivre's formula can be derived from Euler's formula:

$$e^{ix} = (\cos x + i\sin x)$$

Also $\left(e^{ix}\right)^n = e^{i(nx)} = \cos(nx) + i\sin(nx)$

Solved Examples

1. Evaluate $\left[6\left(\cos 30^0 + i\sin 30^0\right)\right]^2$ using De Moivre's formula

Solution: $\left[6\left(\cos 30^0 + i\sin 30^0\right)\right]^2 = 36\left[\cos\left(2\times 30^0\right) + i\sin\left(2\times 30^0\right)\right]$

$$= 36\left[\cos 60^0 + i\sin 60^0\right] = 36\left[\frac{1}{2} + i\frac{\sqrt{3}}{2}\right]$$

$$= \frac{36}{2}\left[1 + i\sqrt{3}\right] = 18\left[1 + \sqrt{3}\right]$$

2. Simplify $3\left(\cos 3x + i\sin 3x\right)^6$

Solution: $3\left(\cos 3x + i\sin 3x\right)^6 = 3\left[\cos 6.3\theta + i\sin 6 \times 3\theta\right]$

$$= 3\left[\cos 18\theta + i\sin 18\theta\right]$$

3. Simplify $\dfrac{\left(\cos 3\theta + i\sin 3\theta\right)}{\left(\cos\theta + i\sin\theta\right)}$

Solution: $\dfrac{\left(\cos 3\theta + i\sin 3\theta\right)}{\left(\cos\theta + i\sin\theta\right)} = \dfrac{\left(\cos\theta + i\sin\theta\right)^3}{\left(\cos\theta + i\sin\theta\right)^1}$

$$= \left(\cos\theta + i\sin\theta\right)^{3-1} = \left(\cos\theta + i\sin\theta\right)^2$$

4. Prove that - $\dfrac{\left(\cos 5\theta - i\sin 5\theta\right)^2 .\left(\cos 7\theta + i\sin 7\theta\right)^{-3}}{\left(\cos 4\theta - i\sin 4\theta\right)^9 .\left(\cos\theta + i\sin\theta\right)^5} = 1$

Solution: LHS $= \dfrac{\left(\cos 5\theta - i\sin 5\theta\right)^2 .\left(\cos 7\theta + i\sin 7\theta\right)^{-3}}{\left(\cos 4\theta - i\sin 4\theta\right)^9 .\left(\cos\theta + i\sin\theta\right)^5}$

$$= \dfrac{\left(\cos 2\times 5\theta - i\sin 2\times 5\theta\right).\left(\cos-3\times 7\theta + i\sin-3\times 7\theta\right)}{\left(\cos 9\times 4\theta - i\sin 9\times 4\theta\right).\left(\cos 5\times\theta + i\sin 5\times\theta\right)}$$

$$= \dfrac{\left(\cos\theta - i\sin\theta\right)^{10} .\left(\cos\theta + i\sin\theta\right)^{-21}}{\left(\cos\theta - i\sin\theta\right)^{36} .\left(\cos\theta + i\sin\theta\right)^5}$$

$$= \left(\cos\theta - i\sin\theta\right)^{10-36} .\left(\cos\theta + i\sin\theta\right)^{-21-5}$$

$$= \left(\cos\theta - i\sin\theta\right)^{-26} .\left(\cos\theta + i\sin\theta\right)^{-26}$$

$$= \left(\cos\theta + i\sin\theta\right)^{26}.\left(\cos\theta + i\sin\theta\right)^{-26} = \left(\cos\theta + i\sin\theta\right)^{26-26}$$

$$= \left(\cos\theta + i\sin\theta\right)^{0} = 1$$

The complex number $z = a + ib$ can also be represented in polar form as shown below:

$$z = r\left(\cos\theta + i\sin\theta\right)s$$

Where $r = |z| = \sqrt{\left(a^2 + b^2\right)}$

And $\tan\theta = \dfrac{b}{a}$. We also have $a = r\cos\theta$ and $b = r\sin\theta$

5. Use De Moivre's formula to write $\left(1 - i\right)^4$ in the standard form $a + ib$

Solution: Comparing $1 - i$ with $a + ib$

We get $a = 1$, $b = -1$

$$r = a^2 + b^2 = \left(1\right)^2 + \left(-1\right)^2 = 1 + 1 = 2$$

$$\sin\theta = \frac{b}{r} = \frac{-1}{2} \Rightarrow \sin\theta = \sin\left(-45^0\right) \Rightarrow \theta = -45^0$$

$$z = r\left(\cos\theta + i\sin\theta\right) = 2\left[\cos(-45^0) + i\sin(-45^0)\right]$$

$$\left(1 - i\right)^4 = z^4 = \left[2\left\{\cos\left(45\right) - i\sin\left(45\right)\right\}\right]^4$$

$$= \left[2\left\{\cos\left(45\right) - i\sin\left(45\right)\right\}\right]^4$$

$$= \left[2^4\left\{\cos 4\times\left(45\right) - i\sin 4\times\left(45\right)\right\}\right]$$

$$= \left[16\left\{\cos\left(180\right) - i\sin\left(180\right)\right\}\right]$$

$$= \left[16\left\{-1 - 0\right\}\right] = -16$$

6. Use De Moivre's Theorem to find the value of $\left(1 + i\right)^4$

Solution: First write the given complex number in polar form.

$$\left(1 + i\right)^4 = \left[\sqrt{2}\left(\cos\pi/4 + i\sin\pi/4\right)\right]^4$$

$$= \left[\left(\sqrt{2}\right)^4\left(\cos\pi/4 + i\sin\pi/4\right)^4\right]$$

$$= \left[4\left(\cos 4\pi/4 + i\sin 4\pi/4\right)\right]$$

$$= \left[4\left(\cos\pi + i\sin\pi\right)\right] = \left[4\left(-1 + i.0\right)\right] = -4$$

7. Prove that: $(1+i)^4 + (1-i)^4 = 2^{\frac{5}{2}}$

Solution: $z = 1+i$

$\therefore$ Modulus $r = \sqrt{a^2 + b^2} = \sqrt{1+1} = \sqrt{2}$

$\because$ Arguement $\tan \theta = \dfrac{b}{a} = \dfrac{1}{1} = 1 \Rightarrow \theta = 45^0 = \pi/4$

Writing z in polar form $z = r(\cos \theta + i\sin \theta) = \sqrt{2}(\cos \pi/4 + i\sin \pi/4)$

$\therefore z^4 = (1+i)^4 = \left[\sqrt{2}(\cos \pi/4 + i\sin \pi/4)\right]^4 = \left[(\sqrt{2})^4 (\cos \pi/4 + i\sin \pi/4)^4\right]$

$= \left[2^2 (\cos 4\pi/4 + i\sin 4\pi/4)\right]$

Similarly we can show $(1-i)^4 = 2^2 (cos4\pi/4 + \sin 4\pi/4)$

Now $(1+i)^4 + (1-i)^4 = \left[2^2 (\cos 4\pi/4 + i\sin 4\pi/4)\right] + \left[2^2 (\cos 4\pi/4 - i\sin 4\pi/4)\right]$

$= 2^2 [2\cos 4\pi/4] = 2^3 \cos \pi = -8$

$\because \cos \pi = -1$

Practice Problems: (De Moivre's Theorem)

1. **Find** $(\cos 30^0 + i\sin 30^0)^3$,

2. **Evaluate:** $\dfrac{(\cos 3\theta + i\sin 3\theta)^2}{(\cos 2\theta + i\sin 2\theta)^3}$,

3. **Evaluate** $\left[6(\cos 30 + i\sin 30)\right]^2$

4. **Prove:** $\left(\dfrac{\cos \theta + i\sin \theta}{\sin \theta + i\cos \theta}\right)^6 = (\cos 12\theta + i\sin 12\theta)$

5. Use De Moivre's Theorem to find the value of $(\sqrt{3} + i)^6$

6. **Simplify (i)** $(\cos \theta + i\sin \theta)^5 (cos\theta - i\sin \theta)^2$ **(ii)** $(\sin \theta + i\cos \theta)^2$

Answers:

1. i 2. 1   3. $18(1 + i\sqrt{3})$ 4. -2^6 5. (i) $(\cos 3\theta + i\sin 3\theta)$,

6. **(ii)** $\cos n(\pi/2 - \theta) + i\sin n(\pi/2 - \theta)$

Objective Questions (On Complex Numbers)

1. The value of $\sqrt{-144}$ is:
   - (a) $-12i$
   - (b) $12i$
   - (c) $-i\sqrt{12}$
   - (d) ± 12

2. The value of $\sqrt{-64}.\sqrt{-36}$ is:
   - (a) $-48i$
   - (b) -46
   - (c) -48
   - (d) $48i$

3. The value of $8i(9-8i)$ is:
   - (a) $64+72i$
   - (b) $72i-64i^2$
   - (c) $72i+64i^2$
   - (d) $72i-64$

4. The value of $(9+8i)-(2+8i)+(9+2i)$ is:
   - (a) $16+8i$
   - (b) $16+2i$
   - (c) $-2-2i$
   - (d) $2-2i$

5. $\left(8+\sqrt{-8}\right)+\left(10-\sqrt{-72}\right)$ is equal to:
   - (a) $18-4i\sqrt{5}$
   - (b) $18+4\sqrt{2}$
   - (c) $18+4i\sqrt{5}$
   - (d) $18-4i\sqrt{2}$

6. $\dfrac{1+2i+3i^2}{1-2i+3i^2}$ is equal to:
   - (a) i
   - (b) -1
   - (c) $-i$
   - (d) 4

7. $\dfrac{i^5+i^6+i^7+i^8+i^9}{(1+i)}$ is equal to:
   - (a) $\dfrac{1}{2}(1+i)$
   - (b) $\dfrac{1}{2}(1-i)$
   - (c) 1
   - (d) -1

8. i^{42} is equivalent to:
   - (a) i
   - (b) $-i$
   - (c) 1
   - (d) -1

9. Simplify $\frac{3}{4}\sqrt{-80}$:

   (a) $\left(12\sqrt{5}\right)i$ (b) $\left(6\sqrt{5}\right)i$

   (c) $\left(3\sqrt{5}\right)i$ (d) $\left(-6\sqrt{5}\right)$

10. The expression $\frac{\sqrt{-36}}{-\sqrt{36}}$ is equivalent to:

   (a) $6i$ (b) $-i$

   (c) i (d) -1

11. Simplify $\frac{8-3i}{i}$:

   (a) $3+8i$ (b) $3-8i$

   (c) $-3-8i$ (d) $-8+3i$

12. The product of i^7 and i^5 is equal to:

   (a) -1 (b) $-i$

   (c) i (d) 1

13. Simplify $\left(-\sqrt{-1}\right)^{4n+3}$:

   (a) 1 (b) $-i$

   (c) i (d) -1

14. Solve $\frac{i^{4n+1}-i^{4n-1}}{2}$:

   (a) 1 (b) $(-1-i)$

   (c) i (d) -1

15. If $(x+iy)(1+i)=1-5i$:

   (a) $x=2, y=3$ (b) $x=-2, y=-3$

   (c) $x=-2, y=3$ (d) $x=2, y=-3$

16. If $\frac{1-ix}{1+ix}=a+ib$, then a^2+b^2 is equal to:

   (a) -1 (b) 0

   (c) i (d) 1

17. $\left(\cos\theta - i\sin\theta\right)^2 = x+iy$, then x^2+y^2 is equal to:

   (a) -1 (b) 0

   (c) i (d) 1

18. If $z = \dfrac{1}{(1-i)(2+3i)}$, then $|z|$ is equal to:

(a) 1

(b) $\dfrac{1}{\sqrt{26}}$

(c) $\dfrac{5}{\sqrt{26}}$

(d) $\dfrac{1}{26}$

19. If $a = 1 + i$, then a^2 is:

(a) $i - 1$

(b) $2i$

(c) $(1 + i)(1 - i)$

(d) $1 - i$

20. The amplitude of $\dfrac{x + iy}{x - iy}$ is equal to:

(a) $\tan^{-1}\left(\dfrac{2ab}{a^2 - b^2}\right)$

(b) $\tan^{-1}\left(\dfrac{ab}{a^2 + b^2}\right)$

(c) $\tan^{-1}\left(\dfrac{a^2 - b^2}{a^2 + b^2}\right)$

(d) 45^0

21. If $z_1 = 3 + 2i$ and $z_2 = 2 - 3i$, then $z_1 z_2$ is:

(a) $5 - 12i$

(b) $(12 - 5i)$

(c) $(5 + 12i)$

(d) $(12 + 5i)$

22. The standard form of $\dfrac{1}{3 - 4i}$ is:

(a) $\dfrac{3}{25} + \dfrac{4}{25}i$

(b) $\dfrac{3}{25} - \dfrac{4}{25}i$

(c) $\dfrac{4}{25} + \dfrac{3}{25}i$

(d) $\dfrac{4}{25} - \dfrac{3}{25}i$

23. Multiplication of $3 - 2i$ by its conjugate is equal to:

(a) 13

(b) $12i$

(c) $13i$

(d) $1 - i$

24. $z = \dfrac{1}{3 + 4i}$, then $\overline{z}$ is:

(a) $\dfrac{3}{25} + \dfrac{4}{25}i$

(b) $\dfrac{3}{25} - \dfrac{4}{25}i$

(c) $\dfrac{4}{25} + \dfrac{3}{25}i$

(d) $\dfrac{4}{25} - \dfrac{3}{25}i$

25. The modulus and argument of $1 + i\sqrt{3}$ are:

(a) $2,\ \tan^{-1}\left|\dfrac{\sqrt{3}}{1}\right|$

(b) $2,\ \tan^{-1}\left|\dfrac{1}{\sqrt{3}}\right|$

(c) $3,\ \tan^{-1}\left|\dfrac{1}{\sqrt{3}}\right|$

(d) $3,\ \tan^{-1}\left|\dfrac{\sqrt{3}}{1}\right|$

26. By DeMoivre's theorem, $(\cos\theta + i\sin\theta)^n$ is equal to:

(a) $n(\cos\theta + i\sin\theta)$

(b) $(\cos n\theta + i\sin n\theta)$

(c) $n(\sin\theta + i\cos\theta)$

(d) $(\sin n\theta + i\cos n\theta)$

27. By DeMoivre's theorem, $(\cos\theta - i\sin\theta)^n$:

(a) $n(\cos\theta - i\sin\theta)$

(b) $(\cos n\theta - i\sin n\theta)$

(c) $n(\sin\theta - i\cos\theta)$

(d) $(\sin n\theta + i\cos n\theta)$

Answers

1b	2c	3a	4c	5d	6c	7b
8b	9c	10b	11c	12b	13c	14b
15b	16d	17d	18b	19b	20a	21b
22a	23a	24a	25	26b	27b	

Functions, Limits and Continuity

1.0 FUNCTIONS

1.0.1 Brief History

The concept of *function* is one of the most important in mathematics. **Galileo** (1564-1642) in his first statement of function conceived it as dependency of one quantity on another. According to Dieudonné and Ponte, the concept of a function emerged in the 17th century as a result of the development of analytic geometry and the infinitesimal calculus.

As a mathematical term, **"function"** was coined by **Gottfried Leibniz.** In 1673 , he described a function as a quantity related to a curve, such as a curve's slope at a specific point. The functions which Leibniz considered are today called differentiable functions..

By 1718, **Johann Bernoulli** regarded a function as "any expression made up of a variable and some constants", while **Alexis Claude Clairaut** (in approximately 1734) and Leonhard Euler introduced the familiar notation "f(x)" for the value of a function.

In 1755, however, in his *Institutiones Calculi Differentialis,* Euler gave a more general concept of a function.

In his *Théorie Analytique de la Chaleur,* Fourier claimed that an arbitrary function could be represented by a Fourier series. Fourier had a general conception of a function, which included functions that were neither continuous nor defined by an analytical expression.

During the 19th century, Cauchy made some general remarks about functions in Chapter I, Section 1 of his *Analyse algébrique* (1821)

What was new and what was to be essential for the whole of mathematics was the entirely general conception of a *function*

1.0.2 Definition of Function

A function is a relationship between a set of inputs (denoted by x) and a set of outputs (denoted by y) with the property that each input value is associated with exactly one output value under a rule 'f'. We can write: $f: x \to y$ or $f(x) = y$

One – one function: If different elements of set $\{x\}$ have different images shown by the set $\{y\}$, then the function 'f'. is a one – one function

For example let $y = f(x) = 2x$. Here different elements in $\{x\}$ will have different images. If $x = 1 \Rightarrow y = 2$. Similarly if x = 2 then y = 4 and so on.

Many – one function: If two or more elements in $\{x\}$ have same images in $\{y\}$, then the function f is said to be the many – one function.

For example let $y = f(x) = x^2$

If $x = -1$ then $y = 1$

If $x = 1$ then $y = 1$

It is seen that two values in $\{x\}$ have the same image in $\{y\}$. Hence the function $f(x) = x^2$ is a **many – one** function.

1.0.3 Some Important Functions

1. **Constant Function**: Let **c** be a fixed real number. Then the function $f(x)$ = c is called a constant function.

2. **Identity Function**: If the function returns the same value then it is the identity function. The function defined by $y = f(x) = x$ is called the identity function.

3. **Modulus Function**: The function defined by $f(x) = |x|$ is called a modulus function. It returns positive value. $|x|$ is either 0 or a positive real number.

4. **Reciprocal Function:** The function defined by $f(x) = \dfrac{1}{x}$ is called a reciprocal function. It may also be written as $f(x) = x^{-1}$.

   Clearly $f(x)$ can not be defined when x = 0. Hence $f(x)$ is defined for all real numbers except x = 0

5. **Square Root Function:** It is defined by $f(x) = \sqrt{x}$. Since negative real numbers do not have real square roots. Hence function $f(x)$ can not be defined when x is negative real number.

6. **Exponential Function:** The function of the form $f(x) = e^x$ is called the exponential function.

7. **Logarithmic Function:** The function of the form $f(x) = \log x$ is known as the logarithmic function.

8. **Even and Odd Functions:** If $f(x)$ is a function of x such that $f(-x) = -f(x)$ .Then $f(x)$ is called an odd function. But if $f(-x) = f(x)$

9. **Undefined Functions:** If putting a value say **a** in the function $f(x)$, we get $f(a)$ in the form of $\dfrac{0}{0}$, $\dfrac{\infty}{\infty}$, (∞,∞) or $(0\times\infty)$, then we say $f(x)$ is not defined when x = a

10. **Equal Functions:** Two functions f and g are said to be equal if $f(x) = g(x)$ for all

$$x \in R$$

Evaluation of a function

1. If $f(x) = x^2 - 3x + 3$, then evaluate $f(0)$, $f(1)$, $f(4)$

Solution: $$f(x) = x^2 - 3x + 3$$

$\therefore f(0) = 0^2 - 3 \times 0 + 3 = 3$

$$f(1) = 1^2 - 3 \times 1 + 3 = 1$$

$\therefore f(4) = 4^2 - 3 \times 4 + 3 = 16 - 12 + 3 = 7$

2. Evaluate: $f(x) = 1 - x + x^2$, if $x = a - 4$

Solution: $f(x) = 1 - x + x^2$

$$f(a-4) = 1 - (a-4) + (a-4)^2$$
$$= 1 - a + 4 + a^2 - 8a + 16 = 21 - 9a + a^2$$

3. Find the value of a in $f(x) = 3x^2 + ax - 1$, if $f(3) = 8$

Solution: $$f(x) = 3x^2 + ax - 1$$

$\therefore f(3) = 3.3^2 + a.3 - 1 = 27 + 3a - 1 = 26 + 3a$

But $f(3) = 8$

$\therefore 26 + 3a = 8 \Rightarrow 3a = -18 \Rightarrow a = -6$

4. If $f(x) = (x^2 + 1)(3x - 5)$, find $f(5)$.

Solution: $f(x) = (x^2 + 1)(3x - 5)$

$$f(5) = (5^2 + 1)(3 \times 5 - 5) = 26 \times 10 = 260$$

5. Evaluate $g(x) = \dfrac{x}{\sqrt{x-3}}$, for x = 12

Solution: $g(x) = \dfrac{x}{\sqrt{x-3}}$

$$\therefore g(12) = \frac{12}{\sqrt{12-3}} = \frac{12}{3} = 4$$

6. Evaluate: $f(x) = |x-7|$, for x = 6

Solution: $f(x) = |x-7|$

$$f(6) = |6-7| = |-1| = 1$$

1.0.4 Domain and Co-domain of a function

Set of inputs (x) is called as the **domain** and the set of outputs (y) is called the **co-domain** under the function 'f'.

Let y is the function of x such that y = 2x

If we put x = 1, 2, 3, 4…….we get y = 2,4,6,8,

Then the input set { 1, 2, 3, 4 ….} is called the **Domain** and the output set { 2, 4, 6, 8,….} is called the **Co-domain**

Image: Co-domain is also known as the Image. In the above example the elements 2, 4, 6, … are the images of 1, 2, 3, …….

Range: The range of a function refers to either the co-domain or the image of the function. The set {f(x)} may be termed as the range or co-domain of the input set {x}

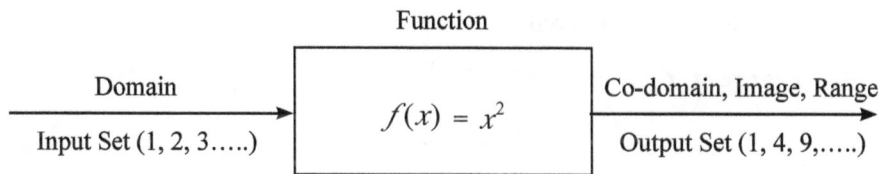

Function

Domain	$f(x) = x^2$	Co-domain, Image, Range
Input Set (1, 2, 3…..)		Output Set (1, 4, 9,…..)

Thus the domain and co-domain are associated with the function 'f'. For each value of a domain there exists a co-domain. Co-domain also gives the range of the function

Solved Examples

Example 1 Find the domain and the range of the functions: **(i)** $f(x) = \dfrac{1}{x}$,

(ii) $f(x) = x^2$, **(iii)** $f(x) = \dfrac{x}{1-x}$ **(iv)** $f(x) = \dfrac{x}{x-7}$ **(v)** $f(x) = \dfrac{x}{1+x^2}$ **(vi)** $f(x) = \dfrac{x^2}{1+x^2}$

(vii) $f(x) = \sqrt{9 - x^2}$, **(viii)** $f(x) = \sin x$, **(ix)** $f(x) = \dfrac{x^2}{x^2 + 7}$

Solution: (i) $f(x) = \dfrac{1}{x}$. It is seen that $f(x)$ can not be defined when x =0. Hence the domain

of $f(x)$ is the set of real numbers except 0. **Hence domain of** $f(x) = R - \{0\}$

Now express given function as $y = \dfrac{1}{x} \Rightarrow x = \dfrac{1}{y}$. We see x is defined for all real number

given to y except 0. **Hence range of** $f(x) = R - \{0\}$

(ii) $f(x) = x^2$. It is seen that $f(x)$ can be defined for all real numbers i.e. for each $x \, \varepsilon \, R$

x^2 gives a unique real number. **Hence domain** of $f(x) = R$ i.e. set of all real numbers.

Write the function as $y = x^2 \Rightarrow x = \sqrt{y}$. Now if y is negative x becomes imaginary

which is not possible. This shows that x is defined for all positive real numbers given to y.

Hence range of $f(x)$ = set of all positive real numbers

(iii) $f(x) = \dfrac{x}{1 - x}$. It may be seen that $f(x)$ can not be defined when $x = 1$ because

$f(x)$ becomes infinite. Hence $f(x)$ is defined for all real number except 1

Therefore domain of $f(x) = R - \{1\}$

We can write $y = \dfrac{x}{1 - x} \Rightarrow y(1 - x) = x$. Solving this we get $x = \dfrac{y}{1 + y}$.

x exits for all values of y except y = -1.

Hence range of $f(x) = R - \{-1\}$

(iv) $f(x) = \dfrac{x}{x - 7}$.

Solution: $f(x)$ is defined if $x - 7 \neq 0$ i.e. $x \neq 7$

So domain of $f(x)$ is all real number except 7 i.e. $\{R - (7)\}$

$$y = \dfrac{x}{x - 7} \Rightarrow yx - 7y = x$$

$$yx - x = 7y$$

$$x = \dfrac{7y}{y - 1}$$

x is defined if $y - 1 \neq 0$

or $\qquad y \neq 1$

Hence, range of $f(x)$ is all real number except 1 i.e. $\{R-(1)\}$

(v) $f(x) = \dfrac{x}{x-7}$. It may be seen that $f(x)$ is defined for all real numbers.

Hence Domain of $f(x) = \{R\}$

Now write the function as $y = \dfrac{x^2}{1+x^2} \Rightarrow x^2 y - x + y = o$. This is the quadratic equation

in x. Its roots can be found as $x = \dfrac{1 \pm \sqrt{1-4y^2}}{2}$ (applying the formula $x = \dfrac{-b \pm \sqrt{a-4ac}}{2}$)

x will be real if $1 - 4y^2 \geq 0$ This implies $1 \geq 4y^2$ or $y \leq \left(\pm \dfrac{1}{2} \right)$

Hence range of $f(x) = \left[\dfrac{-1}{2}, \dfrac{1}{2} \right]$

(vi) $f(x) = \dfrac{x^2}{1+x^2}$. It may be seen that $f(x)$ can be defined for all real numbers.

Hence domain of $f(x) = \{R\}$

Now express $f(x)$ as $y = \dfrac{x^2}{1+x^2}$. We can write this as $y = \dfrac{1+x^2-1}{1+x^2} = 1 - \dfrac{1}{1+x^2}$

If we put x = 0 we get y = 0

If we put $x \geq 1$ we get y < 1. Hence y can take values between 0 and 1

Hence range of $f(x) = [0, 1]$

(vii) $f(x) = \sqrt{9-x^2}$

$f(x)$ is not defined when $\sqrt{9-x^2} < 0$, since it leads to imaginary number. Hence

$f(x)$ is defined only when $\sqrt{9-x^2} \geq 0 \Rightarrow 9 \geq x^2$ **or x ≤ ± 3.**

Hence domain of $f(x) = [-3, 3]$

Now $y = \sqrt{9-x^2} \Rightarrow y^2 = 9 - x^2$. We see y is defined if **x ≤ ± 3.**

Hence range of $f(x) = [-3, 3]$

(viii) $f(x) = \sin x$.

Sin x is defined for all values of x belonging to set of real number, i.e. $x \in R$.

Hence domain of $f(x) = \{R\}$

For any real number $-1 \le \sin x \le 1$ i.e. limits of sin x is $(-1, 1)$

Hence range of $f(x) = [-1, 1]$

(ix) $f(x) = \dfrac{x^2}{x^2 + 7}$.

$f(x)$ is defined for all $x \in R$.. **Hence domain** of $f(x) = \{R\}$

$\dfrac{x^2}{x^2 + 7}$ is a positive fraction > 0 but < 1. **Hence range** of $f(x) = [0, 1]$

1.0.5 Even and Odd Functions

(i) A function is said to be **even** if $f(-x) = f(x)$ .for example $f(x) = \cos x$ is an even function for all values of x since $f(-x) = \cos(-x) = \cos x = f(x)$

(ii) A function is said to be **odd** if $f(-x) = -f(x)$ .for example $f(x) = \sin x$ is an odd function for all values of x since $f(-x) = \sin(-x) = -\sin x = -f(x)$

Solved Examples

1. Show that following functions are even functions:

(i) $f(x) = \dfrac{1}{x^2}$ **(ii)** $f(x) = \dfrac{|x|}{x^2 + 1}$ **(iii)** $f(x) = 5 - x^2$

Solution: (i) $f(x) = \dfrac{1}{x^2}$. Putting $-x$ in place of x in the given function we get –

$f(-x) = \dfrac{1}{(-x)^2} = \dfrac{1}{x^2} = f(x)$. Hence the given function is an even function.

(ii) $f(x) = \dfrac{|x|}{x^2 + 1}$. Putting $-x$ in place of x in the given function we get –

$f(-x) = \dfrac{|-x|}{(-x)^2 + 1} = \dfrac{|x|}{x^2 + 1} = f(x)$. Hence the given function is an even function.

(iii) $f(x) = 5 - x^2$. Putting $-x$ in place of x in the given function we get –

$$f(-x) = 5 - (-x)^2 = 5 - x^2 = f(x).$$ Hence the given function is an even function.

2. Show that following functions are odd functions:

(i) $f(x) = x^3 - x$, **(ii)** $f(x) = x + \dfrac{1}{x}$, **(iii)** $f(x) = \dfrac{|x|}{x}$ **(iv)** $f(x) = -x|x|$

Solution: **(i)** $f(x) = x^3 - x$. Putting $-x$ in place of x in the given function we get –

$$f(-x) = (-x)^3 - (-x) = -x^3 + x = -(x^2 - x) = -f(x).$$

Hence the given function is an odd function.

(ii) $f(x) = x + \dfrac{1}{x} \Rightarrow f(-x) = (-x) + \dfrac{1}{(-x)} = -x - \dfrac{1}{x} = -(x + \dfrac{1}{x}) = -f(x).$

Since $f(x) = -f(x)$, Hence the given function is an odd function.

(iii) $f(x) = \dfrac{|x|}{x} \Rightarrow f(-x) = \dfrac{|-x|}{-x} = -\dfrac{|x|}{x} = -f(x).$ Since $f(x) = -f(x)$, Hence the given

function is an odd function.

(iv) $f(x) = -x|x| \Rightarrow f(-x) = -(-x)|-x| = x|x| = -\{-x|x|\} = -f(x).$ Since $f(x) = -f(x)$

, Hence the given function is an odd function.

3. Find out which of the following functions are even, odd or neither even nor odd.

(i) $f(x) = x^2$ **(ii)** $f(x) = x^3$ **(iii)** $f(x) = x$ **(iv)** $f(x) = -|x| + 2$ **(v)** $f(x) = x^2 + 2$

(vi) $f(x) = \cos x$ **(vii)** $f(x) = \sin x$ **(viii)** $f(x) = -x^3$ **(ix)** $f(x) = 1 - x^2 - 2x^4 + x^6$

(x) $f(x) = x + 2x^3 - x^5$ **(xi)** $f(x) = 2x^3 - 4x$ **(xii)** $f(x) = 2x^3 - 3x^2 - 4x + 4$

(xiii) $f(x) = x\cos x$ **(xiv)** $f(x) = x\sin x$ **(xv)** $f(x) = |x - 2|$.

Solutions:

(i) $f(x) = x^2$. Replacing x by $-x$, we get $f(-x) = (-x)^2 = x^2 = f(x).$

Hence $f(x)$ is **an even function**

(ii) $f(x) = x^3$. Replacing x by $-x$, we get $f(-x) = (-x)^3 = -x^3 = -f(x).$

Hence $f(x)$ is **an odd function**

(iii) $f(x) = x$. Replacing x by $-x$, we get $f(-x) = -x = -f(x)$.

Hence $f(x)$ is **an odd function**

(iv) $f(x) = -|x| + 2$. Replacing x by $-x$, we get $f(-x) = -|-x| + 2 = -|x| + 2 = f(x)$.

Hence $f(x)$ is **an even function.**

(v) $f(x) = x^2 + 2$. Replacing x by $-x$, we get $f(-x) = (-x)^2 + 2 = f(x)$

Hence $f(x)$ is **an even function.**

(vi) $f(x) = \cos x$. Replacing x by $-x$, we get $f(-x) = \cos(-x) = \cos x = f(x)$.

Hence $f(x)$ is **an even function.**

(vii) $f(x) = \sin x$. Replacing x by $-x$, we get $f(-x) = \sin(-x) = -\sin x = -f(x)$.

Hence $f(x)$ is **an odd function**

(viii) $f(x) = -x^3$. Replacing x by $-x$, we get $f(-x) = -(-x)^3 = x^3 = f(x)$.

Hence $f(x)$ is **an even function**

(ix) $f(x) = 1 - x^2 - 2x^4 + x^6$.

Replacing x by $-x$, we get

$$f(-x) = 1 - (-x)^2 - 2(-x)^4 + (-x)^6 = 1 - x^2 - 2x^4 + x^6 = f(x).$$

Hence $f(x)$ is **an even function**

(x) $f(x) = x + 2x^3 - x^5$.

Replacing x by $-x$, we get

$$f(-x) = (-x) + 2(-x)^3 - (-x)^5 = -x - 2x^3 + x^5 = -\{x + 2x^3 - x^5\} = -f(x)$$

Hence $f(x)$ is **an odd function**

(xi) $f(x) = 2x^3 - 4x$.

Replacing x by $-x$, we get

$$f(-x) = 2(-x)^3 - 4(-x) = -2x^3 + 4x = -\{2x^3 - 4x\} = -f(x)$$

Hence $f(x)$ is **an odd function.**

(xii) $f(x) = 2x^3 - 3x^2 - 4x + 4$. Replacing x by $-x$, we get

$$f(-x) = 2(-x)^3 - 3(-x)^2 - 4(-x) + 4 = -2x^3 - 3x^2 + 4x + 4.$$

The function is neither even nor odd.

(xiii) $f(x) = x\cos x$

$\Rightarrow f(-x) = -x\cos(-x) = -x\cos x = -f(x)$.

The function is odd

(xiv) $f(x) = x\sin x$.

Replacing x by $-x$, we get

$$f(-x) = (-x)\{\sin(-x)\} = (-x)\{-\sin x\} = x\sin x = f(x).$$

Hence $f(x)$ is **an even function.**

(xv) $f(x) = |x-2|. \Rightarrow f(-x) = |-x-2| = |x+2|$.

The function $f(x)$ is neither even nor odd

1.0.6 Miscellaneous Problems

1. If $f(x) = x^3 + (k-2)x^2 + 2$, **for all** x **and** $f(x)$ **is an odd function. Find the value of k.**

Solution: $f(-x) = (-x)^3 + (k-2)(-x)^2 + 2(-x)$

Since $f(x)$ is odd function i.e. $f(-x) = -f(x)$

Hence $\quad (-x)^3 + (k-2)(-x)^2 + 2(-x) = -\{x^3 + (k-2)x^2 + 2x\}$

$-x^3 + (k-2)x^2 - 2x = -x^3 - (k-2)x^2 - 2x$

$(k-2)x^2 = -(k-2)x^2$

$2(k-2)x^2 = 0 \quad \Rightarrow k-2 = 0 \Rightarrow k = 2$

2. If $f(x) = x^2 + 3x + 2$. **, find the value of x such that** $f(x+1) = f(x)$

Solution: $f(x+1) = (x+1)^2 + 3(x+1) + 2 = x^2 + 2x + 1 + 3x + 3 + 2 = x^2 + 5x + 6$

Since $f(x+1) = f(x)$

Hence $x^2 + 5x + 6 = x^2 + 3x + 2 \Rightarrow 2x = -4 \Rightarrow x = -2$

3. If $f(x) = x^2 - \dfrac{1}{x^2}$. **Show that** $f(x) + f\left(\dfrac{1}{x}\right) = 0$

$$f\left(\frac{1}{x}\right) = \left(\frac{1}{x}\right)^2 - \left(\frac{1}{1/x}\right)^2 = \left(\frac{1}{x}\right)^2 - x^2$$

$$f(x) + f\left(\frac{1}{x}\right) = x^2 - \frac{1}{x^2} + \frac{1}{x^2} - x^2 = 0 \text{ Proved}$$

4. If $f(x) = \dfrac{2x}{1+x^2}$ **. Prove that** $f(\tan\theta) = \sin 2\theta$

Solution: Putting $\tan\theta$ in place of x in the given function we get-

$$f(\tan\theta) = \frac{2\tan\theta}{1+\tan^2\theta} = \frac{2\tan\theta}{\sec^2\theta} = \frac{2\sin\theta}{\cos\theta}.\cos^2\theta = 2\sin\theta\cos\theta = \sin 2\theta \text{ Proved.}$$

5. If $f(x) = \log\left(\dfrac{1+x}{1-x}\right)$ **. Show that** $f\left(\dfrac{2x}{1+x^2}\right) = 2f(x)$

Solution: Putting $\left(\dfrac{2x}{1+x^2}\right)$ in place of x in the given function we get-

$$f\left(\frac{2x}{1+x^2}\right) = \log\left(\frac{1+\frac{2x}{1+x^2}}{1-2x/1+x^2}\right) = \log\left(\frac{\frac{1+x^2+2x}{1+x^2}}{\frac{1+x^2-2x}{1+x^2}}\right) = \log\left(\frac{(1+x)^2}{(1-x)^2}\right)$$

$$= \log\left(\frac{1+x}{1-x}\right)^2 = 2\log\left(\frac{1+x}{1-x}\right)$$

Hence $f\left(\dfrac{2x}{1+x^2}\right) = 2f(x)$

6. If $f(x) = \begin{cases} 2x+1 & \text{when } x \geq 2 \\ x & \text{when } x < 2 \end{cases}$ **, find** $f(1), f(2) \ and \ f(5)$

Solution: Since $1 < 2$, hence $f(x) = x \Rightarrow f(1) = 1$

Put $x = 2$, in the function applicable is $f(x) = 2x+1$, we get $f(2) = 2\times 2 +1 = 5$

Put $x = 5 > 2$ in the function applicable is $f(x) = 2x+1$, we get $f(5) = 2\times 5 +1 = 11$

7. If $f(x) = \left(x^2 - 3x + 1\right)$ **. Find** $f(0), f(-1), f(-\sqrt{2})$

Solution: $f(0) = (0-0+1) = 1$, $f(-1) = (1+3+1) = 5$,

$$f(-\sqrt{2}) = (-\sqrt{2})^2 - 3(-\sqrt{2}) + 1 = 2 + 3\sqrt{2} + 1 = 3(1+\sqrt{2})$$

8. If $f(x) = \sin^{-1}(\log x)$, **find** $f(1)$.

Solution: $f(1) = \sin^{-1}(\log 1) = \sin^{-1}(0) = 0$

9. If $f(x) = \dfrac{1-x^2}{1+x^2}$. **Show** $f(\tan\theta) = \cos 2\theta$

Solution: Putting $x = \tan\theta$ we get

$$f(\tan\theta) = \frac{1-\tan^2\theta}{1+\tan^2\theta} = \frac{1 - \dfrac{\sin^2\theta}{\cos^2\theta}}{\sec^2\theta} = \frac{\cos^2\theta - \sin^2\theta}{\cos^2\theta.\sec^2\theta} = \cos 2\theta$$

10. If $f(x) = \log\left(\dfrac{1-x}{1+x}\right)$. **Prove that** $f(a) + f(b) = f\left(\dfrac{a+b}{1+ab}\right)$

Solution: Putting $x = a$ in the given function we get $f(a) = \log\left(\dfrac{1-a}{1+a}\right)$

Putting $x = b$ in the given function we get $f(b) = \log\left(\dfrac{1-b}{1+b}\right)$

Hence $f(a) + f(b) = \log\left(\dfrac{1-a}{1+a}\right) + \log\left(\dfrac{1-b}{1+b}\right) = \log\left\{\left(\dfrac{1-a}{1+a}\right)\left(\dfrac{1-b}{1+b}\right)\right\}$,

Since $\log(x) + \log(y) = \log(xy)$

$$= \log\left(\frac{1-a-b+ab}{1+a+b+ab}\right) \qquad\qquad ...(\mathrm{I})$$

$$f\left(\frac{a+b}{1+ab}\right) = \log\left(\frac{1 - \dfrac{a+b}{1+ab}}{1 + \dfrac{a+b}{1+ab}}\right) = \log\left(\frac{1+ab-a-b}{1+ab+a+b}\right) \qquad\qquad ...(\mathrm{II})$$

Comparing (I) and (II) $f(a) + f(b) = \log\left(\dfrac{1-a-b+ab}{1+a+b+ab}\right) = f\left(\dfrac{a+b}{1+ab}\right)$ Hence Proved.

Practice Problems (On Functions)

1. If $f(x) = \dfrac{x-1}{x+1}$. **Show that** $f\left(\dfrac{x-1}{x+1}\right) = \dfrac{-1}{x}$

2. If $y = f(x) = \dfrac{2x-3}{5x-2}$. **Show that** $f(y) = x$

3. If $f(x) = \dfrac{\sin x}{1+\cos x}$. **Find the value of** $f\left(\dfrac{\pi}{2}\right)$.

4. If $f(\theta) = \dfrac{1-2\tan\theta}{1+2\tan\theta}$. **Find the value of** $f\left(\dfrac{\pi}{4}\right)$.

5. If $f(x) = x^2 + kx + 2$, find the value of k if $f(x)$ **is an odd function.**

6. If $f(x) = a^x$, show that $f(x+y) = f(x).f(y)$

7. If $f(x) = 2x\sqrt{1-x^2}$, then show that $f\left(\sin\dfrac{\theta}{2}\right) = \sin\theta$

8. Find the domain and range of the following functions:

(i) $f(x) = \dfrac{1}{x}$, (ii) $f(x) = -x|x|$, (iii) $f(x)$ $\dfrac{}{\sqrt{}}$ (iv) $f(x) = \dfrac{x}{x^2+7}$

9. If $f(x) = x^2 + kx + 1$ for all x and $f(x)$ is an even function. Find the value of k.

Answers:

3. 1 , **4.** $\dfrac{-1}{3}$, **5.** 0 , **8.** (i) $\{(R)-0\}$ (ii) $\{R\}$ (iii) $x \le 0$,(iv) All real numbers

9. k = 0

1.0.7 Operations on Functions

Let f and g be the real functions of x, then their sum, difference, product and division may be defined as follows:

(i) $(f+g)x = f(x) + g(x)$

(ii) $(f-g)x = f(x) - g(x)$

(iii) $(fg)x = f(x).g(x)$

(iv) $\left(\dfrac{f}{g}\right)x = \dfrac{f(x)}{g(x)}$

(v) $(f+k)x = f(x) + k$

Solved Examples

1. If $f(x) = 3\sin x$ and $g(x) = \cos^2 x$, **find** *(i)* $(f+g)\dfrac{\pi}{3}$, *(ii)* $(f-g)\dfrac{\pi}{3}$,

(iii) $(fg)\dfrac{\pi}{3}$, *(iv)* $f(x) = 1 + x^2$ **and** $g(x) = 1-x$, **find** $\left(\dfrac{f}{g}\right)x$

Solutions

(i) $(f+g)x = f(x) + g(x) = 3\sin x + \cos^2 x = 3\sin\dfrac{\pi}{3} + \cos^2\dfrac{\pi}{3}$

$\therefore (f+g)\dfrac{\pi}{3} = f\left(\dfrac{\pi}{3}\right) + g\left(\dfrac{\pi}{3}\right) = 3\sin x + \cos^2 x$

$$= 3\sin\frac{\pi}{3} + \cos^2\frac{\pi}{3} = 3.\frac{\sqrt{3}}{2} + \left(\frac{1}{2}\right)^2$$

$$= \frac{2\times 3\sqrt{3}+1}{4} = \frac{6\sqrt{3}+1}{4}$$

(ii) $(f-g)x = f(x) - g(x) = 3\sin x - \cos^2 x$

$$\therefore (f-g)\frac{\pi}{3} = f\left(\frac{\pi}{3}\right) - g\left(\frac{\pi}{3}\right) = 3\sin\frac{\pi}{3} - \cos^2\frac{\pi}{3} = 3.\frac{\sqrt{3}}{2} - \left(\frac{1}{2}\right)^2$$

$$= \frac{2\times 3\sqrt{3}-1}{4} = \frac{6\sqrt{3}-1}{4}$$

(iii) $(f.g)x = f(x).g(x) = 3\sin x.\cos^2 x$

$$\therefore (f+g)\frac{\pi}{3} = f\left(\frac{\pi}{3}\right).g\left(\frac{\pi}{3}\right) = 3\sin\frac{\pi}{3}.\cos^2\frac{\pi}{3} = 3.\frac{\sqrt{3}}{2}.\left(\frac{1}{2}\right)^2 = \frac{3\sqrt{3}}{8}$$

(iv) If $f(x) = 1 + x^2$ and $g(x) = 1 - x$, then find $\left(\dfrac{f}{g}\right)x$

$$\left(\frac{f}{g}\right)x = \frac{f(x)}{g(x)} = \frac{1+x^2}{1-x}$$

Practice Problems (Operation on Functions)

1. Let $f(x) = 5\sin(5x)$ and $g(x) = x^2$.

Find **(i)** $(f+g)x$, **(ii)** $(f.g)x$, **(iii)** $(f-g)x$, **(iv)** $(f/g)x$

Answers:

(i) $5\sin 5x + x^2$, **(ii)** $5x^2 \sin 5x$, **(iii)** $5\sin 5x - x^2$, **(iv)** $\dfrac{5\sin 5x}{x^2}$

2. Let $f(x) = \dfrac{1}{x}$ and $g(x) = (x^2 + 3)$. Find the following functions:

(i) $(f+g)x$, **(ii)** $(f.g)x$, **(iii)** $(f-g)x$ **(iv)** $(f/g)x$

Answers:

(i) $\dfrac{1}{x} + x^2 + 3$, **(ii)** $x(x^2 + 3)$, **(iii)** $\dfrac{1}{x} - (x^2 + 3)$, **(iv)** $\dfrac{1}{x(x^2 + 3)}$

3. Find $(f+g)x$ and $(f-g)x$ for the following functions.

    (i) $f(x) = 5x-4$ and $g(x) = 4x-11$

    (ii) $f(x) = 4x^2 - 2x + 3$ and $g(x) = -11x^2 + 8x - 1 \ \ 15x^2 - 10x + 4$

Answers:

    (i) $(9x-15)$, $(x+7)$, **(ii)** $-7x^2 + 6x + 2,$

4. Find $(fg)x$ and $(f/g)x$ for the following functions;

    (i) $f(x) = x+3$ and $g(x) = x-7$

    (ii) $f(x) = x+8$ and $g(x) = x-2$

Answers:

    (i) $x^2 - 4x - 21,\ \dfrac{x+3}{x-7}$, **(ii)** $x^2 - 6x - 16,\ \dfrac{x+8}{x-2}$

1.0.8 Composite Functions

Let $f: A \to B$ **and** $g: B \to C$ be two functions. Then a function $gof: A \to C$ is defined by $(gof)x = g(f(x))$, for all values of x. The function of a function is known as a composite function.

Solved Examples

1. $f(x) = x^2 - 1$ and $g(x) = 3x + 1$.

    Find **(i)** $(gof)x$, **(ii)** $(fog)x$, **(iii)** $(fof)x$, **(iv)** $(gog)x$

Solution: (i) $(gof)x = g\{f(x)\} = g\{1-x^2\} = g(x^2) - g(1)$

$$= 3(x^2) + 1 - 1 = 3x^2 - 3 + 1 = 3x^2 - 2$$

(ii) $(fog)x = f\{g(x)\} = f\{3x+1\} = (3x+1)^2 - 1 = 9x^2 + 6x + 1 - 1 = 9x^2 + 6x$

(iii) $(fof)x = f\{f(x)\} = f\{x^2 - 1\} = (x^2 - 1)^2 - 1 = x^4 - 2x^2 + 1 - 1 = x^4 - 2x^2$

(iv) $(gog)x = g\{g(x)\} = g(3x+1) = 3g(x) + g(1) = 3(3x+1) + (3.1+1) = 9x + 7$

2. If $f(x) = \dfrac{1-x}{1+x},\ \ 0 \le x \le 1$ **and** $g(x) = 4x(1-x)\ \ 0 \le x \le 1$.

   Find $(i)\,(fog)\,x,\ (ii)\,(gof)\,x$

Solution: $(i)\,(fog)\,x = f\{g(x)\} = f\{4x(1-x)\}$

   Now replacing x by $\{4x(1-x)\}$ in $f(x) = \dfrac{1-x}{1+x},$ **we get**

$$f\{4x(1-x)\} = \dfrac{1-4x(1-x)}{1+4x(1-x)} = \dfrac{4x^4 - 4x + 1}{1 + 4x - 4x^2} \quad \text{Ans}$$

$(ii)\,(gof)\,x = g\{f(x)\} = g\!\left(\dfrac{1-x}{1-x}\right)$

   Now replacing x by $\dfrac{1-x}{1+x}$ in $g(x) = 4x(1-x)$

$$g\!\left(\dfrac{1-x}{1-x}\right) = 4\!\left(\dfrac{1-x}{1+x}\right)\!\left\{1 - \left(\dfrac{1-x}{1+x}\right)\right\} = 4\!\left(\dfrac{1-x}{1+x}\right)\!\left\{\dfrac{(1+x)-(1-x)}{1+x}\right\}$$

$$= 4\!\left(\dfrac{1-x}{1+x}\right)\!\left\{\dfrac{2x}{1+x}\right\} = \dfrac{8x(1-x)}{(1+x)^2} \quad \text{Ans}$$

3. If $f(x) = \sin x$, $g(x) = 2x$ **and** $h(x) = \cos x$,

   show that $(fog)\,x = \{go(fh)\}\,x$

Solution: $(fog)\,x = f\{g(x)\} = f(2x)$.

   Replacing x by $2x$ in $f(x) = \sin x$ **we get** $f(2x) = \sin 2x$

   Hence $(fog)\,x = f(2x) = \sin 2x$

   Now $(fh)\,x = f(x).h(x) = \sin x.\cos x$

   Hence $\{go(fh)\}\,x = g\{(fh)\,x\} = g\{f(x)h(x)\} = g\{\sin x.\cos x\}$

$$= 2\sin x.\cos x$$

$$= \sin 2x$$

   Thus $(fog)\,x = \{go(fh)\}\,x$

1.0.9 Evaluation of Composite Function

1. Given $f(x) = -3x$, $g(x) = \sqrt{2x}$, **and** $h(x) = |4x| - 12$.

   Find $f\big[h\{g(18)\}\big]$

Solution: $g(x) = \sqrt{2x}$ $\therefore h\{g(x)\} = h\{\sqrt{2x}\} = h\{\sqrt{2 \times 18}\} = h(6)$

$$h(6) = |4 \times 6| - 12 = 24 - 12 = 12$$

$$f\{h(x)\} = f(12) = -3 \times 12 = -36$$

2. $g(x) = 4x - 1$ **and** $f(x) = 2x + 3.g(x)$**. Find the value of** $g\{f(3)\}$

Solution: $f(x) = 2x + 3.g(x)$

$$\therefore f(3) = 2 \times 3 + 3.g(3) = 6 + 3 \times 11 = 39$$

$$g\{f(3)\} = g\{39\}$$

$\because g(x) = 4x - 1$

$$\therefore g(39) = 4 \times 39 - 1 = 136 - 1 = 135$$

$$g\{f(3)\} = g\{39\} = 135$$

3. $f(x) = 3x^3 + x^2 + 2.h(x)$ **and** $h(x) = -4x$**. Find the value of** $h\{f(-2)\}$ **.**

Solution: $f(x) = 3x^3 + x^2 + 2.h(x)$

$$\therefore f(-2) = 3(-2)^3 + (-2)^2 + 2.h(-2)$$

$$= -24 + 4 + 2.8 = -24 + 20 = -4$$

$$h\{f(-2)\} = h\{-4\}$$

$$h(x) = -4x$$

$$h(-4) = -4 \times -4 = 16$$

$$\therefore h\{f(-2)\} = h\{-4\} = 16$$

Practice Problems (On Evaluation of Composite Functions)

1. Let $f(x) = x - 2$ and $g(x)5x + 3$. Evaluate the following functions:

(i) $f\{g(2)\}$, **(ii)** $g\{f(-4)\}$, **(iii)** $f\{f(1)\}$, **(iv)** $f\{g(x)\}$

2. If $f(x) = x^2$ and $g(x) = x + 1$, then find the followings: **(i)** $(fog)x$, **(ii)** $(gof)x$

3. Let $f(x) = x^2$, $g(x) = -2x$ and $h(x) = x - 2$ .Find the following functions:

(i) $\{(fog)oh\}x$ and **(ii)** $\{fo(goh)\}x$

4. Given $f(x) = 3x + k$ and $g(x) = 4x - 3$. Find value of k if $f\{g(x)\} = g\{f(x)\}$

5. If $f(x) = \dfrac{x-1}{4}$ and $g(x) = 4x - 3$, then find **(i)** $f\{g(a)\}$ and **(ii)** $g\{f(-b)\}$

6. Find $(fog)x$ and $(gof)x$, if $f(x) = 2x + 1$ and $g(x) = -x^2 + 6$

7. Find $f\{g(4)\}$, if $f(x) = 6x + 5$, and $g(x) = 3x^2$.

8. If $f(x) = x - 3$, find **(i)** $(fof)(6)$ **(ii)** $(fofof)(6)$

Answers:

1. (i) (11), **(ii)** (-27), **(iii)** (-3), **(iv)** (5x +1) **2. (i)** $x^2 + 2x + 1$, **(ii)** $x^2 + 1$

3. (i) $4x^2$, **(ii)** $4x^2 - 16x + 16$ **4.** k = 4 **5. (i)** $(a-1)$, **(ii)** $-(b+4)$

6. $13 - 2x^2$, $-2\left(2x^2 + x + 2\right)$ **7.** 293 **8. (i)** 0, **(ii)** -3

1.1 LINEAR FUNCTIONS

The functions of the forms:

$$y = mx + c \qquad \qquad \text{...(1)}$$

or $$(y - y_1) = m(x - x_1) \qquad \qquad \text{...(2)}$$

or $$Ax + By + c = 0 \qquad \qquad \text{...(3)}$$

are known as *linear functions*. In above functions, x is the independent variable and y is the dependent variable. The independent variable x is the input variable. The power of both the variables x, y are 1. The above are the three main forms of linear functions. All the three forms of functions yield a *straight line* when plotted on a graph.

1.1.1 Forms of Linear functions.

1. **Slope- Intercept Form:** A very common way to express a linear function is the slope-intercept form:

$$y = mx + c$$

where m is the slope or inclination of the straight line and c is the intercept it cuts from y-axis. If *m* and *c* are known, then for each value of x, y values may be obtained. If on a graph –sheet each point (x, y) is marked and then joined, a straight line will be obtained.

In the formal function definition the above expression may be expressed as:

$$f(x) = mx + c$$

2. **Point Slope Form:** If the coordinates of the point through which the line passes, say (x_1, y_1) and the slope or inclination of the straight line are known, then the linear function may be obtained as:

$$(y - y_1) = m(x - x_1)$$

The above form may also be written as

$$y = mx + (y_1 - mx_1)$$

As a formal function form it may written as:

$$f(x) = mx + (y_1 - mx_1)$$

$(y_1 - mx_1)$ is a constant term because it is the difference of two constants. It is known as y-intercept c.

If the values of x_1, y_1 and m are entered in the above form, an equation of a straight line will be obtained

3. **General Form:** The general form of linear function is as given below:

$$Ax + By + c = 0$$

Using algebra, the general term can be converted into slope intercept form. This will help to find the slope as well as the intercept of the straight line.

1.1.2 Properties of Linear Function

Let m_1, m_2 are the slopes of two straight lines.

The lines are parallel if $m_1 = m_2$

The lines are perpendicular if $m_1.m_2 = -1$

The lines are horizontal if $m_1 = 0$ i.e. the line is parallel to x-axis.

Solved Examples

1. Find the slope and intercept of the line whose function f satisfies : $f(3) = -2$ and $f(5) = 7$.

Solution: Putting x=3 and 5 respectively in the function $f(x) = mx + c$ we get

$$f(3) = 3m + c = -2 \quad\quad\quad ...(1)$$

$$f(5) = 5m + c = 7 \quad\quad\quad ...(2)$$

Subtracting eqn.(1) from eqn. (2) we get

$$5m - 2m = 7 - (-2) = 9 \Rightarrow 2m = 9 \Rightarrow m = \frac{9}{2}$$

Substituting value of m in (2) we get

$$5\left(\frac{9}{2}\right)+c=7\Rightarrow c=7-\frac{45}{2}=\frac{14-45}{2}=\frac{-31}{2}$$

Hence slope $m=\frac{9}{2}=4.5$ and y-intercept $c=\frac{-31}{2}=-15.5$

2. Find the form of the linear function $f(x)$ **if** $f(5)=-2$ **and** $f(1)=6$

Solution: Let the form of the linear function be $f(x)=mx+c$.

Substitute x = 5 and 1 respectively in the above form of linear function:

$$f(5)=5m+c=-2 \qquad\qquad\qquad ...(1)$$

$$f(1)=1m+c=6 \qquad\qquad\qquad ...(2)$$

Subtracting (2) from (1) we get 4m = –2–6 = –8

$\therefore m=-2$

Put $m=-2$ in eqn (1) then 5(–2) + c = – 2

$\therefore c=-2+10=8$

Hence the form of the linear function is $f(x)=-2x+8$

3. Find the linear function whose slope is – 2/7 and passes through the point (6, –1).

Solution: Since point and slope are given, the point–slope form may be used.

The point-slope form is:

$$\left(y-y_1\right)=m\left(x-x_1\right) \qquad\qquad\qquad ...(1)$$

We have **m = – 2/7, x_1 = 6 and y_1 = –1**

Substituting the given values in (1) we get $\left(y+1\right)=-\frac{2}{7}\left(x-6\right)$

$$7\left(y+1\right)=-2\left(x-6\right)$$

$$7y+7=-2x+12$$

$$7y=-2x+5$$

$$y=-\frac{2}{7}x+\frac{5}{7}$$

Hence in **linear equation form** we have $y=-\frac{2}{7}x+\frac{5}{7}$

In **linear function form** we can write : $f(x)=-\frac{2}{7}x+\frac{5}{7}$

4. Find the Linear function which passes through the points (7,–1) and (4,5)

Solution: The line passing through the points (x_1, y_1) and (x_2, y_2) is given by:

$$\left(y - y_1\right) = \frac{\left(y_2 - y_1\right)}{\left(x_2 - x_1\right)}\left(x - x_1\right)$$

We are given: $x_1 = 7, y_1 = -1, x_2 = 4, y_2 = 5$. Substituting these values in above formula

We get

$$\left(y + 1\right) = \frac{(5+1)}{(4-7)}\left(x - 7\right)$$

$$\left(y + 1\right) = \frac{6}{-3}\left(x - 7\right)$$

$$\left(y + 1\right) = -2\left(x - 7\right) = -2x + 14$$

$$y = -2x + 13$$

5. Find whether the following equations are parallel or perpendicular:

(i) $x - 3y = 5$ **and** $-2x + 6y = 8$ **(ii)** $y = -x+$ **and** $5x - 2y = -4$

Solution: (i) $x - 3y = 5$ **or** $3y = x + 5$ **or** $y = \frac{1}{3}x + \frac{5}{3}$ **. Hence slope m$_1$** $= \frac{1}{3}$

$-2x + 6y = 8$ **or** $6y = 2x + 8$ **or** $y = \frac{2}{6}x + \frac{8}{6}$ **.** Hence slope m$_2$ $= \frac{2}{6} = \frac{1}{3}$

Since $m_1 = m_2$, hence given lines are parallel.

(ii) First equation is $y = \frac{2}{5}x + 2$. **Hence slope m$_1$** $= \frac{2}{5}$

The second equation is: $5x + 2y = -4$ **or** $2y = -5x - 4$ **or** $y = -\frac{5}{2}x - \frac{4}{2}$

Hence slope m$_2$ $= -\frac{5}{2}$

It is seen that $m_1 \times m_2 = \frac{2}{5} \times -\frac{5}{2} = -1$

Hence the given lines are perpendicular to each other.

6. Find the point of intersection of the following linear equations:

(i) $2x - y = 10$ **and** $x + y = -1$ **(ii)** $y = \frac{2}{3}x + 5$ **and** $2x - 3y = -15$

Solution: (i)
$$2x - y = 10 \qquad \text{...(i)}$$
$$x + y = -1 \qquad \text{...(ii)}$$

Adding the above equations, we get $3x = 9 \Rightarrow x = 3$

Put x = 3 in equation (i) $3 + y = -1 \Rightarrow y = -3 - 1 = -4$

Hence point of intersection is (3, -4)

(ii) Equations are
$$y = \frac{2}{3}x + 5 \qquad \text{...(i)}$$

and
$$x + 3y = -15 \qquad \text{...(ii)}$$

Put $y = \frac{2}{3}x + 5$ in equation (2) we get $x + 3\left(\frac{2}{3}x + 5\right) = -15$

$$1x + 3 \times \frac{2}{3}x + 3 \times 5 = -15$$

$$3x + 15 = -15 \Rightarrow x = \frac{-30}{3} = -10$$

From equation (1) $y = \frac{2}{3}(-10) + 5 = \frac{-20 + 15}{3} = \frac{-5}{3}$

Hence point of intersection is (−10, −5/3).

7. If the point (3, 4) lies on the equation 3y = mx + 7, find the value of m?

Solution. The given equation is

$$3y = ax + 7 \text{ equation (1)}$$

According to problem, point (3, 4) lie on it.

So, putting the value $x = 3$ and $y = 4$ in equation (1)

$$3 \times 4 = m \times 3 + 7$$
$$12 = m^3 + 7$$
$$12 - 7 = 3m$$
$$5 = 3m$$
$$\therefore \qquad m = 5/3$$

1.2 LIMITS

Let $f(x)$ be an algebraic function defined on an interval that contain a point x = a then limit

of $f(x)$ at x = a is denoted by L and we can express the limit as $\lim_{x \to a} f(x) = L$

The **limit of a function** is a fundamental concept in calculus and analysis concerning the behavior of that function near a particular input.

Although implicit in the development of calculus of the 17th and 18th centuries, the modern idea of the limit of a function goes back to Bolzano who, in 1817, introduced the basics of the epsilon-delta technique to define continuous functions. However, his work was not known during his lifetime (Fleischer 2000). Cauchy discussed limits in his *Cours d'analyse* (1821) and gave essentially the modern definition, but this is not often recognized because he only gave a verbal definition (Grabiner 1983). Weierstrass first introduced the epsilon-delta definition of limit in the form it is usually written today. He also introduced the notations **lim** and $\textbf{lim}_{x \to x0}$

The modern notation of placing the arrow below the limit symbol is due to Hardy in his book *A Course of Pure Mathematics* in 1908 (Miller 2004).

To say that $\lim_{x \to p} f(x) = L$ means that $f(x)$ can be made as close as desired to L by making x close enough, but not equal, to p.

1.2.1 Different Methods of Finding Limits

A. Limits by direct substitution

If by direct substitution of the value of the given point in the given expression, a finite number is obtained than it is the limit of the expression at the given point. .

Solved Examples

Evaluate the following limits

1. $\lim\limits_{x \to 2} \left(\dfrac{x^2 + 1}{x + 5} \right)$

Solution: By direct substitution i.e. putting x = 2

$$\lim_{x \to 2} \left(\frac{x^2 + 1}{x + 5} \right) = \frac{2^2 + 1}{2 + 5} = \frac{5}{7}$$

2. $\lim\limits_{x \to 5} \left(\dfrac{x - 5}{x^2 + 1} \right)$

Solution: By direct substitution: i.e. putting $x = 5$

$$\lim_{x \to 5} \left(\frac{x - 5}{x^2 + 1} \right) = \frac{5 - 5}{5^2 + 1} = \frac{0}{26} = 0$$

3. $\lim\limits_{x \to -4} |3x| - 4 = |3(-4)| - 4 = 12 - 4 = 8$

4. $\lim\limits_{x\to-2} \dfrac{x^3+2x^2-1}{5-3x}$

Solution: By direct substitution:

$$\lim\limits_{x\to-2} \frac{x^3+2x^2-1}{5-3x} = \frac{(-2)^3+2(-2)^2-1}{5-(-2)} = \frac{-8+8-1}{5+6} = \frac{-1}{11}$$

5. $\lim\limits_{x\to8} \log_2 x = \log_2 8 = \log_2 2^3 = 3\log_2 2 = 3 \quad \because \log_2 2 = 1$

Practice Problems (On Limits by direct substitution)

1. Find the limits by direct substitution method.

(*i*) $\lim\limits_{x\to2}(2x+1)$, (*ii*) $\lim\limits_{x\to c}(2x^2)$, (*iii*) $\lim\limits_{x\to-2}\left(\dfrac{4}{x^2}\right)$, (*iv*) $\lim\limits_{x\to3} x^2(2-x)$,

(*v*) $\lim\limits_{x\to2}\dfrac{\left(x^2+2x+4\right)}{x+2}$ (*vi*) $\lim\limits_{x\to-4}\left|3x\right|-4$, (*vii*) $\lim\limits_{x\to-3} x^3-3x-2$,

(*viii*) $\lim\limits_{x\to3}\left(x^2-\dfrac{3}{5}x+9\right)$ (*ix*) $\lim\limits_{x\to0}\dfrac{2x^2+3x+4}{x^2+3x+2}$

Answers:

(*i*) 5, (*ii*) $2c^2$, (*iii*) 1, (*iv*) -9, (*v*) 3. (*vi*) 8 (*vii*) -20 (*viii*) 16.2 (ix) 2

B. Limits by factorization

If by direct substitution we get indeterminate form viz. $\dfrac{0}{0}$ or $\dfrac{\infty}{\infty}$ then factorize the numerator and denominator and cancel out the common factors. If now by direct substitution we get a definite number than that is the limit of given expression, otherwise we may repeat the process till a definite number is obtained.

1. Find the limits of following functions.

(*i*) $\lim\limits_{x\to2}\dfrac{x^3-8}{x^2-4}$, (*ii*) $\lim\limits_{x\to5}\dfrac{x^2-25}{3(x-5)}$, (*iii*) $\lim\limits_{x\to36}\dfrac{x-36}{\sqrt{x}-6}$, (*iv*) $\lim\limits_{x\to1}\dfrac{x^2-1}{x-1}$

(*v*) $\lim\limits_{x\to1}\dfrac{x-2}{\sqrt{x\sqrt{2}}}$ (*vi*) if $\lim\limits_{x\to1}\dfrac{x^4-1}{x-1}=\lim\limits_{x\to k}\dfrac{x^3-k^3}{x^2-k^2}$, find k (*vii*) $\lim\limits_{x\to3}(x^2-9)\left[\dfrac{1}{x+3}+\dfrac{1}{x-3}\right]$

(*viii*) $\lim\limits_{x\to4}\dfrac{x^2+x-20}{x-4}$

Solutin: (*i*) $\lim\limits_{x \to 2} \dfrac{x^3 - 8}{x^2 - 4}$

If we put x = 2, then $\dfrac{x^3 - 8}{x^2 - 4} = \dfrac{0}{0}$ which is an indeterminate form. Hence by factorization we get:

$$\lim_{x \to 2} \frac{x^3 - 8}{x^2 - 4} = \lim_{x \to 2} \frac{(x-2)(x^2 + x + 4)}{(x-2)} = \lim_{x \to 2} \frac{(x^2 + 2x + 4)}{(x+2)} = \frac{2^2 + 2.2 + 4}{(2+2)} = \frac{12}{4} = 3$$

(*ii*) $\lim\limits_{x \to 5} \dfrac{x^2 - 25}{3(x-5)}$

If we put x = 5, then $\dfrac{x^2 - 25}{3(x-5)} = \dfrac{0}{0}$ which is an indeterminate form. Hence direct

substitution can be applied after factorization.

$$\lim_{x \to 5} \frac{(x^2 - 25)}{3(x-5)} = \lim_{x \to 5} \frac{(x-5)(x+5)}{3(x-5)} = \lim_{x \to 5} \frac{(x+5)}{3}$$

Now by direct substitution: $\lim\limits_{x \to 5} \dfrac{(x+5)}{3} = \dfrac{5+5}{3} = \dfrac{10}{3}$

(*iii*) $\lim\limits_{x \to 36} \dfrac{x - 36}{\sqrt{x} - 6}$

Solution: If we put $\sqrt{x} = 6$, then $\dfrac{x - 36}{\sqrt{x} - 6} = \dfrac{0}{0}$. This is an indeterminate form. Hence direct substitution can be applied after factorization.

$$\lim_{x \to 36} \frac{x - 36}{\sqrt{x} - 6} = \lim_{x \to 36} \frac{\left(\sqrt{x}\right)^2 - (6)^2}{\left(\sqrt{x} - 6\right)} = \lim_{x \to 36} \frac{\left(\sqrt{x} - 6\right)\left(\sqrt{x} + 6\right)}{\left(\sqrt{x} - 6\right)}$$
$$= \lim_{x \to 36} \left(\sqrt{x} + 6\right) = \sqrt{36} + 6 = 6 + 6 = 12$$

(*iv*) $\lim\limits_{x \to 1} \dfrac{x^2 - 1}{x - 1}$

Solution: If we put $x = 1$, then $\dfrac{x^2 - 1}{x - 1} = \dfrac{0}{0}$. This is an indeterminate form. Hence direct

substitution can be applied after factorization.

$$\therefore \lim_{x \to 1} \frac{x^2 - 1}{x - 1} = \lim_{x \to 1} \frac{(x-1)(x+1)}{(x-1)} = \lim_{x \to 1} (x+1) = 1 + 1 = 2$$

(v) $\lim\limits_{x\to 2} \dfrac{x-2}{\sqrt{x}-\sqrt{2}}$

Solution: If we put $x=2$, then $\dfrac{x-2}{\sqrt{x}-\sqrt{2}}=\dfrac{0}{0}$. This is an indeterminate form. Hence direct substitution can be applied after factorization.

$$\therefore \lim_{x\to 2}\frac{x-2}{\sqrt{x}-\sqrt{2}}=\lim_{x\to 2}\frac{\left(\sqrt{x}\right)^2-\left(\sqrt{2}\right)^2}{\sqrt{x}-\sqrt{2}}$$

$$=\lim_{x\to 2}\frac{\left(\sqrt{x}-\sqrt{2}\right)\left(\sqrt{x}+\sqrt{2}\right)}{\left(\sqrt{x}-\sqrt{2}\right)}$$

$$=\lim_{x\to 2}\left(\sqrt{x}+\sqrt{2}\right)=\sqrt{2}+\sqrt{2}=2\sqrt{2}$$

(vi) If $\lim\limits_{x\to 1}\dfrac{x^4-1}{x-1}=\lim\limits_{x\to k}\dfrac{x^3-k^3}{x^2-k^2}$, **find the value of k**

Solution: $\lim\limits_{x\to 1}\dfrac{x^4-1}{x-1}=\lim\limits_{x\to k}\dfrac{x^3-k^3}{x^2-k^2}$,

Taking the limits after factorization of both sides:

$$\lim_{x\to 1}\frac{\left(x^2-1\right)\left(x^2+1\right)}{\left(x-1\right)}=\lim_{x\to k}\frac{\left(x-k\right)\left(x^2+kx+k^2\right)}{\left(x-k\right)\left(x+k\right)}$$

$$\lim_{x\to 1}\frac{\left(x-1\right)\left(x+k\right)\left(x^2+1\right)}{\left(x-1\right)}=\lim_{x\to k}\frac{\left(x-k\right)\left(x^2+kx+k^2\right)}{\left(x-k\right)\left(x+k\right)}$$

$$\lim_{x\to 1}\left(x+k\right)\left(x^2+1\right)=\lim_{x\to k}\frac{\left(x^2+kx+k^2\right)}{\left(x+k\right)}$$

$$(1+1)(1+1)=\frac{\left(k^2+kk+k^2\right)}{\left(k+k\right)}$$

$$4.2k=3k^2\Rightarrow\frac{8}{3}=k$$

$$\therefore k=\frac{8}{3}$$

(vii) $\lim\limits_{x \to 3} \left(x^2 - 9\right)\left[\dfrac{1}{x+3} + \dfrac{1}{x-3}\right]$

Solution: $\lim\limits_{x \to 3}\left(x^2-9\right)\left[\dfrac{1}{x+3}+\dfrac{1}{x-3}\right] = \lim\limits_{x \to 3}\left(x^2-9\right)\left[\dfrac{(x-3)+(x+3)}{(x+3)(x-3)}\right]$

$$= \lim_{x \to 3}\left(x^2-9\right)\left[\dfrac{2x}{\left(x^2-9\right)}\right]$$

$$= \lim_{x \to 3}\left[2x\right] = 2.3 = 6$$

(viii) $\lim\limits_{x \to 4} \dfrac{x^2 + x - 20}{x - 4}$

Solution: $\lim\limits_{x \to 4} \dfrac{x^2 + x - 20}{x - 4} = \lim\limits_{x \to 4} \dfrac{x^2 + 5x - 4x - 20}{x - 4} = \lim\limits_{x \to 4} \dfrac{(x+5)(x-4)}{(x-4)} = \lim\limits_{x \to 4}(x+5) = 9$

Practice Problems (On Limits by Factorization)

1. $\lim\limits_{x \to 1} \dfrac{9x^2 - 9}{5x - 5}$ **2.** $\lim\limits_{x \to \frac{-1}{2}} \dfrac{8x^3 + 1}{2x + 1}$, **3.** $\lim\limits_{x \to 1} \dfrac{(a+x)^2 - a^2}{x}$,

4. $\lim\limits_{x \to 3}\left(\dfrac{1}{x-3} - \dfrac{3}{x^2 - 3x}\right)$ **5** $\lim\limits_{x \to 2} \dfrac{x^2\left(x^2 - 4\right)}{(x-4)}$ **6.** $\lim\limits_{x \to 1} \dfrac{x^2 + x - 2}{x^2 - 1}$

7. $\lim\limits_{\to}$ ─── **8.** $\lim\limits_{x \to 1} \dfrac{x^3 + x - 2}{x - 1}$ **9.** $\lim\limits_{x \to 2} \dfrac{x^2 + 6x - 16}{x^2 - 2x}$

Answers:

1. $\dfrac{8}{5}$ **2.** 3   **3.** 2a   **4.** $\dfrac{1}{3}$ **5.** 24   **6.** $\dfrac{3}{2}$ **7.** 4   **8.** 4   **9.** 5

C. Limits by rationalization

Rationalization is a process of converting an irrational number into a rational number, which can be expressed as the ratio of two integers. This process is applied when numerator or denominator or both consist of square roots. Then numerator or denominator or both are multiplied by their conjugate as shown in solved examples.

Solved Examples

Evaluate the following limits

1. $\displaystyle\lim_{x\to a}\frac{x-a}{\sqrt{x}-\sqrt{a}}$

Solution: If we put $x=a$, then $\dfrac{x-a}{\sqrt{x}-\sqrt{a}}=\dfrac{0}{0}$. This is an indeterminate form. Hence direct

substitution can be applied after rationalization of the given irrational number.

$$\lim_{x\to a}\frac{x-a}{\sqrt{x}-\sqrt{a}}\times\frac{\sqrt{x}+\sqrt{a}}{\sqrt{x}+\sqrt{a}}=\lim_{x\to a}\frac{x-a}{\sqrt{x}-\sqrt{a}}\times\frac{\sqrt{x}+\sqrt{a}}{\sqrt{x}+\sqrt{a}}$$

$$=\lim_{x\to a}\frac{x-a}{\left(\sqrt{x}\right)-\left(\sqrt{a}\right)}\times\frac{\sqrt{x}+\sqrt{a}}{\sqrt{x}+\sqrt{a}}$$

$$=\lim_{x\to a}\frac{(x-a)\left(\sqrt{x}+\sqrt{a}\right)}{\left(\sqrt{x}\right)^2-\left(\sqrt{a}\right)^2}$$

$$=\lim_{x\to a}\frac{(x-a)\left(\sqrt{x}+\sqrt{a}\right)}{(x)-(a)}$$

$$=\lim_{x\to a}\left(\sqrt{x}+\sqrt{a}\right)=\sqrt{a}+\sqrt{a}=2\sqrt{a}$$

2. $\displaystyle\lim_{x\to 0}\frac{\sqrt{2-x}-\sqrt{2+x}}{x}$

Solution: If we put $x=0$, then $\dfrac{\sqrt{2-x}-\sqrt{2+x}}{x}=\dfrac{0}{0}$. This is an indeterminate form. Hence

direct substitution can be applied after rationalization of the given irrational number.

$$\lim_{x\to 0}\frac{\left(\sqrt{2-x}\right)-\left(\sqrt{2+x}\right)}{x}\times\frac{\left(\sqrt{2-x}\right)+\left(\sqrt{2+x}\right)}{\left[\left(\sqrt{2-x}\right)+\left(\sqrt{2+x}\right)\right]}$$

$$=\lim_{x\to 0}\frac{\left(\sqrt{2-x}\right)^2-\left(\sqrt{2+x}\right)^2}{x\left[\left(\sqrt{2-x}\right)+\left(\sqrt{2+x}\right)\right]}$$

$$=\lim_{x\to 0}\frac{(2-x)-(2+x)}{x\left[\left(\sqrt{2-x}\right)+\left(\sqrt{2+x}\right)\right]}$$

$$= \lim_{x \to 0} \frac{-2x}{x\left[\left(\sqrt{2-x}\right)+\left(\sqrt{2+x}\right)\right]}$$

$$= \lim_{x \to 0} \frac{-2}{\left[\left(\sqrt{2-x}\right)+\left(\sqrt{2+x}\right)\right]} = \frac{-2}{2\sqrt{2}} = \frac{-1}{\sqrt{2}}$$

3. $\lim\limits_{x \to 0} \dfrac{\sqrt{1+x}-1}{x}$

Solution: Direct substitution i.e. $x = 0$ will make the given expression as $\dfrac{0}{0}$ which is an indeterminate form. Factorization is not possible. The only way is rationalization of given irrational expression.

$$\lim_{x \to 0} \frac{\left(\sqrt{1+x}\right)-1}{x} \times \frac{\left(\sqrt{1+x}\right)+1}{\sqrt{1+x}+1} = \lim_{x \to 0} \frac{\left(\sqrt{1+x}\right)^2 - 1}{x\left[\sqrt{1+x}+1\right]}$$

$$= \lim_{x \to 0} \frac{(1+x)-1}{x\left[\sqrt{1+x}+1\right]}$$

$$= \lim_{x \to 0} \frac{x}{x\left[\sqrt{1+x}+1\right]}$$

$$= \lim_{x \to 0} \frac{1}{\left[\sqrt{1+x}+1\right]} = \frac{1}{\left[\sqrt{1+0}+1\right]} = \frac{1}{2}$$

4. $\lim\limits_{x \to 16} \dfrac{\sqrt{x}-4}{x-16}$

Solution: Direct substitution i.e. $x = 16$ will make the given expression as $\dfrac{0}{0}$ which is an indeterminate form. Factorization is not possible. The only way is rationalization of given irrational expression.

$$\lim_{x \to 16} \frac{\sqrt{x}-4}{x-16} \times \frac{\sqrt{x}+4}{\sqrt{x}+4} = \lim_{x \to 16} \frac{x-16}{x-16} \times \frac{1}{\sqrt{x}+4} \qquad \because \sqrt{x}-4 \times \sqrt{x}+4 = x-16$$

$$= \lim_{x \to 16} \frac{1}{\sqrt{x}+4} = \frac{1}{\sqrt{16}+4} = \frac{1}{4+4} = \frac{1}{8}$$

5. $\lim\limits_{x \to 25} \dfrac{x-25}{\sqrt{x}-5}$

Solution: Direct substitution i.e. $x = 25$ will make the given expression as $\dfrac{0}{0}$ which is an indeterminate form. Hence by rationalizing the given expression:

$$\lim_{x\to 25}\frac{x-25}{\sqrt{x}-5}\times\frac{\sqrt{x}+5}{\sqrt{x}+5}=\lim_{x\to 25}\frac{x-25}{x-25}\times\left(\sqrt{x}+5\right)=\lim_{x\to 25}\left(\sqrt{x}+5\right)=\sqrt{25}+5=5+5=10$$

Practice Problems(Limits by Rationalization)

Evaluate the following limits:

1. $\lim_{x\to 0}\dfrac{\sqrt{1+x}-\sqrt{1-x}}{2x}$ **2.** $\lim_{x\to 0}\dfrac{\sqrt{2+x}-\sqrt{2}}{x}$ **3.** $\lim_{x\to 1}\dfrac{\sqrt{5x-4}-\sqrt{x}}{x-1}$

4. $\lim_{x\to 4}\dfrac{2-\sqrt{x}}{4-x}$ **5.** $\lim_{x\to 0}\dfrac{\sqrt{1+x}-\sqrt{1}}{x}$ **6.** $\lim_{x\to a}\dfrac{\sqrt{x}-\sqrt{a}}{x-a}$

Answers:

1. $\dfrac{1}{2}$ **2.** $\dfrac{1}{2\sqrt{2}}$ **3.** 2 **4.** $\dfrac{1}{4}$ **5.** $\dfrac{1}{2}$ **6.** $\dfrac{1}{2\sqrt{a}}$

D. Limits using standard results

The limits can be evaluated by using if the given expression resembles the standard form as given below:

Standard Results are:

1. $\lim_{x\to a}\dfrac{x^n-a^n}{x-a}=na^{n-1}$

2. $\lim_{x\to a}\dfrac{x^m-a^m}{x^n-a^n}=\dfrac{m}{n}a^{m-n}$

Solved Examples

1. Evaluate $\lim_{x\to 2}\dfrac{x^6-2^6}{x-2}$

Solution: Direct substitution i.e. $x=2$ gives $\dfrac{x^6-2^6}{x-2}=\dfrac{0}{0}$, which is an indeterminate form.

Hence applying standard result $\lim_{x\to a}\dfrac{x^n-a^n}{x-a}=na^{n-1}$

We get $\lim\limits_{x \to 2} \dfrac{x^6 - 2^6}{x - 2} = 6.2^{6-1} = 6.2^5 = 6.32 = 192$

This can also be solved by method of factorization.

$$\lim_{x \to 2} \frac{x^6 - 2^6}{x - 2} = \lim_{x \to 2} \frac{\left(x^3\right)^2 - \left(2^3\right)^2}{x - 2}$$

$$= \lim_{x \to 2} \frac{\left(x^3 - 2^3\right)\left(x^3 + 2^3\right)}{(x - 2)}$$

$$= \lim_{x \to 2} \frac{(x - 2)\left(x^2 + 2^2 + 2.x\right)\left(x^3 + 2^3\right)}{(x - 2)}$$

$$\because \left(x^3 - a^3\right) = (x - a)\left(x^2 + a^2 + ax\right)$$

$$= \lim_{x \to 2} \left(x^2 + 2^2 + 2.x\right)\left(x^3 + 2^3\right)$$

$$= \left(2^2 + 2^2 + 2.2\right)\left(2^3 + 2^3\right) \text{ (by direct substitution)}$$

$$= (4 + 4 + 4)(8 + 8) = 12 \times 16 = 192$$

2. Evaluate $\lim\limits_{x \to 2} \dfrac{x^{5/7} - 2^{5/7}}{x^{2/7} - 2^{2/7}}$

Solution: We can write $\lim\limits_{x \to 2} \dfrac{x^{5/7} - 2^{5/7}}{x^{2/7} - 2^{2/7}} = \lim\limits_{x \to 2} \dfrac{\left\{ \dfrac{x^{5/7} - 2^{5/7}}{x - 2} \right\}}{\left\{ \dfrac{x^{2/7} - 2^{2/7}}{x - 2} \right\}}$

By limits of standard result-

$$\lim_{x \to 2} \frac{\left\{ \dfrac{x^{5/7} - 2^{5/7}}{x - 2} \right\}}{\left\{ \dfrac{x^{2/7} - 2^{2/7}}{x - 2} \right\}} = \frac{\left(\dfrac{5}{7} 2^{\frac{5}{7} - 1} \right)}{\left(\dfrac{2}{7} 2^{\frac{2}{7} - 1} \right)} = \frac{5.2^{-2/7}}{2.2^{-5/7}} = \frac{5}{2}.2^{-2/7 + 5/7} = \frac{5}{2}.2^{3/7}$$

3. If $\lim\limits_{x \to a} \dfrac{x^9 - a^9}{x - a} = 9$, find the value of a

Solution: $\lim\limits_{x \to a} \dfrac{x^9 - a^9}{x - a} = 9$

$$9a^{9-1} = 9(1)^{9-1} \Rightarrow a^8 = 1 \Rightarrow a = \pm 1$$

4. Prove: $\lim\limits_{x\to a}\dfrac{x^9-a^9}{x-a}=9a^8$

Solution: From standard result (2) $\lim\limits_{x\to a}\dfrac{x^m-a^m}{x^n-a^n}=\dfrac{m}{n}a^{m-n}$

$$\lim_{x\to a}\frac{x^9-a^9}{x-a}=\frac{9}{1}a^{9-1}=9a^8$$

Hence Proved.

5. Evaluate $\lim\limits_{x\to a}\dfrac{x^5-243}{x^2-9}$

The given limit can be written as:

$$\lim_{x\to 3}\frac{x^5-243}{x^2-9}=\lim_{x\to 3}\frac{x^5-3^5}{x^2-3^2}=\frac{5}{3}3^{5-2}=\frac{5}{3}3^3=5.3^2=45$$

$$\because \lim_{x\to a}\frac{x^m-a^m}{x^n-a^n}=\frac{m}{n}a^{m-n}$$

Practice Problems (On Limits by Using Standard Results)

1. If $\lim\limits_{x\to 3}\dfrac{x^n-3^n}{x-3}=81$, find the value of n .

2. Evaluate $\lim\limits_{x\to 3}\dfrac{x^{2/7}-3^{2/7}}{x-3}$

3. If $\lim\limits_{x\to a}\dfrac{x^7+a^7}{x+a}=7$, find the value of **a**

4. Evaluate $\lim\limits_{\to}\dfrac{625}{3\quad 3}$

Answers: 1. n = 4 **2.** $\dfrac{2}{7}.3^{\frac{-5}{7}}$ **3.** a = ± 1 **4.** $\dfrac{20}{3}$

E. Limit when $x\to\infty$

In such problems the direct substitution gives an inderminate form $\dfrac{\infty}{\infty}$ ..

Hence highest power of x is taken as common so that inside we get a constant term

with terms in reciprocal of x viz. $\dfrac{1}{x}$ which becomes zero as $x\to\infty$

Solved Examples

1. Evaluate $\lim\limits_{x \to \infty} \dfrac{1000}{x^2 + 100}$

Solution: The numerator remain 1000 but denominator $\to \infty$ as $x \to \infty$

$$\lim_{x \to \infty} \frac{1000}{x^2 + 100} = \frac{1000}{\infty} = 0$$

2. Evaluate $\lim\limits_{x \to \infty} \dfrac{\left(5x^2 + 1\right)}{\left(3x^2 - x\right)}$

Solution: $\lim\limits_{x \to \infty} \dfrac{\left(5x^2 + 1\right)}{\left(3x^2 - x\right)} = \lim\limits_{x \to \infty} \dfrac{x^2\left(5 + 1/x^2\right)}{x^2\left(3 - 1/x\right)} = \lim\limits_{x \to \infty} \dfrac{\left(5 + 1/x^2\right)}{\left(3 - 1/x\right)} = \dfrac{5}{3}$ $\quad \because \dfrac{1}{\infty} = 0$

3. Evaluate $\lim\limits_{x \to \infty} \dfrac{\left(2x^4 - 3x^2 + 1\right)}{\left(3x^4 - x + 3\right)}$

Solution: $\lim\limits_{x \to \infty} \dfrac{\left(2x^4 - 3x^2 + 1\right)}{\left(3x^4 - x + 3\right)} = \lim\limits_{x \to \infty} \dfrac{x^4\left(2 - 3/x + 1/x^4\right)}{x^4\left(3 - 1/x^3 + 3/x^4\right)}$

$$= \lim_{x \to \infty} \frac{\left(2 - 3/x + 1/x^4\right)}{\left(3 - 1/x^3 + 3/x^4\right)} = \frac{2}{3}$$

4. Evaluate $\lim\limits_{n \to \infty} \dfrac{1 + 2 + 3 + \dots n}{n^2}$

Solution : $1 + 2 + 3 + \dots n$ is an AP

The sum of the AP $\dfrac{n}{2}\left[1 + n\right]$

$$1 + 2 + 3 + \dots n = \frac{n}{2}\left[1 + n\right]$$

$$\lim_{n \to \infty} \frac{1 + 2 + 3 + \dots n}{n^2} = \lim_{n \to \infty} \frac{1}{n^2}\left[\frac{n}{2}(1 + n)\right]$$

$$= \lim_{n \to \infty} \frac{1}{2n}(1 + n) = \frac{1}{2} \lim_{n \to \infty}\left(\frac{1}{n} + 1\right) = \frac{1}{2}(0 + 1) = \frac{1}{2}$$

Practice Problems(On limits when x → ∞)

Evaluate the following limits

1. $\lim\limits_{x\to\infty}\dfrac{4x^3+x}{x^2-x^3}$ **2.** $\lim\limits_{x\to\infty}\dfrac{x}{x+1}$ **3.** $\lim\limits_{x\to\infty}\dfrac{2x+1}{x-1}$ **4.** $\lim\limits_{x\to\infty}\dfrac{3x^{-1}+4x^{-1}}{5x^{-1}+6x^{-2}}$

Answers:

1. −4 **2.** 1 **3.** 2 **4.** $\dfrac{3}{5}$

1.2.2 Limits of Trigonometric Functions

Some Trigonometric Limits

1. $\lim\limits_{\theta\to0}\sin\theta=0$

2. $\lim\limits_{\theta\to0}\cos\theta=1$

3. $\lim\limits_{\theta\to0}\dfrac{\sin\theta}{\theta}=1$

4. $\lim\limits_{\theta\to0}\dfrac{\tan\theta}{\theta}=1$

5. $\lim\limits_{\theta\to a}\dfrac{\sin(\theta-a)}{(\theta-a)}=1$

6. $\lim\limits_{\theta\to a}\dfrac{\tan(\theta-a)}{(\theta-a)}=1$

Solved Examples

Evaluate the following limits

1. $\lim\limits_{x\to0}\dfrac{\sin3x}{5x}=\lim\limits_{x\to0}\left(\dfrac{\sin3x}{3x}\right)\dfrac{3}{5}=1.\dfrac{3}{5}$

2. $\lim\limits_{x\to0}\dfrac{\tan8x}{\sin2x}=\lim\limits_{x\to0}4.\left(\dfrac{\tan8x}{8x}\right)\left(\dfrac{2x}{\sin2x}\right)=4.1.1=4$

3. $\lim\limits_{x\to0}\dfrac{\tan mx}{\tan nx}=\lim\limits_{x\to0}\left(\dfrac{\tan mx}{mx}\right)\left(\dfrac{nx}{\tan nx}\right).\dfrac{m}{n}=1.1.\dfrac{m}{n}=\dfrac{m}{n}$

4. $\lim\limits_{x\to\frac{\pi}{4}}\dfrac{\cos ec^2x-2}{\cot x-1}=\lim\limits_{x\to\frac{\pi}{4}}\left(\dfrac{1+\cot^2x-2}{\cot x-1}\right)$ $\because\cos ec^2x=1+\cot^2x$

$$= \lim_{x \to \frac{\pi}{4}} \left(\frac{\cot^2 x - 1}{\cot x - 1} \right)$$

$$\lim_{\to -} \left(\frac{(\cot x - 1)(\cot x + 1)}{(\cot \quad 1)} \right)$$

$$= \lim_{x \to \frac{\pi}{4}} (\cot x + 1) = \cot \frac{\pi}{4} + 1 = 1 + 1 = 2$$

5. $\displaystyle \lim_{x \to 0} \frac{\sin^2 \left(\frac{x}{2} \right)}{4x^2} = \frac{1}{4} \lim_{x \to 0} \frac{\sin^2 \left(\frac{x}{2} \right)}{4 \cdot \frac{x^2}{4}}$

$$= \frac{1}{4.4} \lim_{x \to 0} \frac{\sin^2 \left(\frac{x}{2} \right)}{\left(\frac{x}{2} \right)^2}$$

$$= \frac{1}{4.4} \lim_{x \to 0} \left[\frac{\sin \left(\frac{x}{2} \right)}{\left(\frac{x}{2} \right)} \right]^2$$

$$= \frac{1}{4.4} = \frac{1}{16}$$

6. $\displaystyle \lim_{x \to 0} \frac{\sin 3x . \sin 5x}{7x^2} = \frac{1}{7} \lim_{x \to 0} \left(3 . \frac{\sin 3x}{3x} \right) \left(5 . \frac{\sin 5x}{5x} \right)$

$$= \frac{15}{7} \left(\lim_{x \to 0} \frac{\sin 3x}{3x} \right) \left(\lim_{x \to 0} \frac{\sin 5x}{5x} \right)$$

$$= \frac{15}{7} . 1.1 = \frac{15}{7}$$

Practice Problems (On limits of Trigonometric Functions)

1. $\displaystyle \lim_{x \to 0} \frac{\sin 8x}{\sin 5x}$ **2.** $\displaystyle \lim_{x \to 0} \frac{\sin 9x}{x}$ **3.** $\displaystyle \lim_{x \to 0} \frac{\tan 8x}{\sin 4x}$ **4.** $\displaystyle \lim_{x \to \frac{\pi}{2}} \frac{\sin 2x}{\cos x}$

5. $\lim\limits_{x\to 0}\dfrac{\cos^2 x}{1-\sin x}$ **6.** $\lim\limits_{x\to\frac{\pi}{2}}\dfrac{1-\sin x}{\cos^2 x}$ **7.** $\lim\limits_{x\to 0}\dfrac{1-\cos 2x}{x^2}$ **(hint:** $\cos 2x = 1-2\sin^2 x$

8. $\lim\limits_{x\to 0}\dfrac{\sin px}{\sin qx}$ **9.** $\lim\limits_{x\to 0}\dfrac{\sin^2 ax}{\sin^2 bx}$ **10.** $\lim\limits_{x\to 0}\dfrac{\cot 2x - \cos ec2x}{x}$

Answers

1. $\dfrac{2}{3}$ **2.** 9   **3.** 2   **4.** 2   **5.** 2   **6.** $\dfrac{1}{2}$ **7.** 2   **8.** $\dfrac{p}{q}$ **9.** $\dfrac{a^2}{b^2}$ **10.** −1

1.3   CONTINUITY OF A FUNCTION

1. A function is said to be continuous at x = a if $\lim\limits_{x\to a} f(x) = f(a)$

2. Furthermore, a function is said to be continuous on an interval $[a, b]$,if it is continuous at every point in the interval

3. A function is said to be continuous at x = a if $\lim\limits_{x\to a^-} f(x) = \lim\limits_{x\to a^+} f(x) = f(a)$

   i.e. Left Hand Limit (LHM) = Right Hand Limit (RHL)= Value of the function at

   x =a If f(x) is not continuous at x =a , we say that f(x) is discontinuous at x = a

1.3.1   Continuity of Some particular Functions

1. A constant function is a continuous function.
2. A function is continuous if f(x) = x
3. A polynomial function is continuous.
4. The functions $\sin x, \cos x, e^x, a^x$ are continuous every where.
5. The functions $\sin^{-1} x \; and \; \cos^{-1} x$ are continuous in the interval $[-1,1]$

1.3.2   Continuity by Right Hand and Left Hand Limits

For Left Hand limit (LHL) of function $f(x)$ at x = a we write as $\lim\limits_{x\to a^-} f(x)$

Working Rule:

Put x = a − h and replace $x \to a^-$ by h → 0. Thus $\lim_{x \to a^-}$ f(x)=$\lim_{h \to 0}$ f(a-h)

Simplify $\lim_{x \to a^-}$ f(x)=$\lim_{h \to 0}$ f(a-h). This gives the value of LHL

For Right Hand limit (LHL) of function $f(x)$ at x = a we write as $\lim_{x \to a^+}$ f(x)

Working Rule:

Put x = a + h and replace $x \to a^+$ by h → 0. Thus $\lim_{x \to a^+}$ f(x)=$\lim_{h \to 0}$ f(a + h)

Simplify $\lim_{x \to a^-}$ f(x)=$\lim_{h \to 0}$ f(a-h). This gives the value of RHL

If $LHL = RHL$, then $\lim_{x \to a}$ f(x) exists and $f(x)$ is continuous at x = a

If $LHL \neq RHL$, then $\lim_{x \to a}$ f(x) does not exist and $f(x)$ is not continuous **at x = a**

Solved Examples

1. Test the continuity of the following functions: $f(x) = \begin{cases} x^3 - 1 & \text{for x<3} \\ x^2 + 14 & \text{for x} \geq 3 \end{cases}$

Solution: Left Hand Limit (**LHL**) : $\lim_{x \to 3^-} f(x) = \lim_{x \to 3^-} (x^3 - 1) = 3^3 - 1 = 27 - 1 = 26$

Right Hand Limit (**RHL**) : $\lim_{x \to 3^+} f(x) = \lim_{x \to 3^-} (x^2 + 14) = 3^2 + 14 = 9 + 14 = 23$

Since **LHL≠RHL** i.e. $\lim_{x \to 3^-} f(x) \neq \lim_{x \to 3^+} f(x)$, hence the function $f(x)$ is not continuous at x = 3

2. Is the function $f(x) = \dfrac{1}{1-x}$ continuous at x = 1.

Solution: We see that $f(1) = \dfrac{1}{1-1} = \dfrac{1}{0} = \infty$ i.e. $f(x)$ is not defined at x = 1

Hence the function $f(x)$ does not exists at x = 3 and is not continuous at x = 3

3. Test the continuity of the function $f(x) = \begin{cases} x+3 & \text{for x<2} \\ 5 & \text{at x =2} \\ x^4 - 11 & \text{for x>2} \end{cases}$

Solution: Left Hand Limit(LHL) : $\lim\limits_{x \to 2^-} f(x) = \lim\limits_{x \to 2^-} (x+3) = 2+3 = 5$

Right Hand Limit(RHL) : $\lim\limits_{x \to 2^+} f(x) = \lim\limits_{x \to 3^-} (x^4 - 11) = 2^4 - 11 = 16 - 11 = 5$

Since LHL = RHL i.e. $\lim\limits_{x \to 2^-} f(x) = \lim\limits_{x \to 2^+} f(x)$, hence the function $f(x)$ is continuous at x = 2

4. Is the function $f(x) = \dfrac{x^2 - 16}{x - 4}$ is continuous at x = 4

Solution: Left Hand Limit(LHL): $\lim\limits_{x \to 4^-} f(x) = \lim\limits_{x \to 4^-} \dfrac{(4-h)^2 - 16}{(4-h) - 4}$

$$= \lim\limits_{x \to 4^-} \frac{16 - 8h + h^2 - 16}{4 - h - 4}$$

$$= \lim\limits_{x \to 4^-} \frac{-8h + h^2}{-h} = \lim\limits_{x \to 4^-} (8 - h) = 8 - 4 = 4$$

Right Hand Limit(RHL): $\lim\limits_{x \to 4^+} f(x) = \lim\limits_{x \to 4^+} \dfrac{(4+h)^2 - 16}{(4+h) - 4}$

$$= \lim\limits_{x \to 4^-} \frac{16 + 8h + h^2 - 16}{4 + h - 4}$$

$$= \lim\limits_{x \to 4^-} \frac{8h + h^2}{h} = \lim\limits_{x \to 4^-} (8 + h) = 8 + 4 = 12$$

Since LHL ≠ RHL i.e. $\lim\limits_{x \to 4^-} f(x) \neq \lim\limits_{x \to 4^+} f(x)$, hence the function $f(x)$ is not continuous at x = 3

5. Show that the function $f(x) = \begin{cases} 2-x & \text{when } x\text{-}2 \\ 2+x & \text{when } x>2 \end{cases}$ **is continuous.**

Solution: Left Hand Limit(LHL): $\lim\limits_{x \to 2^-} f(x) = \lim\limits_{x \to 2^-} 2 - (2 - h) = \lim\limits_{x \to 2^-} h = 2$

Right Hand Limit(LHL): $\lim\limits_{x \to 2+} f(x) = \lim\limits_{x \to 2^+} 2 - (2 + h) = \lim\limits_{x \to 2^-} -h = -2$

Since LHL ≠ RHL i.e. $\lim\limits_{x \to 2^-} f(x) \neq \lim\limits_{x \to 2^+} f(x)$, hence the function $f(x)$ is not continuous at x = 2

6. Show that the function $f(x) = \dfrac{x+1}{|x+1|}$ is discontinuous at x = -1

Solution: $f(x) = \dfrac{x+1}{|x+1|} = \begin{cases} \dfrac{x+1}{(x+1)} & \text{if } x > \text{-}1 \\ \dfrac{x+1}{-(x+1)} & \text{if } x < \text{-}1 \end{cases}$

Left Hand Limit (LHL): $\lim\limits_{x \to -1^-} f(x) = \lim\limits_{x \to -1^-} \dfrac{x+1}{x+1} = 1$

Right Hand Limit (RHL): $\lim\limits_{x \to -1+} f(x) = \lim\limits_{x \to -1^+} \dfrac{x+1}{-(x+1)} = -1$

Since LHL $\square$ RHL i.e. $\lim\limits_{x \to 1^-} f(x) \ne \lim\limits_{x \to 1^+} f(x)$, hence the function $f(x)$ is not continuous at x = −1

Practice Problems

1. Show that the function $f(x) = \begin{cases} x^2 +1 & \text{if } x \ge 2 \\ x +1 & \text{if } x < 2 \end{cases}$ is discontinuous at x = 2

2. Show that the function $f(x) = \begin{cases} \dfrac{2x}{x-3} & \text{if } x \ne 3 \\ 4 & \text{if } x = 2 \end{cases}$ is discontinuous at x = 3

3. Show that the function $f(x) = \begin{cases} 5x-4 & \text{when } 0 < x < 1 \\ 4x^3 -3x & \text{when } 1 < x, 2 \end{cases}$ is continuous at x = 1

4. Show that the function $f(x) = \begin{cases} \dfrac{x}{|x|} & , x \ne 0 \\ 1 & , x = 0 \end{cases}$ is discontinuous at x = 0

Objective Questions: (On Functions)

1. If $f(x) = \left(x^2 - \dfrac{1}{x^2} \right)$, then $f(x) + f\left(\dfrac{1}{x} \right)$ is equal to:

(a) 1   (b) 0

(c) 2   (d) ½

2. If $f(x) \dfrac{\sin x}{1+\cos x}$, then $f\left(\dfrac{\pi}{2} \right)$ is equal to:

(a) 1   (b) 0

(c) ∞   (d) none of these.

3. If $f(x) = x^2 - 3x + 1$, then value of $f(-1)$ is:

(a) −1

(b) 2

(c) −3

(d) 5

4. Given: $f(x) = \dfrac{1+x}{1-x}$, then $f[f(\tan x)]$ is equivalent to:

(a) $\tan x$

(b) $-\tan x$

(c) $\cot x$

(d) $-\cot x$

5. If $f(x) = \log_a x$ and $g(x) = a^x$, then $f[g(x)]$ is equal to:

(a) x

(b) $\log a$

(c) $-x$

(d) none of these

6. $f(x) = x^2 + kx + 1$. If k = 0, then $f(x)$ will be:

(a) Odd

(b) Even

(c) Complete

(d) None of these

7. If $f(x) = e^x$ and $g(x) = \log x$, then $(f+g)x$ at $x = 1$ is:

(a) 1

(b) 2

(c) e

(d) 0

8. If $f(x) = e^x$ and $g(x) = \log x$, then $(fog)x$ at $x = 1$ is:

(a) 0

(b) 1

(c) e

(d) −1

9. If $f(x) = x + 5$ and $g(x) = x^2 - 3$, then value of $f[g(0)]$

(a) −1

(b) −3

(c) 5

(d) 2

10. If $f(x) = 5$, $g(x) = x^3 - x^2 + x + 4$, then $\left(\dfrac{f}{g}\right)x$ at $x = 2$ is:

(a) 2

(b) 4

(c) 5

(d) 3

11. If $f(x) = \dfrac{2x+1}{3x-2}$, then show that $(fof)(x)$ is equal to:

(a) x

(b) $2x$

(c) $3x$

(d) $x/2$

Answers

1. b	2. a	3. a	4	5. a	6. b
7. c	8. b	9.	10.a	11.a	

Objective Questions (On Limits)

1. The value of $\lim\limits_{x \to 2} \dfrac{x^2 - 4}{x^2 + 4}$ is:

   (a) 1   (b) 0

   (c) −1/2   (d) −1

2. The value of $\lim\limits_{x \to \infty} \dfrac{x^2 - 4}{x^2 + 4}$ is:

   (a) 1   (b) 4

   (c) −4   (d) −1

3. The value of $\lim\limits_{\to \infty} \dfrac{}{x - 2x - 3}$ is :

   (a) 1   (b) 4

   (c) 1/4   (d) −3

4. $\lim\limits_{x \to 0} \dfrac{x}{x}$ is equal to:

   (a) 0   (b) 1

   (c) ∞   (d) unknown

5. $\lim\limits_{x \to \infty} \dfrac{4 - x^2}{4x^2 - x - 3}$ is equal to:

   (a) 0   (b) 1

   (c) −1/4   (d) ∞

6. $\lim\limits_{x \to \infty} \dfrac{3x^2 + 27}{x^3 - 27}$ is equal to:

   (a) 0   (b) 1

   (c) −1/4   (d) ∞

7. The value of $\lim\limits_{x \to 0} \dfrac{2^{-x}}{2^x}$ is equal to:

   (a) 0   (b) 1

   (c) 2   (d) ∞

8. $\lim\limits_{x \to 0} x \sin\left(\dfrac{1}{x}\right)$ is equal to:

   (a) 1   (b) 0

   (c) 2   (d) ∞

9. $\lim\limits_{x \to 0} \dfrac{\sin 3x}{5x}$ is equal to:

(a) 3/5   (b) 5/3

(c) 5   (d) 3

10. $\lim\limits_{x \to 0} \dfrac{x^3}{\sin x^2}$ is equal to:

(a) 1   (b) ∞

(c) 0   (d) none of these.

11. $\lim\limits_{x \to 0} \dfrac{\sin ax}{\sin bx}$ is equal to:

(a) a/b   (b) b/a

(c) a.b   (d) 1

12. $\lim\limits_{n \to \infty} \dfrac{n!}{(n+1)! - n!}$ is equal to:

(a) 1   (b) ∞

(c) 0   (d) none of these.

13. The value of $\lim\limits_{x \to a} \dfrac{x^n - a^n}{x - a}$ is:

(a) a^{n-1} (b) na^{n-1}

(c) 0   (d) $\dfrac{a^{n-1}}{n}$.

14. The value of $\lim\limits_{x \to a} \dfrac{\sqrt{x} - \sqrt{a}}{x - a}$ is:

(a) $\dfrac{1}{2\sqrt{a}}$ (b) $\dfrac{1}{2}$

(c) 0   (d) $\dfrac{\sqrt{a}}{2}$.

15. $\lim\limits_{x \to 0} \dfrac{\sqrt{1+x} - 1}{x}$ is equal to:

(a) 1   (b) $\dfrac{1}{2}$

(c) ∞ (d) none of these.

Answers

1. b	2. a	3. a	4. a	5. c	6. a
7. b	8. a	9. b	10. c	11. a	12. c
13. b	14. a	15. b			

5

Differentiation

1.0 BRIEF HISTORY

The concept of differentiation was familiar to Greek mathematicians viz. Euler (300BC) and Archimedes (287BC) who introduced the thought of infinitesimal (smallest).

The use of infinitesimals to study the rate of change was also known to Indian mathematicians such as Aryabhatt (476-550) AD, who used infinitesimal (Δ) to study the motion of moon. This was latter developed by Bhaskar II (1114-1185AD). It is said that many of the key notations of differentiation can be found in his works.

Persian mathematician such as Sharaf –al –Din –al –Tusi (1135-1213 AD) was first to discover the derivative of cubic polynomials. His work 'Treatise on Equations' developed the concepts of differential calculus.

The modern development of calculus is credited to Isaac Newton and Gottfried Leibnitz. The other significant contributors were: Isaac Barrow, Rene' Descartes, Christian Huygenes, Blasé Pascal, and John Wallis.

Nevertheless, Newton and Leibnitz remain key figures in the history of differentiation. Newton was first to use differentiation in theoretical physics and Leibnitz systematically developed much of the notions still used to-day.

1.1 DIFFERENTIATION

Let $y = f(x)$ be a given function. Value of y depends on x. Hence y will change with change in x i.e. an increase or decrease in x will bring an increase or decrease in y.

Let δx be an increment in x. The corresponding increment in y is δy.

We have $y = f(x)$ and after increment we get $y + \delta y = f(x + \delta x)$

$\therefore (y + \delta y) - y = f(x + \delta x) - f(x)$

or
$$\delta y = f(x + \delta x) - f(x)$$

Dividing by δx we get $\dfrac{\delta y}{\delta x} \quad \dfrac{f(x + \delta x) - f(x)}{\delta x}$

So under the limit $\delta x \to 0$ $\dfrac{\delta y}{\delta x} = \dfrac{dy}{dx}$

The above limit is called the derivative or differential coefficient of $y = f(x)$ with respect to x.

The derivative of $y = f(x)$ is written as $\dfrac{dy}{dx}$ or $\dfrac{df(x)}{dx}$ or $f'(x)$

The process of finding the derivative is known as *'differentiation'*

1.1.1   Some Important Derivatives

A. $\dfrac{d}{dx}(x^n) = nx^{n-1}$

Solved Example.

1. Find the derivative of (*i*) x^9 , (*ii*) x^{-3} , (*iii*) $\sqrt[3]{x}$, (*iv*) $\dfrac{1}{\sqrt{x}}$

Solution:

(*i*) $\dfrac{d}{dx}(x^9) = 9x^{9-1} = 9x^8$

(*ii*) $\dfrac{d}{dx}(x^{-3}) = -3x^{-3-1} = -3x^{-4}$

(*iii*) $\dfrac{d}{dx}(\sqrt[3]{x}) = \dfrac{d}{dx}(x^{\frac{1}{3}}) = \dfrac{1}{3}x^{\frac{1}{3}-1} = \dfrac{1}{3}x^{\frac{-2}{3}}$

(*iv*) $\dfrac{d}{dx}\left(\dfrac{1}{\sqrt{x}}\right) = \dfrac{d}{dx}\left(x^{\frac{-1}{2}}\right) = \dfrac{-1}{2}x^{\frac{-1}{2}-1} = \dfrac{-1}{2}x^{\frac{-3}{2}}$

B. $\dfrac{d}{dx}(a^x) = a^x \log a$

C. $\dfrac{d}{dx}(a^x) = a^x \log a$

Solved example:

(*i*) $\dfrac{d}{dx}(2^x) = 2^x(\log_e a)$

D. $\dfrac{d}{dx}(\log x) = \dfrac{1}{x}$

E. $\dfrac{d}{dx}(\sin x) = \cos x$

F. $\dfrac{d}{dx}(\cos x) = -\sin x$

G. $\dfrac{d}{dx}(\tan x) = \sec^2 x$

H. $\dfrac{d}{dx}(\cot x) = -\text{cosec}^2 x$

I. $\dfrac{d}{dx}\left(\sec x\right) = \sec x . \tan x$

J. $\dfrac{d}{dx}\left(\text{cosec}x\right) = -\text{cosec } x . \cot x$

K. $\dfrac{d}{dx}\{cf(x)\} = c\dfrac{d}{dx}f(x)$

Solved examples

(*i*) $\dfrac{d}{dx}(8x^3) = 8.\dfrac{d}{dx}x^3 = 8.3x^{3-1} = 24x^2$

(*ii*) $\dfrac{d}{dx}(6\sqrt{x}) = 6\dfrac{d}{dx}x^{\frac{1}{2}} = 6.\dfrac{1}{2}x^{\frac{1}{2}-1} = 3.x^{\frac{-1}{2}}$

(*iii*) $\dfrac{d}{dx}(5.e^x) = 5\dfrac{d}{dx}e^x = 5e^x$

(*iv*) $\dfrac{d}{dx}(9.2^x) = 9\dfrac{d}{dx}2^x = 9.2^x$

(*v*) $\dfrac{d}{dx}(\log_{10} x) = \log_e x . \log_{10} e$

(*vi*) $\dfrac{d}{dx}\left(x^3 + e^x + 3^x + \cot x\right) = 3x^2 + e^x + 3^x \log 3 - c\sec^2 x$

(*vii*) $\dfrac{d}{dx}\left(9x^2 + \dfrac{3}{x} + 5\sin x\right) = \dfrac{d}{dx}\left(9x^2 + 3x^{-1}5\sin x\right) = 18x - 3x^{-2} + 5\cos x$

(*viii*) $\dfrac{d}{dx}\left(x^2 + \dfrac{4}{x^2} - \dfrac{2}{3}\tan x + 7\log_e x + 6e\right) = \dfrac{d}{dx}\left(x^2 + 4x^{-2} - \dfrac{2}{3}\tan x + 7\log_e x + 6e\right)$

$= 2x - 4(-2)x^{-3} - \dfrac{2}{3}\sec^2 x + \dfrac{7}{x} + 0 = 2x + \dfrac{8}{x^3} - \dfrac{2}{3}\sec^2 x + \dfrac{7}{x}$

(ix) $\frac{d}{dx}(\log x^3) = \frac{1}{x^3}\frac{d}{dx}x^3 = \frac{1}{x^3}.3x^2 = \frac{3}{x}$

(x) $\frac{d}{dx}\{(x^2-5x+6)(x-3)\} = \frac{d}{dx}\{x^3-8x^2+21x-18\} = 3x^2-16x+21$

(xi) $\frac{d}{dx}\left(\sqrt{x}+\frac{1}{\sqrt{x}}\right) = \frac{d}{dx}\left(x^{\frac{1}{2}}+x^{\frac{-1}{2}}\right) = \frac{1}{2}x^{\frac{1}{2}-1}-\frac{1}{2}x^{\frac{-1}{2}-1}$

$$= \frac{1}{2}x^{\frac{-1}{2}}-\frac{1}{2}x^{\frac{-3}{2}} = \underline{\quad\quad}(\quad\quad)$$

Practice Problems:

Differentiate the following with respect to x

(i) $6.\sqrt[3]{x^2}$, (ii) $\log_e \sqrt{x}$, (iii) $\log_2 x$, (iv) $\log_e 2x$, (v) $\left(ax^3+bx^2+cx+d\right)$

Answers:

(i) $4.x^{-\frac{1}{3}}$, (ii) $\frac{1}{2x}$, (iii) $\frac{1}{x}.\log_2 e$, (iv) $\frac{1}{x}$, (v) $3ax^2+2ab$

1.1.2 Derivatives of Product of Two Functions

(The Product Rule of Differentiation)

Let u and v are two functions of x i.e. $u = f(x)$ and $v = g(x)$

Then the derivative if product is given by:

$$\frac{d}{dx}(uv) = vu' + uv' ,$$

where $u' = \frac{du}{dx}$ and $v' = \frac{du}{dx}$

Solved Examples

Find $\frac{dy}{dx}$ of the following functions:

1. $y = 2x\left(x^2+1\right)$.

Solution: Let $u = 2x$ and $v = \left(x^2+1\right)$

$$u' = 2 \quad \text{and v'} = 2x$$

$$\therefore \frac{dy}{dx} = \frac{d}{dx}(uv) = vu' + uv' = \left(x^2+1\right)+2x(2x) = x^2+1+4x^2 = 5x^2+1$$

2. $y = \left(3x^2 - 5\right)\left(3x + 2\right).$

Solution: Let $u = \left(3x^2 - 5\right)$ and $v = \left(3x + 2\right)$

$$u' = 6x \quad \text{and } v' = 3$$

$$\therefore \frac{dy}{dx} = \frac{d}{dx}(uv) = vu' + uv' = \left(3x + 2\right).6x + \left(3x^2 - 5\right)3 = 18x^2 + 12x + 9x^2 - 15$$

$$= 27x^2 + 12x - 15$$

3. $y = 3x\left(x^2 + 1\right)^2$

Solution: Let $u = 3x$ and $v = \left(x^2 + 1\right)^2$

$$u' = 3 \quad \text{and } v' = 2\left(x^2 + 1\right).2x$$

$$\therefore \frac{dy}{dx} = \frac{d}{dx}(uv) = vu' + uv' = \left(x^2 + 1\right)^2 .3 + 3x.4x\left(x^2 + 1\right) = 3\left(x^4 + 2x^2 + 1\right) + 12x^2\left(x^2 + 1\right)$$

$$= \left(3x^4 + 6x^2 + 3\right) + \left(12x^4 + 12x^2\right)$$

$$= 15x^4 + 12x^2 + 3$$

4. $y = xe^x \sin x$

Solution: Let $u = xe^x$ and $v = \sin x$

$$u' = \left(xe^x + e^x.1\right) \quad \text{and } v' = \cos x$$

$$\therefore \frac{dy}{dx} = \frac{d}{dx}(uv) = vu' + uv' = \sin x\left(xe^x + e^x\right) + \left(xe^x\right)\cos x$$

5. $y = e^x\left(1 + \log x\right)$

Solution: Let $u = e^x$ and $v = \left(1 + \log x\right)$

$$u' = e^x \quad \text{and } v' = \frac{1}{x}$$

$$\therefore \frac{dy}{dx} = \frac{d}{dx}(uv) = vu' + uv' = \left(1 + \log x\right)e^x + e^x.\frac{1}{x} = e^x\left(1 + \log x + \frac{1}{x}\right)$$

6. $y = x^2 \sin x \log x$

Solution: Let $u = x^2 \log x$ and $v = \sin x$

$$u' = \left(x^2.\frac{1}{x} + 2x \log x\right) \quad \text{and } v' = \cos x$$

$$\therefore \frac{dy}{dx} = \frac{d}{dx}(uv) = vu' + uv' = \sin x\left[x + 2x\log x\right] + x^2\log x \cos x$$

7. $y = 3\left(x^2 + 1\right)\sin e^x$

Solution: Let $u = 3(x^2 + 1)$ and $v = \sin e^x$

$$u' = 3(2x) \quad \text{and} \quad v' = \cos e^x.\frac{d}{dx}\left(e^x\right) = \cos e^x.e^x$$

$$\therefore \frac{dy}{dx} = \frac{d}{dx}(uv) = vu' + uv' = \sin e^x\left(3.2x\right) + 3(x^2 + 1)e^x.\cos e^x$$

$$= 6x\sin e^x + 3e^x.\cos e^x(x^2 + 1)$$

8. $y = (x + 8)^2.\sec 3x$

Solution: Let $u = (x + 8)^2$ and $v = \sec 3x$ $\therefore y = uv$

$$u' = 2(x + 8) \text{ and } v' = \sec 3x.\tan 3x.3$$

Differentiating y = uv with respect to x, we get

$$\frac{dy}{dx} = \frac{d}{dx}\{u.v\} = uv' + vu' = (x + 8)^2.(\sec 3x.\tan 3x.3) + \sec 3x.2(x + 8)$$

$$= \sec 3x.(x + 8)\{3(x + 8)(\tan 3x) + 2\}$$

9. $y = x^3.\log x^3$

Solution: Let $u = x^3$ and $v = \log x^3$ $\therefore y = uv$

$$u' = 3x^2 \text{ and } v' = \frac{1}{x^3}.3x^2 = \frac{3}{x}$$

Differentiating y = uv with respect to x, we get

$$\frac{d}{dx}\{u.v\} = uv' + vu'$$

$$= x^3.(\frac{1}{x^3}.3x^2) + \log x^3.(3x^2)$$

$$= (3x^2) + \log x^3.(3x^2) = \left(3x^2\right)\{1 + \log x^3\}$$

Practice Problems (On Product Rule)

Differentiate the followings:

1. $y = e^x . \cos x$ 2. $y = (3 - 2x)(2 - 3x)$ 3. $y = x^2 \left(x^3 - 5x \right)$

4. $y = e^x . \log x$ 5. $y = x.e^{3x}$ 6. $y = x \cos x \log x$

7. $y = \sin x . \cos x$ 8. $y = xe^x$

Answers:

1. $e^x (\cos x - \sin x)$,   2. $12x - 13$,   3. $5x^2 - 15x$,

4. $e^x \left[\log x + \dfrac{1}{x} \right]$,   5. $(1 + 3x)e^{3x}$

6. $\log x . [\cos x - x \sin x] + \cos x$,

7. $\cos^2 x - \sin^2 x$.   8. $e^x (x + 1)$

1.1.3   Derivatives of Quotient of two functions

(Quotient Rule of Differentiation)

Let u and v are two functions of x i.e. $u = f(x)$ and $v = g(x)$

Then the derivative of their quotient is given by –

$$\frac{d}{dx} \left(\frac{u}{v} \right) = \frac{v(u') - u(v')}{(v)^2} \text{ where } u' = \frac{du}{dx} \text{ and } v' = \frac{du}{dx}$$

Solved Examples

Find $\dfrac{dy}{dx}$ of the following functions:

1. $y = \dfrac{x^2}{x^3 + 2}$

Solution: Let $u = x^2$ and $v = x^3 + 3$

$\therefore u' = 2x$ and $v' = 3x^2 + 3$

$$\frac{d}{dx}(y) = \frac{d}{dx} \left(\frac{u}{v} \right) = \frac{v(u') - u(v')}{(v)^2} = \frac{(x^3 + 3)(2x) - x^2 (3x^2)}{(x^3 + 3)^2} = \frac{2x^4 + 6x - 3x^4}{(x^3 + 3)^2}$$

$$= \frac{6x - x^4}{(x^3 + 3)^2}$$

2. $y = \dfrac{2x^2 - 1}{x^3}$

Solution: Let $u = 2x^2 - 1$ and $v = x^3$

$\therefore u' = 2x$ and $v' = 3x^2$

$$\frac{d}{dx}(y) = \frac{d}{dx}\left(\frac{u}{v}\right) = \frac{v(u') - u(v')}{(v)^2} = \frac{\left(x^3\right)(2x) - \left(2x^2 - 1\right)\left(3x^2\right)}{\left(x^3\right)^2}$$

$$= \frac{2x^4 - 6x^4 + 3x^2}{x^6}$$

$$= \frac{3x^2 - 4x^4}{x^6} = \frac{x^2\left(3 - 4x^2\right)}{x^6} = \frac{\left(3 - 4x^2\right)}{x^4}$$

3. $y = \dfrac{x^2 - 2x}{x^3 + x}$

Solution: Let $u = x^2 - 2x$ and $v = x^3 + 1$

$\therefore u' = 2x - 2$ and $v' = 3x^2$

$$\frac{d}{dx}(y) = \frac{d}{dx}\left(\frac{u}{v}\right) = \frac{v(u') - u(v')}{(v)^2} = \frac{\left(x^3 + 1\right)(2x - 2) - \left(x^2 - 2x\right)\left(3x^2\right)}{\left(x^3 + 1\right)^2}$$

$$= \frac{2x^4 - 2x^3 + 2x - 2 - 3x^4 + 2x^3}{\left(x^3 + 1\right)^2}$$

$$= \frac{2x - x^4 - 2}{\left(x^3 + 1\right)^2}$$

4. $y = \dfrac{x \log x}{1 + x^2}$

Solution: Let $u = x \log x$ and $v = 1 + x^2$

$$\therefore u' = \log x . 1 + x\frac{1}{x} = (\log x + 1) \text{ and } v' = 2x$$

$$\frac{d}{dx}(y) = \frac{d}{dx}\left(\frac{u}{v}\right) = \frac{v(u') - u(v')}{(v)^2} = \frac{\left(1 + x^2\right)(1 + \log x) - (x \log x)(2x)}{\left(1 + x^2\right)^2}$$

$$= \frac{\left(1+x^2\right)+\left(\log x + x^2 \log x\right)-\left(2x^2 \log x\right)}{\left(1+x^2\right)^2}$$

$$= \frac{1+x^2+\log x-\left(x^2 \log x\right)}{\left(1+x^2\right)^2}$$

5. $y = \dfrac{2x}{3x^2-1}$

Solution: Let $u = 2x$ and $v = 3x^2 - 1$

$\therefore u' = 2$ and $v' = 6x$

$$\therefore \frac{d}{dx}(y) = \frac{d}{dx}\left(\frac{u}{v}\right) = \frac{v(u')-u(v')}{(v)^2} = \frac{\left(3x^2-1\right)(2)-(2x)(6x)}{\left(3x^2-1\right)^2}$$

$$= \frac{\left(6x^2-2\right)-\left(12x^2\right)}{\left(3x^2-1\right)^2} = \frac{-2-6x^2}{\left(3x^2-1\right)^2}$$

6. $y = \dfrac{e^x}{x}$

Solution: Applying quotient rule:

$$\therefore \frac{dy}{dx} = \frac{d}{dx}\left\{\frac{e^x}{x}\right\} = \frac{x\dfrac{d}{dx}(e^x)-e^x\dfrac{d}{dx}(x)}{x^2} = \frac{xe^x-e^x}{x^2} = \frac{e^x}{x^2}(x-1)$$

7. $y = \dfrac{e^x \sin x}{x^2}$

Solution: By quotient rule:

$$\frac{dy}{dx} = \frac{d}{dx}\left\{\frac{e^x \sin x}{x^2}\right\} = \frac{x^2\dfrac{d}{dx}(e^x \sin x)-e^x \sin x\dfrac{d}{dx}(x^2)}{\left(x^2\right)^2}$$

$$= \frac{x^2\left(e^x\dfrac{d}{dx}\sin x+\sin x\dfrac{d}{dx}e^x\right)-e^x \sin x.2x}{\left(x^2\right)^2}$$

$$= \frac{x^2 e^x\left(\cos x+\sin x\right)-2e^x x\sin x}{\left(x^4\right)}$$

Practice Problems (On Quotients Rule)

Find $\dfrac{dy}{dx}$ of the following functions

1. $y = \dfrac{2}{x+4}$

2. $y = \dfrac{\cos x}{x}$

3. $y = \tan x$

4. $y = \dfrac{8}{\log x}$

5. $y = \dfrac{2\log x}{x}$

6. $y = \dfrac{e^x}{1+x}$

7. $y = \dfrac{\cos x}{1+\sin x}$

8. $y = \dfrac{2x+1}{x-3}$

Answers:

1. $\dfrac{-2}{(x+4)^2}$,

2. $(\cos x - x\sin x)$,

3. $\sec^2 x$,

4. $\dfrac{-8}{x(\log x)^2}$,

5. $\dfrac{2(1-\log x)}{x^2}$

6. $\dfrac{xe^x}{(1+x)^2}$,

7. $\dfrac{\{\cos x + \sin x.\cos x + \cos^2 x\}}{(1+\sin x)^2}$

8. $\dfrac{7}{(x-3)^2}$

1.1.4 Derivative of Composite Function (The Chain Rule)

If $y = f(u)$ and $u = g(x)$ are two functions, then derivative of y with respect to x can be obtained by following relation:

$$\frac{dy}{dx} = \frac{dy}{du} \times \frac{du}{dx}$$

This is known as Chain Rule.

Solved Examples

1. Find $\dfrac{dy}{dx}$ **, if** $y = \sin(3x^2 + 1)$

Solution: Let $u = 3x^2 + 1 \Rightarrow y = \sin u$

$$\frac{du}{dx} = 6x \text{ and } \frac{dy}{du} = \cos u = \cos(3x^2 + 1)$$

Applying chain rule $\dfrac{dy}{dx} = \dfrac{dy}{du} \times \dfrac{du}{dx} = \cos u \times 6x = 6x.\cos(3x^2 + 1)$

2. Find $\dfrac{dy}{dx}$, **if** $y = (3x-2)^3$

Solution: Let $u = 3x-2 \Rightarrow y = u^3$

$$\frac{du}{dx} = 3 \text{ and } \frac{dy}{du} = 3u^2 = 3(3x-2)^2$$

Applying chain rule $\dfrac{dy}{dx} = \dfrac{dy}{du} \times \dfrac{du}{dx} = 3(3x-2)^2 \times 3 = 9(3x-2)^2$

3. Find $\dfrac{dy}{dx}$, **if** $y = \sin mx^2$

Solution: Let $u = mx^2 \Rightarrow y = \sin u$

$$\frac{du}{dx} = 2mx \text{ and } \frac{dy}{du} = \cos(u) = \cos(mx^2)$$

Applying chain rule $\dfrac{dy}{dx} = \dfrac{dy}{du} \times \dfrac{du}{dx} = \cos(mx^2) \times 2mx = 2mx.\cos(mx^2)$

4. Find $\dfrac{dy}{dx}$, **if** $y = e^{x^2}$

Solution: Let $u = x^2 \Rightarrow y = e^u$

$$\frac{du}{dx} \qquad \text{and } \frac{dy}{du} = e^u = e^{x^2}$$

Applying chain rule $\dfrac{dy}{dx} = \dfrac{dy}{du} \times \dfrac{du}{dx} = e^{x^2} \times 2x = 2x.e^{x^2}$

5. Find $\dfrac{dy}{dx}$, **if** $y = \tan(x^2 + \sin x)$

Solution: Let $u = x^2 + \sin x \Rightarrow y = \tan(u)$

$$\frac{du}{dx} = 2x + \cos x \text{ and } \frac{dy}{du} = \sec^2(u) = \sec^2(x^2 + \sin x)$$

Applying chain rule $\dfrac{dy}{dx} = \dfrac{dy}{du} \times \dfrac{du}{dx} = \sec^2(x^2 + \sin x) \times (2x + \cos x)$

6. Find $\dfrac{dy}{dx}$, **if** $y = e^{1+x^2}$

Solution: Let $u = 1 + x^2 \Rightarrow y = e^u$

$$\frac{du}{dx} = 2x \text{ and } \frac{dy}{du} = e^u = e^{1+x^2}$$

Applying chain rule $\dfrac{dy}{dx} = \dfrac{dy}{du} \times \dfrac{du}{dx} = e^{1+x^2} \times 2x = 2x.e^{1+x^2}$

7. Find $\dfrac{dy}{dx}$**, if** $y = \sin(x^2 + e^x)$

Solution: Let $u = x^2 + e^x \Rightarrow y = \sin(u)$

$$\frac{du}{dx} = 2x + e^x \text{ and } \frac{dy}{du} = \cos(u) = \cos(x^2 + e^x)$$

Applying chain rule $\dfrac{dy}{dx} = \dfrac{dy}{du} \times \dfrac{du}{dx} = \cos(x^2 + \sin x) \times (2x + \cos x)$

8. Find $\dfrac{dy}{dx}$**, if** $y = \log(x - \cos x)$

Solution: Let $u = (x - \cos x) \Rightarrow y = \log(u)$

$$\frac{du}{dx} = 1 + \sin x \text{ and } \frac{dy}{du} = \frac{1}{u} = \frac{1}{(x - \cos x)}$$

Applying chain rule $\dfrac{dy}{dx} = \dfrac{dy}{du} \times \dfrac{du}{dx} = \dfrac{1}{(x - \cos x)} \times (1 + \sin x) = \dfrac{(1 + \sin x)}{(x - \cos x)}$

Practice Problems(On Chain Rule)

Find $\dfrac{dy}{dx}$ of the following function:

1. $y = \sin 10x$ 2. $y = \log(2x - 1)$ 3. $y = \tan x^2$

4. $y = \sin(\log x)$ 5. $y = \log(\sin x)$ 6. $y = e^{-\cos x}$

7. $y = \cot(1 - 4x)$ 8. $y = \cos(e^{-nx})$

Answers:

1. $10\cos(10x)$ 2. $\dfrac{2}{2x - 1}$ 3. $2x\sec x^2$

4. $\dfrac{\cos(\log x)}{x}$ 5. $\cot(x)$ 6. $\sin x.e^{-\cos x}$

7. $4\cos ec^2(1 - 4x)$ 8. $ne^{-nx}\sin(e^{-nx})$

1.2 APPLICATION OF DERIVATIVES:

If $y = f(x)$, then $\dfrac{dy}{dx}$ denotes the rate of change of y with respect to x.

Solved Examples

1. **Find the rate of change of the area of a circle, when its radius varies.**

   Solution: The area of a circle is $A = \pi r^2$. This implies: $\dfrac{dA}{dr} = \pi.2r = 2\pi r$

   Hence the rate of change of area of a circle with respect to radius is $2\pi r$

1.2.1   Maxima and Minima

Maxima is the plural of the word Maximum. It is the point where the given function attains the maximum or the largest value.

   Minima is the plural of the word minimum. It is the point where the function attains the minimum or the least value.

   Maxima and Minima are also collectively known as extrema i.e. largest and smallest values.

   Local Maximum: The local maximum of a function is a value that is greater than all the values in its neighborhoods.

   Local Minimum: The local minimum of a function is a value that is less than all the values in its neighborhoods.

   Stationary Point: It is the point on the function graph where the gradient or the slope of function is zero. If the gradient changes sign at this point then it is known as turning point.

   Critical Point: A critical point is the point on a function where either the tangent of the function does not exist, or is a horizontal or vertical line. In the case where it is a horizontal line, that *critical point is called a stationary point*. The value of $f(c)$ at the critical point c is called the *critical value*

Finding the Local Maxima and Minima

Following steps are required to find the *Maxima and Minima*.

   Find the fist derivative of $f(x)$ i.e. $f'(x)$ or $\dfrac{dy}{dx}$

   Find stationary points of $f(x)$ by equating first derivative $f'(x)$ to zero.

   Find second derivative $f''(x)$.

   If $f''(x) > 0$, at the stationary points, then $f(x)$ has a local minima at that point.

   If $f''(x) < 0$, at the stationary points, then $f(x)$ has a local maxima at that point.

Solved Examples

1. Find the stationary points of the function $f(x) = x^3 - 3x$ and determine its nature.

Solution : The first derivative is $f'(x) = 3x^2 - 3$

If $f'(x) = 0$, then $3x^2 - 3 = 0 \Rightarrow x^2 = 1 \Rightarrow x = \pm 1$

Hence stationary points are ± 1

The second derivative $f''(x) = \dfrac{d}{dx}\left(3x^2 - 3\right) = 6x$

At the point $x = +1$ $f''(1) = 6.1 = 6 > 0$

Hence $f(x)$ has a <u>local minima</u> at $x = 1$ (see rule 3 (i) given above)

The value of local minima is $f(+1) = (+1)^3 - 3(+1) = +1 - 3 = -2$

At the point $x = -1$ $f''(-1) = 6.(-1) = -6 < 0$

Hence $f(x)$ has a <u>local maxima</u> at $x = -1$ (see rule 3 (ii) given above)

The value of local maxima is $f(-1) = (-1)^3 - 3(-1) = -1 + 3 = 2$

2. Find the maxima and minima of the function $f(x) = x^3 - 27x + 1$

Solution: Given function is: $f(x) = x^3 - 27x + 1$

Then, $f'(x) = 3x^2 - 27$

Putting $f'(x) = 0$, we get $3x^2 - 27 = 0 \Rightarrow x^2 = 9 \Rightarrow x = \pm 3$

The stationary points are 3,-3

Second derivative is: $f''(x) = 3.2x = 6x$

At $x = 3$, $f''(3) = 6 \times 3 = 18 > 0$

Hence function $f(x)$ has a local minima at the point $x = 3$ (see rule 3 (i) given above)

Value of function at its local minima is $f(3) = 3^3 - 27(3) + 1 = 27 - 81 + 1 = -53$

At $x = -3$, $f''(-3) = 6 \times (-3) = -18 < 0$

Hence function $f(x)$ has a local maxima at the point $x = -3$ (see rule 3 (ii) given above)

Value of function at its local maxima is $f(3) = 3^3 - 27(3) + 1 = 27 - 81 + 1 = -53$

3. Find the maxima and minima of the function $f(x) = \sin 2x \quad 0 < x < \pi$

Solution: Given $f(x) = \sin 2x$

$$\therefore f'(x) = 2\cos 2x$$

$$f'(x) = 0 \Rightarrow 2\cos 2x = 01$$

$$\cos 2x = 0 \Rightarrow 2x = \frac{\pi}{2} or \frac{3\pi}{2} \Rightarrow x = \frac{\pi}{4} or \frac{3\pi}{4}$$

$$f''(x) = -4\sin 2x$$

when $x = \frac{\pi}{4}$ $f''(\frac{\pi}{4}) = -4\sin 2\frac{\pi}{4} = -4\sin\frac{\pi}{2} = -4\sin 90 = -4.1 = -4 < 0$

when $f''(\frac{\pi}{4}) < 0$ i.e. negative , then there is maxima at $x = \frac{\pi}{4}$.(see rule 3 (ii) given above)

Hence maximum value of function $f(x) = \sin 2x$ is

$$f\left(\frac{\pi}{4}\right) = \sin 2\left(\frac{\pi}{4}\right) = \sin\left(\frac{\pi}{2}\right) = \sin 90 = 1$$

When $x = \frac{3\pi}{4}$ $f''(\frac{\pi}{4}) = -4\sin 2\frac{3\pi}{4} = -4\sin\frac{3\pi}{2} = -4.(-1) = 4 > 0$

when $f''(\frac{3\pi}{4}) > 0$ i.e. positive , then there is minima at $x = \frac{3\pi}{4}$.(see rule 3 (ii) given above)

Hence minimum value of function $f(x) = \sin 2x$ is

$$f\left(\frac{3\pi}{4}\right) = \sin 2\left(\frac{3\pi}{4}\right) = \sin\left(\frac{3\pi}{2}\right) = \sin(180 + 90) = -1$$

4. Find local maxima and minima of (i) e^x (ii) $\log x$

Solution: (i) $f(x) = e^x \Rightarrow f'(x) = e^x$

Since e^x can not be 0 , hence $f'(x) \neq 0$.

So the function $f(x) = e^x$ has neither a maxima nor a minima.

(ii) $f(x) = \log x \Rightarrow f'(x) = \frac{1}{x}$

$$f'(x) = 0 \Rightarrow \frac{1}{x} = 0.$$

This indicates x is infinite.

Hence $f(x) = \log x$ has neither a local maxima nor a local minima.

5. $f(x) = \dfrac{x}{2} + \dfrac{2}{x}$ $x>0$

Solution: $f'(x) = \dfrac{1}{2} - \dfrac{2}{x^2}$

When $f'(x) = 0 \Rightarrow \dfrac{1}{2} - \dfrac{2}{x^2} = 0 \Rightarrow x^2 = 4 \Rightarrow x = \pm 2$

Given x>0 so x can not be negative. Hence $x = 2$

By second derivative $\qquad\qquad f''(x) = -\dfrac{4}{x^3}$

$$x = 2, \quad f''(2) = -\dfrac{4}{2^3} = -\dfrac{1}{2} < 0$$

Hence at x = 2 $f(x) = \dfrac{x}{2} + \dfrac{2}{x}$ has a minima.

The minimum value is $f(2) = \dfrac{2}{2} + \dfrac{2}{2} = 2$

Practice Problems: (On Maxima & Minima)

Find the local maxima and local minima of the following functions:

1. $f(x) = (5x-1)^2 + 4$,   **2.** $f(x) = (x^3 - 3x)$,

3. $f(x) = -x^3 + 12x^2 - 8$

Answers:

1. minimum value = 4,
2. Local minimum value= −2 at x =1 and local maximum value = 2 at x = −1
3. Local max. Value = 251 at x = 8 , Local min. value = −5 at x = 0

1.2.2   Equations of Tangent to a Curve

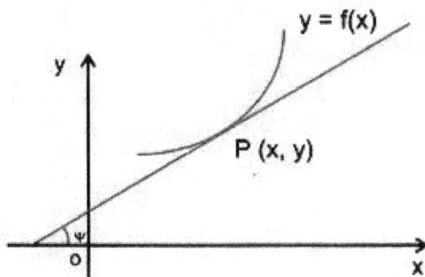

Let $y = f(x)$ be the equation of the curve. Then $\dfrac{dy}{dx}$ at a point $P(x', y')$ on the curve gives the slope or gradient of the tangent line.

The equation of tangent at $P(x', y')$ with slope $\dfrac{dy}{dx}$ is given by:

$$(y - y') = \frac{dy}{dx}(x - x')$$

1.2.3   Equation of Normal to a Curve

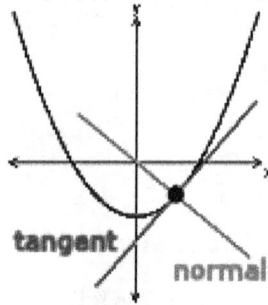

Let $y = f(x)$ be the equation of the curve. Then $\dfrac{dy}{dx}$ is the slope or gradient of tangent at a point $P(x', y')$ on the curve.

A perpendicular line to the tangent at a point $P(x', y')$ on the curve $y = f(x)$ is called the normal. Obviously slope of normal would be $\dfrac{-1}{dy/dx}$.

$$(\because m_1 m_2 = -1 \Rightarrow m_2 = \frac{-1}{m_1})$$

The equation of normal is:

$$(y - y') = \frac{-1}{dy/dx}(x - x')$$

or

$$\frac{dy}{dx}(y - y') + (x - x') = 0$$

Solved Examples

1. Find the equation of tangent to the curve $y = x\sin x$ **at the point** $x = \dfrac{\pi}{3}$.

Solution: When $-., \ y = \dfrac{\pi}{3}\sin\dfrac{\pi}{3} = \dfrac{\pi}{3}\sin 60^0 = \dfrac{\pi}{3}\dfrac{\sqrt{3}}{2} = \dfrac{\pi}{2\sqrt{3}}$

Hence point of touch is $\left(\dfrac{\pi}{3}, \dfrac{\pi}{2\sqrt{3}}\right)$

Now differentiating $y = x\sin x$

$$\dfrac{dy}{dx} = x\cos x + 1.\sin x = x\cos x + \sin x$$

$$\left(\dfrac{dy}{dx}\right)_{at\ x=\frac{\pi}{3}} = \dfrac{\pi}{3}\cos\dfrac{\pi}{3} + \sin\dfrac{\pi}{3} = \dfrac{\pi}{3}\cos 60 + \sin 60 = \dfrac{\pi}{3}\dfrac{1}{2} + \dfrac{\sqrt{3}}{2} = \dfrac{\pi}{6} + \dfrac{\sqrt{3}}{2}$$

The equation of tangent passing through $\left(\dfrac{\pi}{3}, \dfrac{\pi}{2\sqrt{3}}\right)$ and having slope (m) $= \dfrac{\pi}{6} + \dfrac{\sqrt{3}}{2}$ is:

$$(y - y_1) = m(x - x_1)$$

$$\left(y - \dfrac{\pi}{2\sqrt{3}}\right) = \left(\dfrac{\pi}{6} + \dfrac{\sqrt{3}}{2}\right)\left(x - \dfrac{\pi}{3}\right)$$

$$y = \left(\dfrac{\pi}{6} + \dfrac{\sqrt{3}}{2}\right)x - \dfrac{\pi}{3}\left(\dfrac{\pi}{6} + \dfrac{\sqrt{3}}{2}\right) + \dfrac{\pi}{2\sqrt{3}}$$

$$y = \left(\dfrac{\pi}{6} + \dfrac{\sqrt{3}}{2}\right)x - \dfrac{\pi^2}{18} - \dfrac{\sqrt{3}\pi}{3.2} + \dfrac{\pi}{2\sqrt{3}}$$

$$y = \left(\dfrac{\pi}{6} + \dfrac{\sqrt{3}}{2}\right)x - \dfrac{\pi^2}{18} - \dfrac{\pi}{\sqrt{3}.2} + \dfrac{\pi}{2\sqrt{3}}$$

$$y = \left(\dfrac{\pi}{6} + \dfrac{\sqrt{3}}{2}\right)x - \dfrac{\pi^2}{18}$$

2. Find the equation of tangent and normal to the curve $y = x + \log x$ at the point $x = 1$.

Solution: When $x = 1., \ y = 1 + \log 1 = 1 \ \because \log 1 = 0$

Hence the tangent touches the curve at (1, 1)

Differentiating the equation of the curve:

$$\frac{dy}{dx} = 1 + \frac{1}{x}$$

$$\left(\frac{dy}{dx}\right)_{at\ x=1} = 1 + \frac{1}{1} = 2$$

Hence **equation of tangent** at (1, 1) with slope $m = 2$ is:

$$(y-1) = 2(x-1) \Rightarrow y = 2x - 1$$

The slope of the normal would be $\dfrac{1}{m} = \dfrac{-1}{2}$

The **equation of normal** at (1, 1) with slope $1/m = -1/2$ is:

$$(y-1) = \frac{-1}{2}(x-1)$$

$$y = \frac{-1}{2}(x-1) + 1 \Rightarrow y = \frac{-1}{2}x + \frac{1}{2} + 1 = \frac{1}{2}(3-x)$$

Hence the equation of normal is $y = \dfrac{1}{2}(3-x)$

3. Find the equations of tangent and normal to the curve $y = x + \dfrac{1}{x}$ **at the point x = 1**

Solution: When x = 1 then $y = 1 + \dfrac{1}{1} = 2$. Hence point of touch of the tangent is (1, 2)

Now differentiating the equation of curve , we get $-\dfrac{dy}{dx} = 1 - \dfrac{1}{x^2}$. This is the slope of tangent.

The slope at (1, 2) is $\left(\dfrac{dy}{dx}\right)_{at\ (1,2)} = 1 - \dfrac{1}{1^2} = 0$

$$(y-2) = 0(x-1) \Rightarrow y = 2$$

Hence the equation of the tangent is $y = 2$

The slope of the normal is $= \dfrac{-1}{dy/dx} = \dfrac{-1}{0} = \infty$.

Hence equation of normal is indeterminate.

4. Find the equation of the normal to the curve $y = x^2 - 2x + 1$ **at x = 3**

Solution: when x = 3, then $y = (3)^3 - 2(3) + 1 = 9 - 6 + 1 = 4$. Hence point of touch of the tangent is (3, 4)

Differentiating the curve w.r.t . we get:

$$\frac{dy}{dx} = 2x - 2.$$

This is the slope of the tangent.

Slope of the normal is $\dfrac{-1}{dy/dx} = \dfrac{-1}{2x-2}$

Slope of the normal at $x = 3$ is $\left(\dfrac{-1}{dy/dx}\right)_{at\ x=3} = \dfrac{-1}{2.3-2} = \dfrac{-1}{4}$

Hence the equation of the normal is: $(y-4) = \dfrac{-1}{4}(x-3)$ or $4y + x - 19 = 0$

5. Find the equation of the tangent and normal to the curve $x^2 - y^2 = 5$ **at (x, y) = (3, 2)**

Solution: Differentiating the given equation wrt x : $2x - 2y\dfrac{dy}{dx} = 0 \Rightarrow \dfrac{dy}{dx} = \dfrac{x}{y}$

$\therefore \left(\dfrac{dy}{dx}\right)_{at\ (3,2)} = \dfrac{3}{2}$

Hence equation of tangent is: $(y-2) = \dfrac{3}{2}(x-3)$ or $2y - 3x + 5 = 0$

Slope of the normal $= \dfrac{-1}{3/2}$

Hence equation of normal is: $(y-2) = \dfrac{-1}{3/2}(x-3)$ or $2y - 3x + 5 = 0$

$\dfrac{3}{2}(y-2) = -1(x-3)$ or $3(y-2) = -2(x-3)$ or $3y + 6x - 12 = 0$

6. Find the equation of the normal to the curve: $y = \sin 2\theta + 2\cos\theta$ **at** $\theta = \pi/4$

Solution: when $\theta = \pi/4$, then $y = \sin 2(\pi/4) + 2\cos(\pi/4)$

$$= \sin 90 + 2\cos 45$$

$$= 1 + 2\dfrac{1}{\sqrt{2}} = 1 + \sqrt{2}$$

Hence point of touch of the tangent is $\left[(\pi/4), (1+\sqrt{2})\right]$

Differentiating the curve w.r.t . we get : $\dfrac{dy}{dx} = 2\cos 2\theta - 2\sin\theta$.

At $\left[(\pi/4), (1+\sqrt{2})\right]$ $\dfrac{dy}{dx} = 2\cos 2\dfrac{\pi}{4} - 2\sin\dfrac{\pi}{4} = 2\cos 90 - 2\sin 45 = 0 - 2\dfrac{1}{\sqrt{2}} = -\sqrt{2}$

The slope of the tangent is $= -\sqrt{2}$

The slope of the normal is $= -\dfrac{1}{\sqrt{2}}$

Hence the equation of the normal is: $\left[y - \left(1 + \sqrt{2}\right)\right] = \dfrac{-1}{\sqrt{2}}\left[x - \dfrac{\pi}{4}\right]$

or $y + \dfrac{x}{\sqrt{2}} = (1 + \sqrt{2}) + \dfrac{\pi}{4\sqrt{2}}$

7. Find the equation of tangent to the curve: $y = 3x^2 p$ **at the point whose x co-ordinate is 2.**

Solution: Putting x = 2, y = 3.4 = 12. The tangent touches the curve at (2, 12)

Differentiating $y = 3x^2$, **we get** $\dfrac{dy}{dx} = 3.2x = 6x$

$$\left(\dfrac{dy}{dx}\right)_{at\ x=2} = 3.2x = 6.2 = 12$$

Hence slope of the tangent = 12

The equation of tangent at (2, 12) is : $(y - 12) = 12(x - 2) \Rightarrow y = 12x - 12$

The equation of normal at (2, 12) is : $(y - 12) = \dfrac{-1}{12}(x - 2)$

or $\qquad\qquad\qquad 12(y - 12) = -1(x - 2)$

or $\qquad\qquad\qquad 12y + x = 142$

8. Find at what point the tangent to the curve $y = 2x^3 - 9x^2 + 10x - 5$ **is parallel to the straight line** $y + 2x = 12$

Solution: Since the tangent and the straight line are parallel, hence their gradient (slope) will be equal.

Differentiating the equation of the curve, we get the slope of the tangent:

$$\dfrac{dy}{dx} = 6x^2 - 18x + 10$$

The gradient of the line is –2 $\qquad (\because y = -2x)$

$\therefore 6x^2 - 18x + 10 = -2$

$6x^2 - 18x + 12 = 0$

$$6\left(x^2-3x+2\right)=0$$

or $$\left(x^2-3x+2\right)=0$$

or $$\left(x^2-1x-2x+2\right)=0$$

or $$\left(x^2-1x\right)-\left(2x-2\right)=0$$

or $$x(x-1)-2(x-1)=0$$

or $$(x-1)(x-2)=0$$

Hence $x=1$ or 2

Putting $x=1$ in the equation of the curve:

$$y=2.1^3-15.1^2+22.1-15=-6$$

Putting $x=2$ in the equation of the curve:

$$y=2.2^3-9.2^2+10.2-5=16-36+20-5=36-41=-5$$

Hence required point is (1, -6) or (2, -5).

Practice Problems (Tangent & Normal)

1. Find the equation of the tangent to the curve $y=2x^2+3\sin x$ at x = 0.

2. Find the equation of the tangent to the curve $y=2x^2-3x-1$ at (1, 2).

3. Find the equation of tangent to the curve $y=2x^2+7$ which is parallel to the straight line $4x-y+3=0$.

4. Find the equation of normal to the curve $y(x-2)(x-3)-x+7=0$ at the point where it cuts the x- axis(i.e. y = 0)

Answers:

1. $y=3x$

2. $x-y+1=0$ or $x+y-3=0$

3. $y-4x-5=0$

4. $y+19x-133=0$

1.3 PARTIAL DIFFERENTIATION

Let $f(x,y)$ denote a function with two variables viz. $f(x,y)=x^2+2y$ .If we differentiate $f(x,y)$ with respect to x treating y as constant, then it is known as *partial differentiation*. The derivative with respect to x is termed as the *partial derivative* and it is denoted as $\dfrac{\partial f}{\partial x}$ or f_x .

Similarly we can get $\dfrac{\partial f}{\partial y}$ or f_y

Solved Examples

1. Find partial derivatives $\dfrac{\partial f}{\partial x}$ and $\dfrac{\partial f}{\partial y}$ of the function given by:

$$f(x, y) = x^2 + 2xy + y^2$$

Solution: $\dfrac{\partial f}{\partial x} = \dfrac{\partial}{\partial x}\left[x^2 + 2xy + y^2\right] = \dfrac{\partial}{\partial x}(x^2) + 2\dfrac{\partial}{\partial x}(xy) + \dfrac{\partial}{\partial x}(y^2)$

$= 2x + 2y + 0 = 2(x + y)$ (Treating y as constant)

$\dfrac{\partial f}{\partial y} = \dfrac{\partial}{\partial y}\left[x^2 + 2xy + y^2\right] = \dfrac{\partial}{\partial y}(x^2) + 2\dfrac{\partial}{\partial y}(xy) + \dfrac{\partial}{\partial y}(y^2)$

$= 0 + 2x + 2y = 2(x + y)$ (Treating x as constant)

2. Find $\dfrac{\partial f}{\partial x}$ and $\dfrac{\partial f}{\partial y}$ of the function given by: $f(x, y) = 2x^2 + 7y^3$

Solution:

$$\dfrac{\partial f}{\partial x} = \dfrac{\partial}{\partial x}\left[2x^2 + 7y^3\right] = 4x + 0 = 4x$$

$$\dfrac{\partial f}{\partial y} = \dfrac{\partial}{\partial y}\left[2x^2 + 7y^3\right] = 0 + 7.3y^2 = 21y^2$$

3. Find $\dfrac{\partial f}{\partial x}$ and $\dfrac{\partial f}{\partial y}$ of the function given by: $f(x, y) = xe^{xy}$

Solution: Using product rule

$$\dfrac{\partial f}{\partial x} = \dfrac{\partial}{\partial x}\left[xe^{xy}\right] = x\dfrac{\partial}{\partial x}\left[e^{xy}\right] + e^{xy}\dfrac{\partial}{\partial x}[x]$$

$$\left(\text{recall } \dfrac{d}{dx}(uv) = u\dfrac{dv}{dx} + v\dfrac{du}{dx}\right)$$

$$= x.(e^{xy}.y) + e^{xy}.1 = (1 + xy)e^{xy}$$

$$\dfrac{\partial f}{\partial y} = \dfrac{\partial}{\partial y}\left[xe^{xy}\right] = x\dfrac{\partial}{\partial y}\left[e^{xy}\right] + e^{xy}\dfrac{\partial}{\partial y}[x]$$

$$= x.e^{xy}.\dfrac{\partial}{\partial y}(xy) + 0 = xe^{xy}.x = x^2.e^{xy} \qquad \dfrac{\partial}{\partial y}[x] = 0 \because x \text{ is constt.}$$

4. Find $\dfrac{\partial f}{\partial x}$ and $\dfrac{\partial f}{\partial y}$ at the point (2, 3) of the function: $f(x, y) = yx^2 + 2y$

Solution : $\dfrac{\partial f}{\partial x} = y.2x + 0 = 2xy$ $\therefore \dfrac{\partial f}{\partial x}(2,3) = 2.2.3 = 12$

$$\dfrac{\partial f}{\partial y} = 1.x^2 + 2.1 = x^2 + 2 \qquad\qquad \dfrac{\partial f}{\partial y}(2,3) = (2)^2 + 2 = 6$$

5. Find f_x and f_y, if $f(x+y) = \sin(xy) + \cos y$

Solution: Differentiating the given function w.r.t x treating y as constant:

$$f_x = \dfrac{\partial f}{\partial x} = \cos(xy).y - \sin x = y\cos(xy) - \sin x$$

$$f_y = \dfrac{\partial f}{\partial y} = \cos(xy).x + 0 = x\cos(xy)$$

6. Find f_x, f_y, f_{xy}, f_{xx} and f_{yy}, if $f(x, y) = y.e^x$

Solution: $f_x = \dfrac{\partial f}{\partial x} = \dfrac{\partial}{\partial x}(ye^x) = y\dfrac{\partial}{\partial x}(e^x) + e^x.\dfrac{\partial}{\partial x}(y)$ $\qquad$ (recall $\dfrac{d}{dx}(uv) = u\dfrac{dv}{dx} + v\dfrac{du}{dx}$)

$$= y\dfrac{\partial}{\partial x}(e^x) + e^x.\dfrac{\partial}{\partial x}(y)$$

$$= y.e^x + e^x.0 = ye^x$$

$$f_y = \dfrac{\partial f}{\partial y} = \dfrac{\partial}{\partial y}(ye^x) = y\dfrac{\partial}{\partial y}(e^x) + e^x.\dfrac{\partial}{\partial y}(y)$$

$$= y.0 \quad + \quad e^x.1 \ = e^x$$

$$f_{xy} = \dfrac{\partial^2 f}{\partial xy} = \dfrac{\partial}{\partial x}\left[\dfrac{\partial}{\partial y}(ye^x)\right] = \dfrac{\partial}{\partial x}\left[e^x\right] = e^x$$

$$f_{xx} = \dfrac{\partial^2 f}{\partial x^2} = \dfrac{\partial}{\partial x}\left[\dfrac{\partial}{\partial x}(ye^x)\right] = \dfrac{\partial}{\partial x}(ye^x) = ye^x$$

$$f_{yy} = \dfrac{\partial^2 f}{\partial y^2} = \dfrac{\partial}{\partial y}\left[\dfrac{\partial}{\partial y}(ye^x)\right] = \dfrac{\partial}{\partial y}\left[(e^x)\right] = 0$$

(since x is assumed as a constant)

7. Find f_x, and f_y, if $f(x, y) = 2y\sin(xy^2)$

Solution: $f_x = \dfrac{\partial f}{\partial x} = \dfrac{\partial}{\partial x}(2y\sin xy^2) = 2.y\dfrac{\partial}{\partial x}\left[\sin\left(xy^2\right)\right] + \sin\left(xy^2\right).\dfrac{\partial}{\partial x}(2y)$

$$= 2.y\left[\cos\left(xy^2\right)\frac{\partial}{\partial x}\left(xy^2\right)\right] + \sin\left(xy^2\right).(2)$$

$$= 2.y\left[\cos\left(xy^2\right)\left(y^2\right)\right] + \sin\left(xy^2\right).(2)$$

$$= 2.y^3\cos\left(xy^2\right) + 2\sin\left(xy^2\right)$$

$$f_y = \frac{\partial f}{\partial y} = \frac{\partial}{\partial y}(2y\sin xy^2) = 2.y\frac{\partial}{\partial y}\left[\sin\left(xy^2\right)\right] + \sin\left(xy^2\right).\frac{\partial}{\partial y}(2y)$$

(by product rule)

$$= 2.y\left[\cos\left(xy^2\right)\frac{\partial}{\partial y}(xy^2)\right] + \sin\left(xy^2\right).(2)$$

$$= 2y\left[\cos\left(xy^2\right)(x.2y)\right] + 2\sin\left(xy^2\right)$$

$$= 4xy^2\left[\cos\left(xy^2\right)\right] + 2\sin\left(xy^2\right)$$

8. Find f_x , f_y ,and f_z, if $f(x, y, z) = x\cos z + x^2 y^3 e^{4z}$

Solution: $f_x = \dfrac{\partial f}{\partial x} = \dfrac{\partial}{\partial x}(x\cos z + x^2 y^3 e^{4z}) = \cos z + 2xy^3 e^{4z}$

$$f_y = \frac{\partial f}{\partial y} = \frac{\partial}{\partial y}(x\cos z + x^2 y^3 e^{4z}) = 0 + x^2 3y^2 e^{4z} = 3x^2 y^2 e^{4z}$$

$$f_z = \frac{\partial f}{\partial z} = \frac{\partial}{\partial z}(x\cos z + x^2 y^3 e^{4z}) = x\frac{\partial}{\partial z}(\cos z) + x^2 y^3 \frac{\partial}{\partial z}\left(e^{4z}\right) = -x\sin z + 4e^{4z}x^2 y^3$$

Practice Problems (Partial Differentiation)

1. Find $\dfrac{\partial f}{\partial x}$, $\dfrac{\partial f}{\partial y}$ and $\dfrac{\partial^2 f}{\partial y \partial x}$ of the function given by: $f(x, y) = 2x + 3xy - 5y^2$

2. Find $\dfrac{\partial f}{\partial x}$ and $\dfrac{\partial f}{\partial y}$ at (2, 4) of the function by: $f(x, y) = x\sqrt{y} + y$

3. Find f_x and f_y of $f(x, y) = \log(x^2 + 2y)$ at (1, 2)

4. Find f_x , f_y, f_{xy}, f_{yx}, f_{xx} and f_{yy} of $f(x, y) = \left(\dfrac{y}{x}\right)\log x$

Answers:

1. $f_x = 2 + 3y$, $f_y = 3x - 10y$

2. $f_x = \sqrt{y}$, $f_y = \dfrac{x}{2\sqrt{y}} + 1$,

3. $f_x = \dfrac{2}{5}$, $f_y = \dfrac{2}{5}$

4. $f_x = \dfrac{y}{x^2}(1 - \log x)$, $f_y = \dfrac{1}{x} \cdot \log x$, $f_{xy} = \dfrac{1}{x^2}(1 - \log x)$,

   $f_{yx} = \dfrac{1}{x^2}(1 - \log x)$, $f_{xx} = \dfrac{y}{x^3}(2 \log x - 3)$, $f_{yy} = 0$

1.4   SUCCESSIVE DIFFERENTIATION

If $y = f(x)$ is a given function, then its successive derivatives are given by:

(i) $f'(x)$, ...the first derivative.

(ii) $f''(x)$ ....the second derivative.

(iii) $f'''(x)$ ....the third derivative. and

(iv) $f^{(n)}(x)$............the nth derivative.

Also $f', f'', f'''...$ can be written as y', y'', y''' .....

where $y' = \dfrac{dy}{dx}$, $y'' = \dfrac{d^2 y}{dx^2}$, and $y'' = \dfrac{d^3 y}{dx^3}$ .....

The first derivative of $y = f(x)$ is denoted by y' or $\dfrac{dy}{dx}$.

The derivative of first derivative can be found by differentiating it again. This will be called the second derivative denoted by y'' or $\dfrac{d^2 y}{dx^2}$. Similarly we can find derivative of second derivative and this will be called the third derivative of y and denoted as y''' or $\dfrac{d^3 y}{dx^3}$

Thus derivative of higher order can be determined.

Solved Examples

1. Given: $y = e^{kx}$. Find third and nth derivatives.

Solution: we have to find y''' and $y^{(n)}$

$$y = e^{kx}$$

Differentiating the above, we get $y' = \dfrac{d}{dx}(y) = \dfrac{d}{dx}\left(e^{kx}\right) = e^{kx}.k$

Differentiating the y' , we get $y'' = \dfrac{d}{dx}(y') = \dfrac{d}{dx}\left(ke^{kx}\right) = k\left(e^{kx}.k\right) = k^2.e^{kx}$

Differentiating the y'' , we get $y''' = \dfrac{d}{dx}(y'') = \dfrac{d}{dx}\left(k^2 e^{kx}\right) = k^2\left(e^{kx}.k\right) = k^3.e^{kx}$

Hence by induction $y^{(n)} = \dfrac{d}{dx}\left(y^{(n-1)}\right) = \dfrac{d}{dx}\left(k^{n-1}e^{kx}\right) = k^{(n-1)}\left(e^{kx}.k\right) = k^n.e^{kx}$

2. Given $y = \log(kx + c)$. Find y^4 .

Solution: We have to find the 4th derivative of given function.

$$y' = \frac{d}{dx}[\log(kx + c)] = \frac{1}{kx + c}.\frac{d}{dx}(kx + c)$$

$$= \frac{1}{(kx + c)}k$$

$$= k(kx + c)^{-1}$$

$$y'' = \frac{d}{dx}(y') = \frac{d}{dx}\left[k(kx + c)^{-1}\right] = k\frac{d}{dx}(kx + c)^{-1}$$

$$= k.\left[(-1)(kx + c)\quad.k\right]$$

$$= (-1)k^2.\left[(kx + c)^{-2}\right]$$

$$y''' = \frac{d}{dx}(y'') = \frac{d}{dx}\left[(-1)k^2\left(kx + c\right)^{-2}\right]$$

$$= (-1)k^2\frac{d}{dx}(kx + c)^{-2}$$

$$= (-1)k^2\left[(-2)(kx + c)^{-3}.k\right]$$

$$= (-1)(-2)k^3\left[(kx + c)^{-3}\right]$$

$$y^{(4)} = \frac{d}{dx}\left[(-1)(-2)k^3(kx+c)^{-3}\right] = (-1)(-2)k^3\frac{d}{dx}\left[(kx+c)^{-3}\right]$$

$$= (-1)(-2)k^3\left[(-3)(kx+c)^{-4}.k\right]$$

$$= (-1)(-2)(-3)k^4\left[(kx+c)^{-4}\right]$$

3. Given $y = 2x^3 - 3x^2 + 2x + 5$. Find y'''

Solution: we have to determine the second derivative.

$$y' = \frac{d}{dx}(2x^3 - 3x^2 + 2x + 5)$$

$$= 6x^2 - 6x + 2$$

$$y'' = \frac{d}{dx}(y') = \frac{d}{dx}\left(6x^2 - 6x + 2\right) = 12x - 6$$

$$y''' = \frac{d}{dx}(y'') = \frac{d}{dx}(12x - 6) = 12$$

4. Given $y = ax^2 + bx + c$. Show that $y''' = 0$

Solution: $y = ax^2 + bx + c$ ∴ $y' = 2ax + b$

$$y'' = 2a \qquad\qquad \text{(differentiating } y'\text{)}$$

$$y''' = 0 \qquad\qquad \text{(differentiating } y''\text{)}$$

5. $y = a^x$. Find $\dfrac{d^n y}{dx^n}$

Solution: $\dfrac{dy}{dx} = \dfrac{d}{dx}(a^x) = a^x.\log(a)$

$$\frac{d^2 y}{dx^2} = \frac{d}{dx}(a^x \log(a)) = \log(a)\frac{d}{dx}(a^x) = \log(a)\left[a^x.\log a\right] = \left[\log(a)\right]^2\left[a^x\right]$$

$$\frac{d^3 y}{dx^3} = \frac{d}{dx}\left[\frac{d^2 y}{dx^2}\right] = \frac{d}{dx}\left[\{\log(a)\}^2 a^x\right] = \{\log(a)\}^2 .\frac{d}{dx}\left[a^x\right]$$

$$= \{\log(a)\}^2 .\left[\log(a).a^x\right]$$

$$= \{\log(a)\}^3 .a^x$$

By induction we can write $\dfrac{d^n y}{dx^n} = \left[\log(a)\right]^n a^x$

Practice Problems: (Successive differentiation)

1. $y = 4x^3 - 6x^2 + 4x + 6$. Find y'''.

2. $y^2 = 4ax$. Find y''.

3. $y = 3x^4$. Find y'''

4. $y = e^{ax}$. Find y''

5. $y = \log x$. Find y^{iv}

6. $y = 2x^2 + x + 3$. Find y'''.

7. $r = \sin a\theta$. Show that $r'' + a r^2 = 0$

8. $y = 3^x + x^3 + 3^3$. Find y^{iv}.

9. $y = \tan\theta + \sec\theta$. Show that $\dfrac{d^2 y}{d\theta^2} = \dfrac{\cos\theta}{(1 - \sin\theta)^2}$

10. $y = e^{-x}\cos x$. Find y''.

Answers:

1. 24

2. $-4a^2 / y^3$

3. 72x

4. $a^n . e^{ax}$

5. $1.2.3.x^{-4}$

6. 0

8. $(\log 3)^4 .3^x$

10. $2e^{-x}.\sin x$

Objective Questions: Differentiation

1. If $y = xe^x$, then $\dfrac{dy}{dx}$ is:

   (a) e^x

   (b) $(x+1)$

   (c) $x\left(e^x + 1\right)$

   (d) $\left(e^x + 1\right)$

2. If $y = e^x\left(1 + \log x\right)$, then $\dfrac{dy}{dx}$ is:

   (a) $\dfrac{e^x}{x}$

   (b) $\dfrac{e^x}{x} + 1$

   (c) $e^x\left(\dfrac{1}{x} + 1 + \log x\right)$

   (d) $e^x(1 + \log x)$

3. If $y = e^x \cos x$, then $\dfrac{dy}{dx}$ is:

   (a) $-e^x \sin x$

   (b) $-e^x \sin x + \cos x$

   (c) $e^x(\cos x - \sin x)$

4. If $y = \sin x \log x$, then $\dfrac{dy}{dx}$ is:

(a) $(\sin x + \cos x) \log x$

(b) $\dfrac{\sin x}{x} + \cos x \log x$

(c) $\dfrac{\cos x}{x} + \sin x \log x$

5. If $y = \dfrac{e^x}{x}$, then $\dfrac{dy}{dx}$ is:

(a) $\dfrac{e^x(x-1)}{x^2}$

(b) $\dfrac{e^x(x^2)}{x-1}$

(c) $\dfrac{e^x(x)}{x^2-1}$

(d) $\dfrac{e^x(x^2)}{x^2-1}$

6. If $\dfrac{e^x}{(1+x)}$, then $\dfrac{dy}{dx}$ is:

(a) $\dfrac{e^x}{(1+x)^2}$

(b) $\dfrac{xe^x}{(1+x)}$

(c) $\dfrac{xe^x}{(1+x)^2}$

(d) none of there.

7. $y = \sin^3 x$, then $\dfrac{dy}{dx}$ is:

(a) $3 \sin^2 x$

(b) $3 \sin^2 x \cos x$

(c) $3 \sin^2 x + \cos x$

8. $y = \sin x^3$, then $\dfrac{dy}{dx}$ is:

(a) $3 \sin x^2$

(b) $3x^2 \cos x^3$

(c) $3 \sin x^2 \cos x$

(d) $3 \cos x^2 \sin x$

9. $y = x^3 \log x$, then $\dfrac{dy}{dx}$ is:

(a) $3x^2 \dfrac{1}{x}$

(b) $x^2(1 + 3\log x)$

(c) $\left(\dfrac{1}{x} + 3x^2\right)$

(d) $\left(1 + 3x^2 \log x\right)$

10. $y = xe^{x^2}$, then $\dfrac{dy}{dx}$ is

(a) $2x^2 e^{x^2}$

(b) $e^{x^2}(1+2x)$

(c) $e^{x^2}(1+2x^2)$

(d) $\left(1+2x^2 e^{x^2}\right)$

11. $y = x\sin 2x$, then $\dfrac{dy}{dx}$ is:

(a) $2x\cos 2x$

(b) $2x(\cos 2x + \sin 2x)$

(c) $2x\cos 2x + \sin 2x$

(d) $2x\cos 2x\sin 2x$

12. $y = x^2 e^{-2x}$, then $\dfrac{dy}{dx}$ is:

(a) $e^{-2x}(1-2x)$

(b) $2e^{-2x}(1-x)$

(c) $2xe^{-2x}(1-x)$

(d) $xe^{-2x}(1-2x)$

13. $y = x^3 \log(4x)$, then $\dfrac{dy}{dx}$ is:

(a) $3x^2 [1 + 4\log(4x)]$

(b) $3x^2 [1 + \log(4x)]$

(c) $x^2 [1 + 3\log(4x)]$

(d) $3x^2 \left[1 + \dfrac{4}{x}\right]$

14. $y = \log(17 - x)$, then $\dfrac{dy}{dx}$ is:

(a) $\dfrac{-1}{x}$

(b) $\dfrac{1}{17-x}$

(c) $\dfrac{-1}{17-x}$

(d) $\dfrac{-1}{x-17}$

15. $y = \dfrac{\sin^2 x}{\cos^2 x}$, then $\dfrac{dy}{dx}$ is:

(a) $2\tan^2 x\sec^2 x$

(b) $\tan x\sec^2 x$

(c) $2\tan x\sec^2 x$

(d) $\tan^2 x\sec^2 x$

16. $y = x^3 \tan x$, then $\dfrac{dy}{dx}$ is:

(a) $3x^2 \sec^2 x$

(b) $x^2 \left(x \sec^2 x + 3 \tan x \right)$

(c) $3x^2 \left(\tan x + \sec^2 x \right)$

17. $y = \sin 2x + \cos^2 x$, then $\dfrac{dy}{dx}$ is:

(a) $\left(\cos 2x - \sin 2x \right)$

(b) $\left(\cos 2x - 2\sin 2x \right)$

(c) $2 \left(\cos 2x - \sin 2x \right)$

18. $y = \cos 2x + \sin^2 x$, then $\dfrac{dy}{dx}$ is:

(a) $-\sin 2x$

(b) $-\sin 2x . \cos 2x$

(c) $-\cos 2x$

19. $y = 3 \tan \sqrt{x}$, then $\dfrac{dy}{dx}$ is:

(a) $3 \sec^2 \sqrt{x}$

(b) $\dfrac{3 \sec^2 \sqrt{x}}{2\sqrt{x}}$

(c) $\dfrac{\sec^2 \sqrt{x}}{\sqrt{x}}$

(d) $\dfrac{2 \sec^2 \sqrt{x}}{3\sqrt{x}}$

20. $y = x^x$, then $\dfrac{dy}{dx}$ is:

(a) $x^x \log x$

(b) $x^x + \log x$

(c) $x^x (1 + \log x)$

(d) x^x

21. $y = 3xe^{2x}$, then $\dfrac{dy}{dx}$ is:

(a) $6xe^{2x}$

(b) $3e^{2x} (2x + 3)$

(c) $3e^{2x} (3x + 2)$

(d) $12e^{2x}$

22. $\dfrac{x^2}{a^2} + \dfrac{y^2}{b^2} = 1$, then $\dfrac{dy}{dx}$ is:

(a) $\dfrac{b^2 x^2}{a^2 y^2}$

(b) $\dfrac{-b^2 x^2}{a^2 y^2}$

(c) $\dfrac{-b^2 x}{a^2 y}$

(d) $\dfrac{b^2 y^2}{a^2 x^2}$

23. $x^n + y^n = a^n$, then $\dfrac{dy}{dx}$ is:

(a) $\left(\dfrac{x}{y}\right)^{n-1}$

(b) $\left(\dfrac{y}{x}\right)^{n-1}$

(c) $-\left(\dfrac{x}{y}\right)^{n-1}$

(d) $-\left(\dfrac{y}{x}\right)^{n-1}$

24. $x = a\cos\theta$ and $y = b\sin\theta$, then $\dfrac{dy}{dx}$ is:

(a) $\dfrac{a}{b}\cot\theta$

(b) $\dfrac{-a}{b}\cot\theta$

(c) $\dfrac{-b}{a}\cot\theta$

(d) $\dfrac{a}{b}\tan\theta$

25. $x = at^2$ and $y = 2at$, then $\dfrac{dy}{dx}$ is:

(a) $\dfrac{1}{t}$

(b) t

(c) $\dfrac{-1}{t}$

(d) $-t$

26. $y = x^2 + 2$, differentiate w.r.t x^3

(a) $\dfrac{2}{3}$

(b) $\dfrac{2x}{3}$

(c) $\dfrac{2}{3x}$

(d) $\dfrac{2x}{3}$

27. If $y = \dfrac{x+2}{x-2}$, then $\dfrac{dy}{dx}$ at $x = -2$ is:

    (a) $\dfrac{1}{4}$ (b) $\dfrac{-1}{4}$

    (c) $\dfrac{1}{2}$ (d) $\dfrac{-1}{2}$

28. If $y = \dfrac{x+2}{x^2-3}$, then $y'(0)$ is

    (a) $\dfrac{1}{3}$ (b) $\dfrac{-1}{3}$

    (c) $\dfrac{1}{2}$ (d) $\dfrac{-1}{2}$

Answers

1. b	2. c	3. c	4. b	5. a	6. c	7. b
8. b	9. b	10. c	11. c	12. c	13. c	14. c
15. c	16. b	17. c	18. a	19. b	20. c	21. b
22. c	23. c	24. c	25. b	26. c	27. b	28. b

6

Integration

1.0 BRIEF HISTORY

The history of integration can be traced as far back as 'Ancient Egyptica'- 1800BC., demonstrating that the knowledge of a formula for the volume of a "pyramidal frustum" The first documented systematic technique capable of determining integrals is "Method of Exhaustion" of the ancient Greek astronomer Eudoxus(370BC).This was applied to find areas and volumes by breaking them up into an infinite number of shapes for which area and volume was known.. This method was further developed and employed by Archimedes in 3rd century BC. He calculated areas of parabolas and an approximation to the area of circle.

Similar methods were independently developed in China by 'Liu Hui'(300AD).

The next significant development in integral calculus did not appear until the 16th century AD. At this time the work of Cavalier and Fermat began to lay foundation of modern calculus.

Notations

Isaac Newton used small vertical bar above a variable to indicate integration or placed a variable inside a box. But vertical bar was easily confused with $\dot{x}$ or x' as this was used for differentiation and box notation was difficult to print. So these notations were not popular. The modern notation for integration was given by Gottfried Leibniz (1675AD).He adapted the modern integral symbol $\int$ (long s) standing for summa, written as $\int umma$ (Latin for sum or total). The modern notation for definite integral with limits was first used by Joseph Fourier.

The simple way of writing integral of a function $f(x)$ is $\int f(x)dx$. $f(x)$ is called the integrand

1.1 INDEFINITE INTEGRATION

Integration is the reverse process of differentiation. If derivative of F(x) is f(x), then anti-derivative or integral of f(x) is F(x) and it is written as $\int f(x)dx = $ F(x)

In general $\int f(x)dx = $ F(x) + c, where c is a constant called Constant of Integration. The function $f(x)$ is called integrand .Since constant c may take any value; we may get many solutions of a single integrand. That is why it is known as *indefinite integral.*

Properties

1. $\int dx = x + c$ **2.** $\int a\,dx = a\int dx$ **3.** $\int (dx + dy + dz) = \int dx + \int dy + \int dz$

Important Formulae

Indefinite Integration

1. $\int x^n dx = \dfrac{x^{n+1}}{n+1} + c$ **2.** $\int \dfrac{1}{x} dx = \log|x| + c$ **3.** $\int e^x dx = e^x + c$ **4.** $\int a^x dx = \dfrac{a^x}{\log a} + c$

Solved Examples

1. $\int x^6 dx = \dfrac{x^{6+1}}{6+1} + c = \dfrac{x^7}{7} + c$

2. $\int x^{\frac{2}{3}} dx = \dfrac{x^{\frac{2}{3}+1}}{\frac{2}{3}+1} + c = \dfrac{3x^{\frac{5}{3}}}{5} + c$

3. $\int x^{\frac{-3}{4}} dx = \dfrac{x^{\frac{-3}{4}+1}}{\frac{-3}{4}+1} + c = \dfrac{x^{\frac{1}{4}}}{\frac{1}{4}} + c = 4x^{\frac{1}{4}} + c$

4. $\int (3x^2 + 2x + 1) dx = 3\int x^2 dx + 2\int x dx + \int x^0 dx = 3\dfrac{x^3}{3} + 2\dfrac{x^2}{2} + x + c = x^3 + x^2 + x + c$

5. $\int (ax^3 + bx^2 + cx + d) = a\int x^3 dx + b\int x^2 dx + c\int x dx + d\int dx + c$

$$= a\dfrac{x^4}{4} + b\dfrac{x^3}{3} + c\dfrac{x^2}{2} + dx + C$$

6. $\int \left(\dfrac{a}{x^2} + \dfrac{b}{x} + c\right) dx = \int (ax^{-2} + bx^{-1} + c) dx = a\dfrac{x^{-2+1}}{-2+1} + b\log x + cx + C$

$$= \dfrac{-a}{x} + b\log x + cx + C$$

7. $\int \dfrac{(4x^2 + 3x + 2)}{x^3} dx = \int \dfrac{4}{x} dx + \int \dfrac{3}{x^2} dx + \int \dfrac{2}{x^3} dx + C = 4\int \dfrac{1}{x} dx - 3\int x^{-2} dx - 2\int x^{-3} dx + c$

$$= 4\log x - 3\dfrac{x^{-2+1}}{-2+1} + 2\dfrac{x^{-3+1}}{-3+1} + c$$

$$= 4\log x - \dfrac{3}{x} - \dfrac{1}{x} + c$$

8. $\int\left(x^2+8\right)^2 dx = \int\left(x^4+16x^2+64\right)dx = \dfrac{x^5}{5}+16\dfrac{x^3}{3}+64x+c$

9. $\int\left(\dfrac{3^x+4^x}{5^x}\right)dx = \int\left(\dfrac{3}{5}\right)^x dx + \int\left(\dfrac{4}{5}\right)^x dx$

$$= \left(\dfrac{(3/5)^x}{\log(3/5)}\right)+\left(\dfrac{(4/5)^x}{\log(4/5)}\right)+c \text{ (Applying } \int a^x dx = \dfrac{a^x}{\log a}+c\text{)}$$

10. $\int\dfrac{2x^2+7x-3}{x-2}dx = \int\left(2x+11+\dfrac{19}{x-2}\right)dx = \int 2xdx+11\int dx+19\int\dfrac{dx}{x-2}$

$$= x^2+11x+19\log(x-2)+c$$

Practice Problems: (Indefinite Integration)

1. $\int\left(x^2+3x+5\right)dx$,

2. $\int\left(\dfrac{3x^4+7x-11}{x^3}\right)dx$,

3. $\int\left(-2x^{-3}+24x^{-5}\right)dx$

4. $\int\dfrac{a^x+b^x}{c^x}dx$

5. $\int\left(\sqrt{x}-\dfrac{1}{\sqrt{x}}\right)dx$

6. $\int(3x+5)^2 dx$

7. $\int\left(x-\dfrac{1}{x}\right)^2 dx$

8. $\int 2.3^x dx$

9. $\int 3x^{-1}dx$

Answers

1. $\left(\dfrac{x^3}{3}+\dfrac{3}{2}x^2+9x+c\right)$

2. $\dfrac{3}{2}x^2-\dfrac{7}{x}+\dfrac{11}{2x^2}+c$,

3. $\dfrac{-1}{x^2}-\dfrac{6}{x^4}+c$

4. $\dfrac{(a/c)^x}{\log a-\log c}+\dfrac{(b/c)^x}{\log b-\log c}+c$

5. $\left(\dfrac{2x^{\frac{3}{2}}}{3}-2x^{\frac{1}{2}}+c\right)$,

6. $\left(\dfrac{x^3}{3}+\dfrac{3x^2}{2}+2x+c\right)$,

7. $\left(\dfrac{x^3}{3}-\dfrac{1}{x}-2x+c\right)$

8. $2\left(\dfrac{3^x}{\log 3}\right)+c$

9. $3\log x+c$

1.1.2 Integration of Trigonometric Functions

$\int \sin x \, dx = -\cos x + c$ $\qquad$ $\int \cos x \, dx = \sin x + c$

$\int \sec^2 x \, dx = \tan x + c$ $\qquad$ $\int \sec x \tan x \, dx = \sec x + c$

$\int \csc^2 x \, dx = -\cot x + c$ $\qquad$ $\int \csc x \cot x \, dx = -\csc x + c$

Solved Examples

1. $\int (1 + \sin x + \cos x) dx = x - \cos x + \sin x + c$

2. $\int \left(3 \sin x - 2 \sec^2 x \right) dx = 3(-\cos x) - 2 \tan x + c$

3. $\int \sec x (\sec x + \tan x) dx = \int (\sec^2 x + \sec x \tan x) dx = \tan x + \sec x + c$

4. $\int \left(e^x + 2 \sin x - 3 \cos x \right) dx = \int e^x dx + 2 \int \sin x dx - 3 \int \cos x + c = e^x - 2 \cos x - 3 \sin x + c$

5. $\int \left(5 \cos x + 2 \sec^2 x - 10 \right) dx = 5 \int \cos x dx + 2 \int \sec^2 x dx - 10 \int dx + c$

$$= 5 \sin x + 2 \tan x - 10x + c$$

6. $\int \left(\sec x \tan x - 5 \cos ec^2 x \right) dx = \int \sec x \tan x . dx - 5 \int \cos ec^2 x . dx + c$

$$= \sec x - 5(-\cot x) + c = \sec x + 5 \cot x + c$$

7. $\int \tan^2 x dx = \int \left(\sec^2 x - 1 \right) dx$

$\because \sec^2 x = \tan^2 x + 1 = \int \sec^2 x dx - \int 1 . dx = \tan x - x + c$

8. $\int \left(3 \sin x - 4 \cos x + 5 \sec^2 x - 2 c \sec^2 x \right) dx$

$= 3 \int \sin x dx - 4 \int \cos x dx + 5 \int \sec^2 x dx - 2 \int \cos ec^2 x dx$

$= 3(-\cos x) - 4(\sin x) + 5 \tan x - 2(-\cot^2 x) + c$

$= -3 \cos x - 4 \sin x + 5 \tan x + 2 \cot x + c$

9. $\int \left(\dfrac{a + b \sin x}{\cos^2 x} \right) dx = \int \dfrac{a}{\cos^2 x} dx + \int \dfrac{b \sin x}{\cos^2 x} dx = a \int \sec^2 x dx + b \int \dfrac{1}{\cos x} . \dfrac{\sin x}{\cos x} dx$

$= a \tan x + b \int \sec x . \tan x . dx$

$= a \tan x + b \int \sec x . \tan x dx + c$

$= a \tan x + b \sec x + c$

10. $\int \cos ecx (\cos ecx - \cot x) dx = \int \left(\cos ec^2 x - \cos ecx . \cot x \right) dx$

$$= -\cot x - (-\cos ecx) + c = \cos ecx - \cot x + c$$

Key Point

$$\int \sin kx\, dx = \frac{-\sin kx}{k} + c \quad \int \cos kx.dx = \frac{\sin kx}{k} + c$$

Practice Problems: (Integration of Trigonometric Functions)

1. $\int 3\sec^2 x.dx$

2. $\int \dfrac{2}{\sec x}\, dx$

3. $\int \left(\dfrac{\sin^2 x - \cos^2 x}{\sin^2 x.\cos^2 x}\right) dx$,

4. $\int \left(\dfrac{a + b\sin x}{\cos^2 x}\right) dx$,

5. $\int \cos ecx(\cos ecx - \cot x)dx$

6. $\int (\tan x + \cot x)^2\, dx$,

7. $\int \dfrac{1}{1 + \sin x}\, dx$,

8. $\int \dfrac{1}{1 + \cos x}\, dx$

Answers:

1. $3\tan x + c$

2. $2\sin x + c$

3. $(\tan x + \cot x) + c$

4. $(a\tan x + b\sec x) + c$.

5. $(\cos ecx - \cot x + c)$,

6. $(\tan x - \cot x + c)$,

7. $(\tan x - \sec x + c)$,

8. $(-\cot x - \cos ecx + c)$

1.1.3 Indefinite Integration by Substitution

Let the given integral is $\int f\{g(x)\}\, dx$

1. Choose a subsititute t such that $y\,(x) = t$

2. Determine the value of dt = g'(x) dx $\Rightarrow$ dx $= \dfrac{dt}{g(x)}$

3. By substitution the integral becomes $\int f(t) = \dfrac{dt}{g'(x)}$

4. Integrate the new integral

5. Return to initial value

Solved Examples

1. Evaluate $\int (ax + b)^n dx$

Solution: Substitute for $(ax + b) = t$. Differentiating it w.r.t t, we get

$$a\frac{dx}{dt} = 1 \Rightarrow dx = \frac{dt}{a}$$

$$\therefore \int (t)^n \frac{dt}{a} = \frac{1}{a}.\frac{t^{n+1}}{n+1} + c = \frac{(ax + b)^{n+1}}{a(n+1)} + c$$

2. Evaluate $\int (3x+5)^6 dx$. Put $(3x+5) = t \Rightarrow 3dx = dt . \Rightarrow dx = \dfrac{dt}{3}$

Solution: $\int (3x+5)^6 dx = \int t^6 \dfrac{dt}{3} = \dfrac{t^7}{3.7} = \dfrac{(3x+5)^7}{21} + c$

3. Evaluate $\int (4-9x)^5 dx$.

Solution: Put $4-9x = t$. Differentiating it, we get $-9dx = dt \Rightarrow dx = \dfrac{dt}{-9}$

$$\int (4-9x)^5 dx. = \int t^5 \dfrac{dt}{-9} = -\dfrac{1}{9} \dfrac{t^6}{6} + c = -\dfrac{(4-9x)^6}{54} + c$$

4. Evaluate $\int \dfrac{\log x}{x} dx$.

Solution: Put $\log x = t$. Differentiating we get $\dfrac{dx}{x} = dt$

$$\therefore \int \dfrac{\log x}{x} dx = \int t \, dt = \dfrac{t^2}{2} + c = \dfrac{(\log x)^2}{2} + c \ .$$

5. Evaluate $\int \dfrac{3x^2}{x^3 + 1} dx$.

Solution: Put $x^3 + 1 = t$. . Differentiating it we get, $3x^2 dx = dt$

$$\therefore \int \dfrac{3x^2}{x^3 + 1} dx. = \int \dfrac{dt}{t} = \log t + c = \log (x^3 + 1) + c$$

6. Evaluate $\int \dfrac{e^x}{e^x + 1} dx$.

Solution: Put $e^x + 1 = t$. Differentiating it we get, $e^x dx = dt$

$$\therefore \int \dfrac{e^x}{e^x + 1} dx. = \int \dfrac{dt}{t} = \log t + c = \log(e^x + 1) + c$$

7. Evaluate $\int \dfrac{\cos ec^2 x}{1 + \cot x} dx$.

Solution: Put $1 + \cot x = t$. Differentiating it, we get $-\cos ec^2 x.dx = dt$

$$\therefore \int \dfrac{\cos ec^2 x}{1 + \cot x} dx. = -\int \dfrac{dt}{t} = -\log t + c = -\log (1 + \cot x) + c$$

8. Evaluate $\int \dfrac{1}{x\log x}\,dx$

Solution: Put $\log x = t$ .Differentiating it we get $\dfrac{1}{x}\,dx = dt$

$$\therefore \int \dfrac{1}{x\log x}\,dx = \int \dfrac{dt}{t} = \log t + c = \log\left(\log x\right) + c$$

9. Evaluate $\int \dfrac{\cos x}{\sin x}\,dx$

Solution: Put $\sin x = t$. Differentiating it we get $\cos x\,dx = dt$

$$\therefore \int \dfrac{\cos x}{\sin x}\,dx = \int \dfrac{1}{t}\,dt = \log t + c = \log(\sin x) + c$$

10. Evaluate $\int \dfrac{10x}{5x^2 - 8}\,dx$

Solution: Let $5x^2 - 8 = u$. Differentiating it we get $10x\,dx = du$

$$\int \dfrac{10x}{5x^2 - 8}\,dx = \int \dfrac{du}{u} = \log u = \log\left(5x^2 - 8\right) + c$$

Practice Problems: (Indefinite Integration by substitution)

Evaluate the following integerals :

1. $\int (3x+6)^8 \, dx$

2. $\int 4x^3 (x^4 - 1)dx$

3. $\int t^3 (1+t^4)dt$

4. $\int \cos(3x+3)dx$

5. $\int \sin x.\cos x\,dx$

6. $\int \tan x.\sec^2 x\,dx$

7. $\int x\sin x^2 dx$

8. $\int xe^{x^2} dx$

9. $\int 2x\sqrt{\left(x^2 + 1\right)}dx$

10 . $\int \dfrac{4x}{\sqrt{2x^2 + 1}}\,dx$

Answers

1. $\dfrac{(3x+6)^9}{27}$

2. $\dfrac{\left(x^4 - 1\right)^2}{2}$

3. $\dfrac{\left(1+t^4\right)^2}{2}$

4. $\dfrac{-\sin(3x+3)}{3}$

5. $\dfrac{\sin^2 x}{2}$

6. $\dfrac{\tan^2 x}{2}$

7. $-\dfrac{1}{2}\cos x^2 + c$

8. $\dfrac{1}{2}e^{x^2} + c$

9. $\dfrac{2}{3}\left(x^2 + 1\right)^{\frac{3}{2}} + c$

10. $2\left(2x^2 + 1\right)^{\frac{1}{2}} + c$.

1.2 DEFINITE INTEGRATION

The definite integral of function $f(x)$ in the intervals $[a,b]$ is denoted by $\int_a^b f(x)dx$. Its

solution is given by $\int_a^b f(x)dx = \left[F(x)\right]_a^b = F(b)-F(a)$, a and b are known as lower and

upper limits of integral. Note that it does not involve a constant of integration and it gives us a definite value (a number) at the end of the calculation.

The primary difference between definite and indefinite integrals is that the definite integral, if it exists, is a real number value, while the indefinite integral represents an infinite number of functions that differ only by a constant.

Solved Examples

1. $\int_2^4 \frac{1}{x}dx = \left[\log x\right]_2^4 = \log 4 - \log 2 = \log \frac{4}{2} = \log 2$

2. $\int_4^9 \sqrt{x}\,dx = \int_4^9 x^{\frac{1}{2}}\,dx = \left[\frac{x^{\frac{1}{2}+1}}{\frac{1}{2}+1}\right]_4^9 = \frac{2}{3}\left[x^{\frac{3}{2}}\right]_4^9$

$$= \frac{2}{3}\left[\left(\sqrt{x}\right)^3\right]_4^9 = \frac{2}{3}\left[\left(\sqrt{9}\right)^3 - \left(\sqrt{4}\right)^3\right] = \frac{2}{3}\left[(3)^3 - (2)^3\right]$$

$$= \frac{2}{3}[27-8] = \frac{2\times19}{3} = \frac{38}{3}$$

$$= \frac{1}{9}\left[4^3 - 2^3\right] = \frac{64-8}{9} = \frac{56}{9}$$

3. $\int_0^{\frac{\pi}{4}} \sec^2 x\,dx = \left[\tan x\right]_0^{\frac{\pi}{4}} = \tan\frac{\pi}{4} - \tan 0 = 1-0 = 1$

4. $\int_0^{\frac{\pi}{2}} (\cos x - \sin x)dx = \left[\sin x - (-\cos x)\right]_0^{\pi/2} = (\sin x)_0^{\pi/2} + (\cos x)_0^{\pi/2}$

$$= \left(\sin\frac{\pi}{2} - \sin 0\right) + \left(\cos\frac{\pi}{2} - \cos 0\right)$$

$$= (1-0)+(0-1) = 0$$

5. $\int\limits_{2}^{6} f(x)dx = 12$ and $\int\limits_{6}^{9} f(x) = 7$. Evaluate $\int\limits_{2}^{6} f(x)dx$

Solution: $\int\limits_{2}^{9} f(x)dx = \int\limits_{2}^{6} f(x)dx + \int\limits_{6}^{9} f(x)dx$

Hence $\int\limits_{2}^{6} f(x)dx = \int\limits_{2}^{9} f(x)dx - \int\limits_{6}^{9} f(x)dx = 12 - 7 = 5$

6. $\int\limits_{1}^{2}\left(\dfrac{2x^5 - x + 3}{x^2}\right)dx$

Solution: $\int\limits_{1}^{2}\left(\dfrac{2x^3 - x + 3}{x^2}\right)dx = \int\limits_{1}^{2}\left(\dfrac{2x^3}{x^2} - \dfrac{x}{x^2} + \dfrac{3}{x^2}\right)dx$

$$= \int\limits_{1}^{2} \left(2 - \dfrac{1}{x} + 3x^{-2}\right)dx$$

$$= \left[2x - \log x - 3\dfrac{x^{-1}}{-1}\right]_{1}^{2}$$

$$= \left[(2 \times 2) - \log 2 + \dfrac{3}{2}\right] - \left[(2 \times 1) - \log 1 + \dfrac{3}{1}\right]$$

$$= \left[\left(4 - \log 2 + \dfrac{3}{2}\right) - (2 - 0 + 3)\right] = \dfrac{-7}{2} - \log 2$$

Practice Problems: (Definite Integrals)

Evaluate the following integrals.

1. $\int\limits_{0}^{6} 3dx$,

2. Given $\int\limits_{1}^{3} f(x)dx = 10$ and $\int\limits_{0}^{6} g(x)dx = 15$, find $\int\limits_{1}^{3}\left[f(x) + g(x)\right]$

3. $\int\limits_{4}^{9} \sqrt{x}dx = \dfrac{16}{3}$, find $\int\limits_{9}^{4} \sqrt{x}dx$ $\left[H\text{int} \int\limits_{a}^{b} f(x)dx = -\int\limits_{b}^{a} f(x)dx\right]$

4. $\int\limits_{0}^{2}\left(x^2 + 2\right)dx$,   5. $\int\limits_{1}^{2}\left(x^2 + \dfrac{1}{x^2}\right)dx$,   6. $\int\limits_{-3}^{1}\left(6x^2 - 5x + 2\right)dx$

Answers

1. 12,

2. 25

3. $\dfrac{-16}{3}$,

4. $\dfrac{14}{3}$,

5. $\dfrac{17}{6}$,

6. 84

1.2.1   Definite Integral by Substitution

Solved Examples

1. Evaluate $\int\limits_{0}^{2} e^{\frac{x}{2}}\, dx.$

Solution: Put $\dfrac{x}{2} = t \Rightarrow dx = 2\,dt$

New limits are: $t = 0$ when $x = 0$ and $t = e$ when $x = 2$

$$\int\limits_{0}^{2} e^{\frac{x}{2}}\, dx = \int\limits_{0}^{1} e^{t}\, 2\,dt = 2\Big[e^{t} \Big]_{0}^{1} = 2\left(e^{1} - e^{0} \right) = 2(e-1)$$

2. Evaluate $\int\limits_{0}^{2} \sqrt{(6x+4)}\, dx$.

Solution: Put $6x + 4 = t \Rightarrow 6\,dx = dt$

New limits. When $x = 0$ $t = 4$, when $x = 2$ $t = 16$

$$\therefore \int\limits_{0}^{2} \sqrt{(6x+4)}\, dx = \int\limits_{4}^{16} \sqrt{t}\, \frac{dt}{6} = \frac{1}{6} \left[\frac{t^{\frac{3}{2}}}{\frac{3}{2}} \right]_{4}^{16} = \frac{1}{6} \times \frac{2}{3}\left[\left(\sqrt{t} \right)^{3} \right]_{4}^{16} = \frac{1}{9}\left[\left(\sqrt{16} \right)^{3} - \left(\sqrt{4} \right)^{3} \right]$$

$$= \frac{1}{9}[64 - 8] = \frac{56}{9}$$

3. Evaluate $\int\limits_{2}^{4} \dfrac{x}{x^{2}+1}\, dx$

Solution: Put $x^{2} + 1 = t \Rightarrow 2x\,dx = dt \Rightarrow x\,dx = \dfrac{dt}{2}$

New limits are: $t = 5$ when $x = 2$, $t = 17$ when $x = 4$

$$\int\limits_{2}^{4} \frac{x}{x^{2}+1}\, dx = \int\limits_{5}^{17} \frac{1}{t}\frac{dt}{2} = \frac{1}{2}\int\limits_{5}^{17} \log|t| = \frac{1}{2}\left[\log 17 - \log 5 \right] = \frac{1}{2}\log\left| \frac{17}{5} \right|$$

4. Evaluate $\int\limits_{0}^{1}\dfrac{e^x}{1+e^{2x}}dx$

Solution: Put $e^x = t$. Limits of t: $t = 1$ when x=0 and $t = e$ when x=1

$\because e^x = t \therefore e^x dx = dt$

$$\int\limits_{0}^{1}\frac{e^x}{1+e^{2x}}dx = \int\limits_{1}^{e}\frac{dt}{1+t^2} = \left[\tan^{-1}x\right]_{1}^{e} = \tan^{-1}e - \tan^{-1}1$$

5. Evaluate $\int\limits_{1}^{3}\dfrac{\cos(\log x)}{x}dx$

Solution: Put $\log x = t$. Limits of t: $t = 0$ when x=1 and $t = \log 3$ when x=3

$\because \log x = t \therefore \dfrac{1}{x}dx = dt$

$$\therefore \int\limits_{1}^{3}\frac{\cos(\log x)}{x}dx = \int\limits_{0}^{\log 3}\cos t.dt = \left[\sin t\right]_{0}^{\log 3} = \sin\left(\log 3\right) - \sin 0 = \sin\{\log(3)\}$$

6. Evaluate $\int\limits_{1}^{2}\dfrac{3x}{x^2+4}dx$

Solution: Put $x^2+4 = u \Rightarrow 2xdx = du$ Lower limit $= 1 + 4 = 5$ Upper limit $= 2^2 + 4 = 8$

$$\int\limits_{1}^{2}\frac{3x}{x^2+4}dx = \frac{3}{2}\int\limits_{5}^{8}\frac{du}{u} = \frac{3}{2}\left[\log u\right]_{5}^{8} = \frac{3}{2}\left[\log 8 - \log 5\right] = \frac{3}{2}\log\frac{8}{5}$$

7. Evaluate $\int\limits_{0}^{1}\dfrac{e^{5x}}{1+e^{5x}}dx$

Solution : Let $1+e^{5x} = u \Rightarrow 5e^{5x}dx = du$ Lower limit $= \dfrac{1}{1+1} = \dfrac{1}{2}$, Upper limit $= \dfrac{e^5}{1+e^5}$

$$\int\limits_{0}^{1}\frac{e^{5x}}{1+e^{5x}}dx = \frac{1}{5}\int\limits_{\frac{1}{2}}^{\frac{e^5}{1+e^5}}\frac{du}{u} = \frac{1}{5}\left[\log u\right]_{\frac{1}{2}}^{\frac{e^5}{1+e^5}} = \frac{1}{5}\left[\log\left(\frac{e^5}{1+e^5}\right) - \log\left(\frac{1}{2}\right)\right]$$

Practice Problems: (Definite integral by substitution)

Evaluate the following integrals :

1. $\int\limits_{0}^{1}x\sqrt{1-x^2}\,dx$

2. $\int\limits_{0}^{1}\dfrac{5r}{\left(4+5r^2\right)}dr$

3. $\int\limits_{0}^{\frac{\pi}{2}}\sin x.\cos x dx$,

4. $\int_{0}^{\frac{\pi}{2}} \sin^2 x.\cos x dx$

5. $\int_{1}^{4} \frac{3x}{x^2+4} dx$

6. $\int_{-2}^{-1} \frac{x}{\left(x^2-2\right)} dx$

7. $\int_{0}^{\frac{\pi}{2}} \cos x.dx$

Answers

1. $\frac{1}{3}$,

2. $\frac{1}{2}\log\left(\frac{9}{4}\right)$,

3. $\frac{1}{2}$,

4. $\frac{1}{3}$,

5. $\frac{3}{2}\log\left(\frac{8}{5}\right)$,

6. $-\frac{1}{48}$

7. 1

1.3 Area under a Curve

The area under a curve between two points of an axis can be determined just as we determine the definite integral between two limits.

Let $y = f(x)$ be the equation of curve. The area under the curve between the limits

$x = a$ and $x = b$ can be determined as: $A = \int_{a}^{b} f(x)dx$

Solved Examples

1. Find the area between the curve: $y = x^2 - 4$ and the x-axis.

Solution: On x-axis, the value of ordinate is zero i.e. $y = 0$

Hence from the equation of the curve, $x^2 - 4 = 0 \Rightarrow x = \pm 0$

∴ Area under the curve is given by:

$$A = \int_{-2}^{2} y dx = \int_{-2}^{2}\left(x^2-4\right)dx = 2\int_{0}^{2}\left(x^2-4\right)dx \quad \because \int_{-a}^{a} f(x)dx = 2\int_{0}^{a} f(x)dx$$

$$= 2\left[\frac{x^3}{3}-4x\right]_{0}^{2}$$

$$= 2\left[\frac{2^3}{3}-4.2\right]$$

$$= 2\left[\frac{8}{3} - 8\right] = 2.\frac{16}{3} = \frac{32}{3} \text{ sq.units}$$

2. Find the area between the curve $y = x^2$ and $x-axis$ from $x = -2$ and $x = 2$.

Solution: $A = \int_{-2}^{2} y\,dx = \int_{-2}^{2} x^2\,dx = 2\int_{0}^{2} x^2\,dx$ $\because \int_{-a}^{a} f(x)dx = 2\int_{0}^{a} f(x)dx$

$$= 2\left[\frac{x^4}{4}\right]_0^2 = \frac{2}{4}\left[2^4\right] = \frac{1}{2}.16 = 8 \text{ sq.units}$$

3. Find the area between the curve $y = (x-1)^2 + 3$ and the straight line $y = 7$.

Solution: $y = (x-1)^2 + 3 = x^2 - 2x + 1 + 3 = x^2 - 2x + 4$

Also $y = 7$.

$\therefore\ x^2 - 2x + 4 = 7$

$or\ \ x^2 - 2x + 4 - 7 = 0$

$or\ \ x^2 - 2x - 3 = 0$

$or\ \ (x+1)(x-3) = 0$

$\therefore\ x = -1 \text{ or } 3$.

$$A = \int_{-1}^{3} y\,dx = dx = \int_{-1}^{3}\left(x^2 - 2x + 4\right)dx \qquad \because y = x^2 - 2x + 4$$

$$= \left(\frac{x^3}{3} - 2\frac{x^2}{2} - 4x\right)_{-1}^{3}$$

$$= \left(\frac{3^3}{3} - 2\frac{3^2}{2} - 4(3)\right) - \left(\frac{(-1)^3}{3} - 2\frac{(-1)^2}{2} - 4(-1)\right)$$

$$= (9 - 9 - 12) - \left(\frac{-1}{3} - 1 + 4\right)$$

$$= (-12) - \left(\frac{-1}{3} + 3\right) = (-12) - \left(\frac{8}{3}\right) = -\frac{44}{3} = -14.7$$

The area under the curve is 14.7 sq.units.

4. Find the area of the curve $y = e^x$ bound by $x = a$ and $x = b$.

Solution: $A = \int_a^b e^x dx = \left[e^x\right]_a^b = e^b - e^a \quad \left[\int e^x dx = e^x\right]$

5. Find the area bound by the curve $y = \log x$ and the x-axis between $x = a$ and $x = b$.

Solution: $A = \int_a^b \log x dx = \int_a^b \log x . 1 dx$

Recall $\int uv dx = u \int v dx - \int \left[\frac{d}{dx} u . \int v\right] dx$

$$A = \int_a^b \log x dx = \int_a^b \log x . 1 dx = \log x \int_a^b 1 . dx - \int_a^b \left[\frac{d}{dx}\log x \int 1 dx\right] dx$$

$$= \log x [x]_a^b - \int_a^b \left[\frac{1}{x} x\right] dx$$

$$= \log x [x]_a^b - [x]_a^b$$

$$= (b - a)[\log x - 1]$$

1.3.1 Area between two Curves

The area between two curves: $f(x)$ and $g(x)$ between the limits $x = a$ and $x = b$ .is given

by : $A = \int_a^b [f(x) - g(x)] dx$

Solved Examples

1. Find the area between two curves: $y = x^2 + 5x$ and $y = 3 - x^2$

Solution: From the given curves $x^2 + 5x = 3 - x^2$

or $2x^2 + 5x - 3 = 0$

or $2x^2 + 6x - x - 3 = 0$

or $2x(x + 3) - 1(x + 3) = 0$

or $(2x - 1)(x + 3) = 0$

$$x = \frac{1}{2} \text{ or -3}$$

Hence area between two curves: $y = x^2 + 5x$ and $y = 3 - x^2$ in the limits $x = \frac{1}{2}$, x = -3

is given by $A = \int_a^b [f(x) - g(x)]dx = \int_{-3}^{\frac{1}{2}} [(x^2 + 5x) - (3 - x^2)]dx$

$$= \int_{-3}^{\frac{1}{2}} [2x^2 + 5x - 3]dx$$

$$= \left[2\frac{x^3}{3} + 5\frac{x^2}{2} - 3x \right]_{-3}^{\frac{1}{2}}$$

$$= \left[\frac{2}{3}\left(\frac{1}{2}\right)^3 + \frac{5}{2}\left(\frac{1}{2}\right)^2 - 3\left(\frac{1}{2}\right) \right] - \left[\frac{2}{3}(-3)^3 + \frac{5}{2}(-3)^2 - 3(-3) \right]$$

$$= \left[\frac{2}{3} \cdot \frac{1}{8} + \frac{5}{2} \cdot \frac{1}{4} - 3 \cdot \frac{1}{2} \right] - \left[\frac{2}{3}(-27) + \frac{5}{2}(9) + 9 \right]$$

$$= \left[\frac{1}{12} + \frac{5}{8} - \frac{3}{2} \right] - \left[-18 + \frac{45}{2} + 9 \right]$$

$$= \left[\frac{1}{12} + \frac{5}{8} - \frac{3}{2} \right] - \left[-18 + \frac{45}{2} + 9 \right]$$

$$= \left[\frac{2 + 15 - 18}{24} \right] - \left[-9 + \frac{45}{2} \right]$$

$$= \left[\frac{-1}{24} \right] - \left[\frac{-27}{2} \right] = \frac{-1}{24} + \frac{27}{2} = \frac{-1 + 12 \times 27}{24} = \frac{323}{24}$$

2. Find the area between two curves: $y = x^2 - 4x + 5$ and $y = 7 - x^2$, $x = 0$ and $x = 5$

Solution: $A = \int_a^b [f(x) - g(x)]dx = \int_0^5 [(x^2 - 4x + 5) - (7 - x^2)]dx$

$$= \int_0^5 [2x^2 - 4x - 2]dx$$

$$= 2\int_0^5 [x^2 - 2x - 1]dx$$

$$= 2 \left[\frac{x^3}{3} - 2\frac{x^2}{2} - x \right]_0^5$$

$$= 2 \left[\left(\frac{5^3}{3} - 2\frac{5^2}{2} - 5 \right) - 0 \right]$$

$$= 2 \left[\left(\frac{125}{3} - 25 - 5 \right) \right]$$

$$= 2 \left[\left(\frac{125 - 90}{3} \right) \right] = \frac{2 \times 35}{3} = 16.7 \text{ sq.units}$$

3. Find the area between two curves: $x \quad y$ and $x = y + 6$

Solution: $y^2 = y + 6 \Rightarrow y^2 - y - 6 = 0$

By factorization we get $(y + 2)(y - 3) = 0$

$$y = -2, 3$$

Applying $A = \int_a^b [f(x) - g(x)]dx$

$$A = \int_{-2}^{3} [y^2 - y - 6)]dy = \left[\frac{y^3}{3} - \frac{y^2}{2} - 6y \right]_{-2}^{3}$$

$$= \left[\frac{y^3}{3} - \frac{y^2}{2} - 6y \right]_{-2}^{3} = \left[\left(\frac{3^3}{3} - \frac{3^2}{2} - 6.3 \right) - \left(\frac{(-2)^3}{3} - \frac{(-2)^2}{2} - 6.(-2) \right) \right]$$

$$= \left[\left(\frac{27}{3} - \frac{9}{2} - 18 \right) - \left(\frac{-8}{3} - \frac{4}{2} + 12 \right) \right]$$

$$= \left[\left(9 - \frac{9}{2} - 18 \right) - \left(\frac{-8}{3} - 2 + 12 \right) \right]$$

$$= \left[\left(-\frac{9}{2} - 9 \right) - \left(\frac{-8}{3} + 10 \right) \right]$$

$$= \left[\left(\frac{-27}{2} \right) - \left(\frac{22}{3} \right) \right] = -\left(\frac{81 + 44}{6} \right) = -\frac{125}{6} = 20.8 \text{ sq. units}$$

Practice Problems (Area of Curves by Integration)

Find the area of the following curves:

1. $y = 3x^2$, between $x = 1$ and $x = 10$

2. $y = x^2 + 1$, between $x = 0$ and $x = 4$

3. $y = \sqrt{x-1}$, between $y = 1$ and $y = 5$

4. $y = x^2 + 5$, between $x = 2$ and $x = 7$

5. Find the area bound by the curve $y = x^2 - 4$ and the straight line $y = 2x$

6. Find the area bound by the curve $y = x^2$ and the straight line $y = x$

7. Find the area bound by the curves $x = y^2$ and $x = 2 - y^2$

8. Find the area bound by the curve $y = x^3$ and $y = 3x - 2$

Answers

1. 999

2. $\dfrac{76}{3}$ sq. units

3. $45\dfrac{1}{3}$ sq. units

4. $136\dfrac{2}{3}$ sq. units

5. 36 sq. units

6. $\dfrac{1}{6}$ sq. units

7. $\dfrac{8}{3}$ sq. units

8. 6.75 sq. units

Objective Questions: Integration

Indefinite Integrals

1. $\int x^{\frac{1}{3}} dx = \ldots.$

   (a) $\dfrac{1}{3} x^{\frac{2}{3}} + c$

   (b) $\dfrac{3}{4} x^{\frac{4}{3}} + c$

   (c) $\dfrac{4}{3} x^{\frac{3}{4}} + cp$

   (d) $\dfrac{1}{3} x^{\frac{4}{3}} + c$

2. $\int \left(x^2 + 5 \right) dx = \ldots$

   (a) $\left(2x + c \right)$

   (b) $\left(\dfrac{1}{2} x^3 + c \right)$

   (c) $\left(x^3 + c \right)$

   (d) $\left(- + \ x + c \right)$

3. $\int \left(2x^3 + 4\right) dx = \ldots$

   (a) $\left(6x^2 + c\right)$ (b) $\left(\dfrac{x^4}{2} + 4x + c\right)$

   (c) $\left(6x^4 + 4x\right)$ (d) $\left(\dfrac{2x^4}{3} + 4x + c\right)$

4. $\int \dfrac{3}{x^3} dx = \ldots$

   (a) $\left(\dfrac{3}{2x^2} + c\right)$ (b) $\left(\dfrac{-3}{4x^3} + c\right)$

   (c) $\left(\dfrac{-9}{x^4} + c\right)$ (d) $\left(\dfrac{-3}{4x^4} + c\right)$

5. $\int \left(3x^2 - 2x^3\right) dx = \ldots$

   (a) $\left(x^3 - \dfrac{x^2}{2} + c\right)$ (b) $\left(\dfrac{3x^3}{4} - \dfrac{2x^4}{3} + c\right)$

   (c) $\left(x^3 - \dfrac{3x^4}{2} + c\right)$ (d) $\left(\dfrac{2x^3}{3} - \dfrac{3x^3}{4} + c\right)$

6. $\int 18e^{-3x} dx = \ldots$

   (a) $\left(6e^{-3x} + c\right)$ (b) $\left(-6e^{3x} + c\right)$

   (c) $\left(54e^{-3x} + c\right)$ (d) $\left(-6e^{-3x} + c\right)$

7. $\int 6x^{-1} dx = \ldots$

   (a) $\left(\dfrac{6}{x^2} + c\right)$ (b) $\left(-6x + c\right)$

   (c) $\left(6\log x + c\right)$ (d) $\left(\log 6x + c\right)$

8. $\int \dfrac{\log x}{x} dx = \ldots$

   (a) $\left(\dfrac{1}{\log x} + c\right)$ (b) $\left(\dfrac{1}{2}(\log x)^2 + c\right)$

   (c) $\left(2(\log x)^2 + c\right)$ (d) $\left((\log x)^2 + c\right)$

9. $\int \sin x \cos x \, dx = \ldots$

(a) $-\dfrac{1}{4}\cos 2x + c$

(b) $-\dfrac{1}{4}\sin 2x + c$

(c) $\dfrac{1}{2}\cos 2x + c$

(d) $\dfrac{\sin^2 x}{2} + c$

10. $\int \sin^2 x \cos x \, dx = \ldots 1$

(a) $\dfrac{1}{3}\cos^3 x + c$

(b) $\dfrac{1}{2}\cos^2 x + c$

(c) $\dfrac{1}{3}\sin^3 x + c$

(d) $\tan x + c$

11. $\int \sin x \sec^2 x \, dx = \ldots$

(a) $\tan x + c$

(b) $\sec x + c$

(c) $\sec x \tan x + c$

(d) $\sec^2 x + c$

12. $\int 3 \cos ec^2 x \, dx = \ldots$

(a) $\cos ec^3 x + c$

(b) $-\cos ec^3 x + c$

(c) $\cos ec^3 x - \cot x + c$

(d) $-3\cot x + c$

Definite Integrals

13. $\int_{2}^{3} x^3 \, dx = \ldots$

(a) $\dfrac{65}{4}$

(b) $\dfrac{66}{4}$

(c) $\dfrac{65}{3}$

(d $\dfrac{55}{4}$

14. $\int_{2}^{1} \dfrac{1}{x} \, dx = \ldots$

(a) 0

(b) $\log 2$

(c) $-\log 2$

(d) $1 - \log 2$

15. $\int_{1}^{5} \dfrac{1}{e} \, dx = \ldots$

(a) $\dfrac{4}{e}$

(b) $\log 4$

(c) $-\log e$

(d) $\dfrac{1}{4e}$

16. $\int\limits_{2}^{6} 3dx = \dots$

    (a) $\dfrac{3}{4}$ (b) 6

    (c) 9   (d) 12

17. $\int\limits_{0}^{3} -4x^2 dx = \dots$

    (a) –36   (b) 27

    (c) – 18   (d) 13

18. $\int\limits_{0}^{\frac{\pi}{2}} \cos x dx = \dots$

    (a) –1   (b) 1

    (c) $\dfrac{1}{2}$ (d) $\dfrac{1}{\sqrt{2}}$

19. $\int\limits_{1}^{e} \dfrac{1}{x} dx = \dots$

    (a) 1   (b) $\left(\dfrac{1}{e}-1\right)$

    (c) $(e-1)$ (d) e

20. $\int\limits_{0}^{\frac{\pi}{4}} \sec^2 x dx = \dots$

    (a) $\dfrac{1}{2}$ (b) $\dfrac{3}{4}$

    (c) $\dfrac{1}{4}$ (d) 1

Answers

1. b	2. d	3. b	4. d	5. a	6. d	7. c
8. b	9. a, d	10. c	11. b	12. d	13. a	14. c
15. a	16. d	17. a	18. b	19. a	20.d	

CHAPTER 7

Matrices and Determinants

1.0 BRIEF HISTORY

Cardin (1545) gave a rule for solving a system of two linear equations which he termed as *'regula de modo'* i.e. mother of rules. This rule laid the foundation of Cramer's rule for solution of equations. But Cardin could not give the definition of 'determinants'.

The idea of determinant appeared in Japan and Europe almost exactly the same time but Seki in Japan was first to publish it. Using his determinants Seki was able to find the determinants of 2x2, 3x3, 4x4 matrices and applied them to solve equations .Cramer (1750) gave the general rule for n x n systems.

Matrices are used to solve "simultaneous equations" but they are also used to solve problems in:

1. Electronics
2. Statics
3. Robotics
4. Linear Programming
5. Optimisation
6. Genetics
7. Intersection of Planes

Leibnitz worked on mathematical notations of matrices and determinants (1700).

Work on determinants began to appear after 1700.In 1764 Bezout gave methods of calculating determinants as did Vondermonde in 1771.

Actually the term **'determinant'** was first given by Fredric Gauss (1801).

1.1 MATRICES

Matrix: a rectangle array on m n numbers in the form of m rows and n columns is called a matrix of order $m \times n$ or $m \times n$ matrix.

Such an array (of m rows and n columns) is enclosed in [ ] or ( ). There will be $m \times n$ elements. For example 5 rows and 6 columns will constitute a matrix of 5x6 order and it will contain $5 \times 6 = 30$ elements.

The plural of matrix is matrices

Usually a matrix is denoted by a capital letter A, B or C.

Example (i): $A = \begin{bmatrix} 3 & 5 & -4 \\ 1 & 3 & 2 \end{bmatrix}$ is a matrix having 2 rows and 3 columns. It contains $2 \times 3 = 6$ elements. Clearly it is of 2×3 order.

Example (ii): $B = \begin{bmatrix} 8 & 5 & 2 & -1 \\ 1 & 6 & -3 & 2 \\ 6 & 0 & 5 & 7 \end{bmatrix}$. It is a matrix of 3×4 order and has 12 elements.

Description of a matrix

Let A be a matrix of order $m \times n$. It may be defined as follows:

$$A = \begin{bmatrix} c_{11} & c_{12} & c_{13} & \text{........} & c_{1n} \\ c_{21} & c_{22} & c_{23} & \text{........} & c_{2n} \\ \text{........................} \\ c_{m1} & c_{m2} & c_{m3} & \text{........} & c_{mn} \end{bmatrix} = \begin{bmatrix} c_{ij} \end{bmatrix}_{m \times n}$$

An element occurring in i^{th} row and j^{th} column is denoted by c_{ij}. For example c_{12} denotes an element in 1st row and 2nd column.

Consider the matrix $A = \begin{bmatrix} 3 & 5 & -4 \\ 1 & 3 & 2 \end{bmatrix}$.

Its first row elements may be denoted as $c_{11} = 3$, $c_{12} = 4$, $c_{13} = 5$

Its second row elements may be denoted as $c_{21} = 1$, $c_{22} = 3$, $c_{23} = 2$

Examples: Construct a 2×3 matrix whose elements are given by

(1) $c_{ij} = i + j$ (2) $c_{ij} = i - j$ (3) $c_{ij} = \dfrac{i}{j}$

Solution (1) $c_{ij} = i + j$

The first row elements are:

$$c_{11} = 1 + 1 = 2, \ c_{12} = 1 + 2 = 3, \ c_{13} = 1 + 3 = 4$$

The second row elements are:

$$c_{21} = 2 + 1 = 3, \ c_{22} = 2 + 2 = 4, \ c_{23} = 2 + 3 = 5$$

Hence the required matrix $A_{2 \times 3}$ is $A_{2 \times 3} = \begin{bmatrix} 2 & 3 & 4 \\ 3 & 4 & 5 \end{bmatrix}$

(2) $c_{ij} = i - j$

The first row elements are: $c_{11} = 1 - 1 = 0$, $c_{12} = 1 - 2 = -1$, $c_{13} = 1 - 3 = -2$

The second row elements are: $c_{21} = 2 - 1 = 1$, $c_{22} = 2 - 2 = 0$, $c_{23} = 2 - 3 = -1$

Hence the required matrix $A_{2 \times 3}$ is $A_{2 \times 3} = \begin{bmatrix} 0 & -1 & -2 \\ 1 & 0 & -1 \end{bmatrix}$

(3) $c_{ij} = \dfrac{i}{j}$

The first row elements are:

$$c_{11} = 1 \div 1 = 1 \ , c_{12} = 1 \div 2 = \frac{1}{2} \ , c_{13} = 1 \div 3 = \frac{1}{3}$$

The second row elements are:

$$c_{21} = 2 \div 1 = 2 \ , c_{22} = 2 \div 2 = 1 \ , c_{23} = 2 \div 3 = \frac{2}{3}$$

Hence the required matrix $A_{2 \times 3}$ is $A_{2 \times 3} = \begin{bmatrix} 1 & \dfrac{1}{2} & \dfrac{1}{3} \\ 2 & 1 & \dfrac{1}{3} \end{bmatrix}$

1.1.1 Various Types of Matrices

1. **Row Matrix:** It has single row and many columns i.e. $A = \begin{bmatrix} 1 & 3 & 5 & 7 \end{bmatrix}$

2. **Column Matrix:** It has single column and many rows i.e $A = \begin{bmatrix} 1 \\ 3 \\ 5 \\ 7 \end{bmatrix}$

3. **Square Matrix:** It has equal number of rows and column, viz. $A = \begin{bmatrix} 1 & 3 & 5 \\ 2 & 4 & 6 \\ 3 & 5 & 7 \end{bmatrix}$

4. **Zero or Null Matrix:** Its elements are zero: $A = \begin{bmatrix} 0 & 0 & 0 & 0 \end{bmatrix}$

5. **Diagonal Matrix:** A square matrix in which every non-diagonal element is zero is known as Diagonal Matrix, viz.

$$A = \begin{bmatrix} 2 & 0 & 0 \\ 0 & 4 & 0 \\ 0 & 0 & 6 \end{bmatrix}$$

It may be written as $A = diag.\begin{bmatrix} 2 & 4 & 6 \end{bmatrix}$

6. **Square Matrix:** A matrix having same number of rows and columns is called a square matrix.

The matrix $A = \begin{bmatrix} 2 & 3 \\ 4 & 5 \end{bmatrix}$ is a square matrix

7. **Scalar Matrix:** A square matrix in which all non-diagonal elements are zero but diagonal elements are same. viz.

$$A = \begin{bmatrix} 4 & 0 & 0 \\ 0 & 4 & 0 \\ 0 & 0 & 4 \end{bmatrix}$$

8. **Unit Matrix:** A square matrix in which every non-diagonal is zero but every diagonal element is 1. viz.

$$A = \begin{bmatrix} 1 & 0 & 0 \\ 0 & 1 & 0 \\ 0 & 0 & 1 \end{bmatrix}$$

8. **Comparable Matrix:** Two matrices are said to be comparable if they are of same order. Matrices $A_{2 \times 3}$ and $B_{2 \times 3}$ are comparable as they are of same order (2x3).

Matrices: $A_{2 \times 3} = \begin{bmatrix} 0 & -1 & -2 \\ 1 & 0 & -1 \end{bmatrix}$ and $B_{2 \times 3} = \begin{bmatrix} 1 & 2 & 3 \\ 2 & 3 & 4 \end{bmatrix}$ are comparable as they are of same order.

9. **Equal matrices:** Two matrices are said to be equal if they are of same order and their corresponding elements are same. Equality of matrices helps us to detect the unknown variables.

It may be seen that both the given matrices are of same order i.e. 2×2.

If $\qquad \begin{bmatrix} 5 & 3 \\ x & 7 \end{bmatrix} = \begin{bmatrix} y & z \\ 2 & 7 \end{bmatrix} \Rightarrow x = 1, y = 5, z = 3$

10. **Rectangle Matrix:** The matrix where number of rows are different than the columns, is called a rectangle matrix..

The matrix $A = \begin{bmatrix} 1 & 2 & 3 \\ 7 & 6 & 5 \end{bmatrix}$ is a rectangle matrix.

11. **Upper Triangle Matrix:** In upper triangle matrix, the elements located below the diagonal are zeros.

Thus $A = \begin{bmatrix} 1 & 6 & 2 \\ 0 & 3 & 4 \\ 0 & 0 & 3 \end{bmatrix}$ is an upper triangle matrix.

12. **Lower Triangle Matrix:** In lower triangle matrix, the elements located above the diagonal are zeros.

Thus $A = \begin{bmatrix} 2 & 0 & 0 \\ 4 & 3 & 0 \\ 5 & 4 & 3 \end{bmatrix}$ is an upper triangle matrix

13. **Regular Matrix:** A square that has an inverse, is called regular matrix.

14. **Singular Matrix:** A regular matrix is called a singular matrix if it has no inverse.

15. **Idempotent Matrix:** The matrix A is idempotent if : $A^2 = A$

16. **In-volutive Matrix:** A matrix A is involutive if: $A^2 = 1$

17. **Symmetric Matrix:** A square matrix is called a symmetric matrix if it equals its transpose i.e

$$A = A^T$$

18. **Anti-Symmetric Matrix:** A square matrix is called a anti-symmetric matrix if it equals its negative transpose i.e

$$A = -A^T$$

19. **Orthogonal Matrix :** A matrix is orthogonal if it satisfies the following condition:

$$A.A^T = 1$$

1.1.2 Addition of Matrices

If A and B be two matrices, each of order $m \times n$, then the matrix obtained by their addition shall also be of same order i.e. a matrix (A+B) of order $m \times n$. Matrix (A+B) is obtained by adding the corresponding elements of A and B.

Example 1: Add the matrices $A = \begin{bmatrix} 5 & 3 & 2 \\ 3 & 2 & -7 \end{bmatrix}$ and $B = \begin{bmatrix} 4 & -3 & -4 \\ -1 & 0 & 6 \end{bmatrix}$

$$A + B = \begin{bmatrix} 5+4 & 3-3 & 2-4 \\ 3-1 & 2+0 & -7+6 \end{bmatrix} = \begin{bmatrix} 9 & 0 & -2 \\ 2 & 2 & -1 \end{bmatrix}$$

Example 2: Add the matrices $A = \begin{bmatrix} 5 & 3 \\ 3 & 2 \end{bmatrix}$ and $B = \begin{bmatrix} 4 & -3 & -4 \\ -1 & 0 & 6 \end{bmatrix}$

The matrices can't be added since their orders are different. Matrix A is of 2x2 order whereas Matrix B is of 2x3 order, hence they are not comparable.

1.1.3 Properties of Matrix Addition

1. Matrix addition is cumulative i.e. A + B = B + A

Example: Let the matrices are $A = \begin{bmatrix} 2 & 3 & 5 \\ 3 & 2 & -1 \end{bmatrix}$ and $B = \begin{bmatrix} 4 & 3 & -4 \\ -1 & 0 & 6 \end{bmatrix}$

$$A + B = \begin{bmatrix} 2+4 & 3+3 & 5-4 \\ 3-1 & 2+0 & -1+6 \end{bmatrix} = \begin{bmatrix} 6 & 6 & 1 \\ 2 & 2 & 5 \end{bmatrix}$$

$$B+A = \begin{bmatrix} 4+2 & 3+3 & -4+5 \\ -1+3 & 0+2 & 6-1 \end{bmatrix} = \begin{bmatrix} 6 & 6 & 1 \\ 2 & 2 & 5 \end{bmatrix}$$

Hence **A + B = B + A**

2. Matrix Addition is associative i.e. (A + B) + C = A + (B + C)

Example: Let $A = \begin{bmatrix} 2 & 3 & 5 \\ 3 & 2 & -1 \end{bmatrix}$, $B = \begin{bmatrix} 4 & 3 & -4 \\ -1 & 0 & 6 \end{bmatrix}$ and $C = \begin{bmatrix} 5 & 0 & -3 \\ 3 & 2 & 6 \end{bmatrix}$

$$A + B = \begin{bmatrix} 2+4 & 3+3 & 5-4 \\ 3-1 & 2+0 & -1+6 \end{bmatrix} = \begin{bmatrix} 6 & 6 & 1 \\ 2 & 2 & 5 \end{bmatrix}$$

$$(A+B)+C = \begin{bmatrix} 6 & 6 & 1 \\ 2 & 2 & 5 \end{bmatrix} + \begin{bmatrix} 5 & 0 & -3 \\ 3 & 2 & 6 \end{bmatrix} = \begin{bmatrix} 11 & 6 & -2 \\ 5 & 4 & 11 \end{bmatrix} \qquad ...(1)$$

$$(B+C) = \begin{bmatrix} 4 & 3 & -4 \\ -1 & 0 & 6 \end{bmatrix} + \begin{bmatrix} 5 & 0 & -3 \\ 3 & 2 & 6 \end{bmatrix} = \begin{bmatrix} 9 & 3 & -7 \\ 2 & 2 & 12 \end{bmatrix}$$

$$A+(B+C) = \begin{bmatrix} 2 & 3 & 5 \\ 3 & 2 & -1 \end{bmatrix} + \begin{bmatrix} 9 & 3 & -7 \\ 2 & 2 & 12 \end{bmatrix} = \begin{bmatrix} 11 & 6 & -2 \\ 5 & 4 & 11 \end{bmatrix} \qquad ...(2)$$

From (1) and (2): **(A + B) + C = A + (B + C)**

1.1.4 Subtraction of Matrices

If A and B are two matrices of same size i.e. they are comparable, Then their subtraction is defined as: **A − B = A + (−B)**

Example: Let $A = \begin{bmatrix} 2 & 3 & 5 \\ 3 & 2 & -1 \end{bmatrix}$ and $B = \begin{bmatrix} 4 & 3 & -4 \\ -1 & 0 & 6 \end{bmatrix}$ are two comparable matrices, then:

$$-B = \begin{bmatrix} -4 & -3 & +4 \\ +1 & -0 & -6 \end{bmatrix}$$

$$A - B = A + (-B) = \begin{bmatrix} 2 & 3 & 5 \\ 3 & 2 & -1 \end{bmatrix} + \begin{bmatrix} -4 & -3 & +4 \\ +1 & -0 & -6 \end{bmatrix} = \begin{bmatrix} -2 & 0 & 9 \\ 4 & 2 & -7 \end{bmatrix}$$

1.1.5 Scalar Matrix Multiplication

Let A be any matrix and k be a number, then matrix obtained by multiplying each element of A by k is called the scalar multiplication of A by k and denoted by kA.

Example 1: If $A = \begin{bmatrix} 5 & 4 & -2 \\ 6 & -1 & 7 \end{bmatrix}$, find 3A and $\frac{1}{2}A$

$$3A = \begin{bmatrix} 3\times5 & 3\times4 & 3\times-2 \\ 3\times6 & 3\times-1 & 3\times7 \end{bmatrix} = \begin{bmatrix} 15 & 12 & -6 \\ 18 & -3 & 21 \end{bmatrix}$$

$$\frac{1}{2}A = \begin{bmatrix} \dfrac{5}{2} & 2 & -1 \\ 3 & -\dfrac{1}{2} & \dfrac{7}{2} \end{bmatrix}$$

Example2: If $A = \begin{bmatrix} 3 & 5 \\ 7 & -9 \end{bmatrix}$ and $B = \begin{bmatrix} 6 & -4 \\ 2 & 3 \end{bmatrix}$ .Find $4A - 3B$

$$4A - 3b = 4A + (-3B) = 4 \times \begin{bmatrix} 3 & 5 \\ 7 & -9 \end{bmatrix} + (-3) \begin{bmatrix} 6 & -4 \\ 2 & 3 \end{bmatrix}$$

$$= \begin{bmatrix} 4 \times 3 & 4 \times 5 \\ 4 \times 7 & 4 \times -9 \end{bmatrix} + \begin{bmatrix} -3 \times 6 & -3 \times -4 \\ -3 \times 2 & -3 \times 3 \end{bmatrix}$$

$$= \begin{bmatrix} 12 & 20 \\ 28 & -36 \end{bmatrix} + \begin{bmatrix} -18 & 12 \\ -6 & -9 \end{bmatrix}$$

$$= \begin{bmatrix} 12-18 & 20+12 \\ 28-6 & -36-9 \end{bmatrix}$$

$$= \begin{bmatrix} -6 & 32 \\ 22 & -45 \end{bmatrix}$$

Example 3: $A = diag\begin{bmatrix} 3 & -5 & 7 \end{bmatrix}$ and $B = diag\begin{bmatrix} -1 & 2 & 4 \end{bmatrix}$

Find $(A + B)$, $(A - B)$ and $(2A + 3B)$

Solution: $A = \begin{bmatrix} 3 & 0 & 0 \\ 0 & -5 & 0 \\ 0 & 0 & 7 \end{bmatrix}$ and $B = \begin{bmatrix} -1 & 0 & 0 \\ 0 & 2 & 0 \\ 0 & 0 & 4 \end{bmatrix}$

$$(A+B) = \begin{bmatrix} 3 & 0 & 0 \\ 0 & -5 & 0 \\ 0 & 0 & 7 \end{bmatrix} + \begin{bmatrix} -1 & 0 & 0 \\ 0 & 2 & 0 \\ 0 & 0 & 4 \end{bmatrix} = \begin{bmatrix} 3-1 & 0 & 0 \\ 0 & -5+2 & 0 \\ 0 & 0 & 7+4 \end{bmatrix} = \begin{bmatrix} 2 & 0 & 0 \\ 0 & -3 & 0 \\ 0 & 0 & 11 \end{bmatrix}$$

$$(A-B) = \begin{bmatrix} 3 & 0 & 0 \\ 0 & -5 & 0 \\ 0 & 0 & 7 \end{bmatrix} - \begin{bmatrix} -1 & 0 & 0 \\ 0 & 2 & 0 \\ 0 & 0 & 4 \end{bmatrix}$$

$$= \begin{bmatrix} 3 & 0 & 0 \\ 0 & -5 & 0 \\ 0 & 0 & 7 \end{bmatrix} + \begin{bmatrix} 1 & 0 & 0 \\ 0 & -2 & 0 \\ 0 & 0 & -4 \end{bmatrix} = \begin{bmatrix} 4 & 0 & 0 \\ 0 & -7 & 0 \\ 0 & 0 & 3 \end{bmatrix}$$

Example 4: Find the matrices X and Y, if $2\begin{bmatrix} x & 5 \\ 7 & y\text{-}3 \end{bmatrix} + \begin{bmatrix} 3 & 4 \\ 1 & 2 \end{bmatrix} = \begin{bmatrix} 13 & 14 \\ 15 & 16 \end{bmatrix}$

Solution: $\begin{bmatrix} 2.x & 2.5 \\ 2.7 & 2.(y\text{-}3) \end{bmatrix} + \begin{bmatrix} 3 & 4 \\ 1 & 2 \end{bmatrix} = \begin{bmatrix} 7 & 14 \\ 15 & 16 \end{bmatrix}$

$$\begin{bmatrix} 2x & 10 \\ 14 & 2y\text{-}6 \end{bmatrix} + \begin{bmatrix} 3 & 4 \\ 1 & 2 \end{bmatrix} = \begin{bmatrix} 13 & 14 \\ 15 & 16 \end{bmatrix}$$

$$\begin{bmatrix} 2x+3 & 10+4 \\ 14+1 & 2y\text{-}6+2 \end{bmatrix} = \begin{bmatrix} 13 & 14 \\ 15 & 16 \end{bmatrix}$$

$$\begin{bmatrix} 2x+3 & 14 \\ 15 & 2y\text{-}4 \end{bmatrix} = \begin{bmatrix} 13 & 14 \\ 15 & 16 \end{bmatrix}$$

Comparing the corresponding elements:

$$2x + 3 = 13 \Rightarrow 2x = 13 - 3 = 10 \therefore \boldsymbol{x = 5}$$

$$2y - 4 = 16 \Rightarrow 2y = 20 \therefore \boldsymbol{y = 10}$$

Example 5: Find the value of A and B from the following matrices:

$$\left(A+B \right) = \begin{bmatrix} 5 & 2 \\ 3 & 9 \end{bmatrix} \text{ and } \left(A-B \right) = \begin{bmatrix} 7 & 8 \\ 3 & \text{-}1 \end{bmatrix} \backslash$$

Solution: Adding the given matrices:

$$\left(A+B \right) + \left(A-B \right) = \begin{bmatrix} 5 & 2 \\ 3 & 9 \end{bmatrix} + \begin{bmatrix} 7 & 8 \\ 3 & \text{-}1 \end{bmatrix}$$

$$2A = \begin{bmatrix} 5+7 & 2+8 \\ 3+3 & 9\text{-}1 \end{bmatrix} = \begin{bmatrix} 12 & 10 \\ 6 & 8 \end{bmatrix}$$

$$A = \frac{1}{2} \begin{bmatrix} 12 & 10 \\ 6 & 8 \end{bmatrix} = \begin{bmatrix} 6 & 5 \\ 3 & 4 \end{bmatrix} \backslash$$

Subtracting the given matrices:

$$\left(A+B \right) - \left(A-B \right) = \begin{bmatrix} 5 & 2 \\ 3 & 9 \end{bmatrix} - \begin{bmatrix} 7 & 8 \\ 3 & \text{-}1 \end{bmatrix}$$

$$2B = \begin{bmatrix} 5-7 & 2-8 \\ 3\text{-}3 & 9\text{-}(\text{-}1) \end{bmatrix} = \begin{bmatrix} \text{-}2 & \text{-}6 \\ 0 & 10 \end{bmatrix}$$

$$B = \frac{1}{2} \begin{bmatrix} \text{-}2 & \text{-}6 \\ 0 & 10 \end{bmatrix} = \begin{bmatrix} \text{-}1 & \text{-}3 \\ 0 & 5 \end{bmatrix}$$

Exercise 6: Given $A = \begin{bmatrix} 3 & 5 \\ 4 & 6 \end{bmatrix}$ and $B = \begin{bmatrix} 3 & 1 \\ 1 & 5 \end{bmatrix}$. Find X such that $2A + B + X = 0$

Solution: $2A = 2 \times \begin{bmatrix} 3 & 5 \\ 4 & 6 \end{bmatrix} = \begin{bmatrix} 6 & 10 \\ 8 & 12 \end{bmatrix}$

$$2A + B = \begin{bmatrix} 6 & 10 \\ 8 & 12 \end{bmatrix} + \begin{bmatrix} 3 & 1 \\ 1 & 5 \end{bmatrix} = \begin{bmatrix} 6+3 & 10+1 \\ 8+1 & 12+5 \end{bmatrix} = \begin{bmatrix} 9 & 11 \\ 9 & 17 \end{bmatrix}$$

Since $2A + B + X = 0$

$$\begin{bmatrix} 9 & 11 \\ 9 & 17 \end{bmatrix} + X = 0$$

Adding the matrix on both sides $-\begin{bmatrix} 9 & 11 \\ 9 & 17 \end{bmatrix}$

$$-\begin{bmatrix} 9 & 11 \\ 9 & 17 \end{bmatrix} + \begin{bmatrix} 9 & 11 \\ 9 & 17 \end{bmatrix} + X = -\begin{bmatrix} 9 & 11 \\ 9 & 17 \end{bmatrix} \Rightarrow X = \begin{bmatrix} -9 & -11 \\ -9 & -17 \end{bmatrix}$$

1.1.6 Matrix Multiplication

Let A and B are two matrices, then they can be multiplied if number of column of A is equal to the number of rows of B. The order of product matrix would be rows of first matrix and column of second matrix

If $A = \begin{bmatrix} a_{ij} \end{bmatrix}_{m \times n}$ and $B = \begin{bmatrix} b_{ij} \end{bmatrix}_{m \times p}$ are two matrices, then their product is defined as

$AB = \begin{bmatrix} c_{ij} \end{bmatrix}_{m \times p}$ i.e. the order of AB would be m x p

The product matrix's dimensions are: (rows of first matrix) x (columns of second matrix)

Steps for matrix multiplication:

Step1. Make sure that the **number of columns in first matrix** is equal to the **number of rows in the second matrix.**

Step 2. Multiply the elements of **first row** of first matrix by elements of **first column** of second matrix and add. Next multiply the elements of **second row** of first matrix by elements of **first column of second matrix** and so on until the last row.

Illustration:

A is a matrix of order 2×3 such that $A = \begin{bmatrix} r_{11} & r_{12} & r_{13} \\ r_{21} & r_{22} & r_{23} \end{bmatrix}$

B is a matrix of order 3×1 such that $B = \begin{bmatrix} c_{11} \\ c_{21} \\ c_{31} \end{bmatrix}$

It may be seen that number of columns of 1st matrix = number of rows of 2nd matrix, Hence matrix multiplication is possible. We will get a product matrix of order 2×1

$$\therefore AB = \begin{bmatrix} r_{11} & r_{12} & r_{13} \\ r_{21} & r_{22} & r_{33} \end{bmatrix} \begin{bmatrix} c_{11} \\ c_{21} \\ c_{31} \end{bmatrix} = \begin{bmatrix} P_{11} \\ P_{12} \end{bmatrix}$$

Where $P_{11} = r_{11} \times c_{11} + r_{12} \times c_{21} + r_{13} \times c_{31}$

$P_{12} = r_{21} \times c_{11} + r_{22} \times c_{21} + r_{23} \times c_{31}$

Example 1: $A = \begin{bmatrix} 2 & -1 \\ 3 & 4 \\ 1 & 4 \end{bmatrix}$ and $B = \begin{bmatrix} -1 & 3 \\ 2 & 1 \end{bmatrix}$. Find the product AB. Does BA exist?

Solution: A is a 3×2 matrix and B is a 2×2 matrix.

Since columns in A = Rows in B, multiplication is possible i.e. AB exists and its order would be 3×2.

$$AB = \begin{bmatrix} 2 & -1 \\ 3 & 4 \\ 1 & 4 \end{bmatrix} \begin{bmatrix} -1 & 3 \\ 2 & 1 \end{bmatrix}.$$

Now multiply the row elements of A by column elements of B and add as shown below;

$$= \begin{bmatrix} 2.(-1)+(-1).2 & 2.3 +(-1).1 \\ 3.(-1) + (4).2 & 3.3 + 4.1 \\ 1.(-1) + 5.2 & 1.3 + 5.1 \end{bmatrix}$$

$$= \begin{bmatrix} -2-2 & 6-1 \\ -3+8 & 9+4 \\ -1+10 & 3+5 \end{bmatrix} = \begin{bmatrix} -4 & 5 \\ 5 & 13 \\ 9 & 8 \end{bmatrix}$$

BA is not possible since number of columns of B ≠ number of rows of A.

Hence BA does not exist.

Example 2: If $A = \begin{bmatrix} 5 & 3 \\ 1 & 2 \end{bmatrix}$, $B = \begin{bmatrix} 1 & 3 & -2 \\ 0 & 2 & 4 \end{bmatrix}$ and $c = \begin{bmatrix} 2 & 3 & -4 \\ 2 & -5 & 0 \end{bmatrix}$ **Show that A(B+C)=AB+AC**

Solution: $B + C = \begin{bmatrix} 1 & 3 & -2 \\ 0 & 2 & 4 \end{bmatrix} + \begin{bmatrix} 2 & 3 & -4 \\ 2 & -5 & 0 \end{bmatrix}$

$$= \begin{bmatrix} 3 & 6 & -6 \\ 2 & -3 & 4 \end{bmatrix}$$

$$A(B+C) = \begin{bmatrix} 5 & 3 \\ 1 & 2 \end{bmatrix} \begin{bmatrix} 3 & 6 & -6 \\ 2 & -3 & 4 \end{bmatrix}$$

$$= \begin{bmatrix} 5\times3+3\times2 & 5\times6+3\times-3 & 5\times-6+3\times4 \\ 1\times3+2\times2 & 1\times6+2\times-3 & 1\times-6+2\times4 \end{bmatrix}$$

$$\begin{bmatrix} 15 & 6 & 30\text{-}9 & \text{-}30\text{+}12 \\ 3\text{+}4 & 6\text{ - }6 & \text{-}6\text{ + }8 \end{bmatrix}$$

$$= \begin{bmatrix} 21 & 21 & \text{-}18 \\ 7 & 0 & 2 \end{bmatrix} \qquad \qquad ...(1)$$

$$AB = \begin{bmatrix} 5 & 3 \\ 1 & 2 \end{bmatrix}\begin{bmatrix} 1 & 3 & \text{-}2 \\ 0 & 2 & 4 \end{bmatrix}$$

$$= \begin{bmatrix} 5\times1+3\times0 & 5\times3+3\times2 & 5\times-2+3\times4 \\ 1\times1+2\times0 & 1\times3+2\times2 & 1\times-2+2\times4 \end{bmatrix}$$

$$= \begin{bmatrix} 5+0 & 15+6 & -10+12 \\ 1+0 & 3+4 & -2+8 \end{bmatrix} = \begin{bmatrix} 5 & 21 & 2 \\ 1 & 7 & 6 \end{bmatrix}$$

$$AC = \begin{bmatrix} 5 & 3 \\ 1 & 2 \end{bmatrix}\begin{bmatrix} 2 & 3 & \text{-}4 \\ 2 & \text{-}5 & 0 \end{bmatrix}$$

$$= \begin{bmatrix} 5\times2+3\times2 & 5\times3+3\times-5 & 5\times-4+3\times0 \\ 1\times2+2\times2 & 1\times3+2\times-5 & 1\times-4+2\times0 \end{bmatrix}$$

$$= \begin{bmatrix} 10+6 & 15-15 & -20+0 \\ 2+4 & 3-10 & -4+0 \end{bmatrix}$$

$$= \begin{bmatrix} 16 & 0 & -20 \\ 6 & -7 & -4 \end{bmatrix}$$

$$AB + AC = \begin{bmatrix} 5 & 21 & 2 \\ 1 & 7 & 6 \end{bmatrix} + \begin{bmatrix} 16 & 0 & -20 \\ 6 & -7 & -4 \end{bmatrix}$$

$$= \begin{bmatrix} 5+16 & 21+0 & 2\text{-}20 \\ 1+6 & 7\text{-}7 & 6\text{-}4 \end{bmatrix} = \begin{bmatrix} 21 & 21 & \text{-}18 \\ 7 & 0 & 2 \end{bmatrix} \qquad ...(2)$$

From (1) and (2) we find- A (B+C) = AB +AC

Example 3: Find the value of x such that: $[x \ 1 \ 2]\begin{bmatrix} 1 & 0 & 1 \\ 0 & 2 & 2 \\ 0 & 0 & 3 \end{bmatrix}\begin{bmatrix} 1 \\ 4 \\ x \end{bmatrix} = 0$

Solution: First matrix is of size 1 × 3 and second matrix is of size 3 × 3 ,hence multiplication will give us a matrix of size 1 × 3 (row of first x column of second) i.e. a single row matrix with three elements.

$$[x+0+0 \quad 0+2+0 \quad x+2+6]\begin{bmatrix} 1 \\ 4 \\ x \end{bmatrix} = 0$$

$$[x \ 2 \ x+8]\begin{bmatrix} 1 \\ 4 \\ x \end{bmatrix} = 0$$

$$x + 8 + (x + 8)x = 0$$

$$x^2 + 9x + 8 = 0$$

$$(x + 1)(x + 8) = 0$$

$$\therefore x = -1 \text{ or} - 8$$

1.1.7 Transpose of a Matrix

Let A be a $m \times n$ .matrix. If rows and columns are interchanged, i.e. rows become columns and columns become rows, then the matrix so formed is called the *transpose of matrix A* and its order would be $n \times m$. Transpose of matrix A is denoted by A^T .

Example 1: If $A = \begin{bmatrix} 2 & 4 & 5 \\ 3 & 1 & 6 \end{bmatrix}$, then $A^T = \begin{bmatrix} 2 & 3 \\ 4 & 1 \\ 5 & 6 \end{bmatrix}$

Example 2: If $B = \begin{bmatrix} 1 & 2 & 3 & 4 \\ 0 & 1 & 2 & 3 \\ 4 & 5 & 6 & 0 \end{bmatrix}$, then $B^T = \begin{bmatrix} 1 & 0 & 4 \\ 2 & 1 & 5 \\ 3 & 2 & 6 \\ 4 & 3 & 0 \end{bmatrix}$

Rules of Transpose of Matrices:

Rule 1. For any matrix $\left(A^T \right)^T = A$ i.e. *Transpose of a transposed matrix = Original matrix*

Example : If $A = \begin{bmatrix} 2 & 3 & -1 \\ 0 & -5 & 7 \end{bmatrix}$, then show that $\left(A^T \right)^T = A$

$$A^T = \begin{bmatrix} 2 & 0 \\ 3 & -5 \\ -1 & 7 \end{bmatrix} \quad \text{(rows of A have become as colums)}$$

$$\left(A^T \right)^T = \begin{bmatrix} 2 & 3 & -1 \\ 0 & -5 & 7 \end{bmatrix} = A$$

Rule 2. If A and B are two matrices of same order, then $(A + B)^T = A^T + B^T$

i.e. *Transpose of sum of matrices = sum of transposed matrices*

Example: Let $A = \begin{bmatrix} 3 & 4 \\ -2 & 0 \\ 7 & -5 \end{bmatrix}$ and $B = \begin{bmatrix} 2 & -3 \\ 5 & 6 \\ -1 & 8 \end{bmatrix}$

$$A + B = \begin{bmatrix} 3+2 & 4-3 \\ -2+5 & 0+6 \\ 7-1 & -5+8 \end{bmatrix} = \begin{bmatrix} 5 & 1 \\ 3 & 6 \\ 6 & 3 \end{bmatrix}$$

$$(A+B)^T = \begin{bmatrix} 5 & 1 \\ 3 & 6 \\ 6 & 3 \end{bmatrix}^T = \begin{bmatrix} 5 & 3 & 6 \\ 1 & 6 & 3 \end{bmatrix}$$

$$A^T = \begin{bmatrix} 3 & 4 \\ -2 & 0 \\ 7 & -5 \end{bmatrix}^T = \begin{bmatrix} 3 & -2 & 7 \\ 4 & 0 & -5 \end{bmatrix}$$

$$B^T = \begin{bmatrix} 2 & -3 \\ 5 & 6 \\ -1 & 8 \end{bmatrix}^T = \begin{bmatrix} 2 & 5 & -1 \\ -3 & 6 & 8 \end{bmatrix}$$

$$A^T + B^T = \begin{bmatrix} 3 & -2 & 7 \\ 4 & 0 & -5 \end{bmatrix} + \begin{bmatrix} 2 & 5 & -1 \\ -3 & 6 & 8 \end{bmatrix}$$

$$= \begin{bmatrix} 3+2 & -2+5 & 7-1 \\ 4-3 & 0+6 & -5+8 \end{bmatrix} = \begin{bmatrix} 5 & 3 & 6 \\ 1 & 6 & 3 \end{bmatrix} = (A+B)^T$$

Hence $(A+B)^T = A^T \ B^T$

Rule 3: If A is a matrix, and k is a scalar then $(kA)^T = kA^T$

Transpose of a scalar multiplication of matrix = scalar multiplication of transposed matrix

Example: If $A = \begin{bmatrix} 2 & 3 & -5 \\ 0 & -1 & 4 \end{bmatrix}$ then $(3A)^T = 3A^T$

Solution: $3A = \begin{bmatrix} 2 & 3 & -5 \\ 0 & -1 & 4 \end{bmatrix} = \begin{bmatrix} 6 & 9 & -15 \\ 0 & -3 & 12 \end{bmatrix}$

$$(3A)^T = \begin{bmatrix} 6 & 9 & -15 \\ 0 & -3 & 12 \end{bmatrix}^T = \begin{bmatrix} 6 & 0 \\ 9 & -3 \\ -15 & 12 \end{bmatrix}$$

$$A^T = \begin{bmatrix} 2 & 3 & -5 \\ 0 & -1 & 4 \end{bmatrix}^T = \begin{bmatrix} 2 & 0 \\ 3 & -1 \\ -5 & 4 \end{bmatrix}$$

$$3A^T = 3 \times \begin{bmatrix} 2 & 0 \\ 3 & -1 \\ -5 & 4 \end{bmatrix} = \begin{bmatrix} 6 & 0 \\ 9 & -3 \\ -15 & 12 \end{bmatrix}$$

Hence $(3A)^T = 3A^T$

1.1.8 Symmetric and Skewed Symmetric Matrix

Symmetric Matrix: If A is a square matrix then $A^T = A$

Example: $A = \begin{bmatrix} 1 & 3 & 5 \\ 3 & 5 & 7 \\ 5 & 7 & 9 \end{bmatrix}$ then $A^T = \begin{bmatrix} 1 & 3 & 5 \\ 3 & 5 & 7 \\ 5 & 7 & 9 \end{bmatrix}$

Since $A = A^T$, hence A is a symmetric matrix.

Skewed symmetric Matrix: A square matrix is said to be skewed matrix if $A^T = -A$

Example 1: $A = \begin{bmatrix} 0 & 3 & -5 \\ -3 & 0 & 1 \\ 5 & -1 & 0 \end{bmatrix}$. Find whether A is symmetric or skewed symmetric matrix?

Solution: $A^T = \begin{bmatrix} 0 & 3 & -5 \\ -3 & 0 & 1 \\ 5 & -1 & 0 \end{bmatrix}^T = \begin{bmatrix} 0 & -3 & 5 \\ 3 & 0 & -1 \\ -5 & 1 & 0 \end{bmatrix} = (-) \begin{bmatrix} 0 & 3 & -5 \\ -3 & 0 & 1 \\ 5 & -1 & 0 \end{bmatrix} = -A$

Hence A is skewed symmetric matrix.

Example 2: $A = \begin{bmatrix} 4 & 1 \\ 5 & 8 \end{bmatrix}$. Show that $\left(A + A^T \right)$ is symmetric matrix.

Solution: $A^T = \begin{bmatrix} 4 & 1 \\ 5 & 8 \end{bmatrix}^T = \begin{bmatrix} 4 & 5 \\ 1 & 8 \end{bmatrix}$

$$\left(A + A^T \right) = \begin{bmatrix} 4 & 1 \\ 5 & 8 \end{bmatrix} + \begin{bmatrix} 4 & 5 \\ 1 & 8 \end{bmatrix} = \begin{bmatrix} 8 & 6 \\ 6 & 16 \end{bmatrix}.$$

$$\left(A + A^T \right)^T = \begin{bmatrix} 8 & 6 \\ 6 & 16 \end{bmatrix}^T = \begin{bmatrix} 8 & 6 \\ 6 & 16 \end{bmatrix} = \left(A + A^T \right).$$

Hence $\left(A + A^T \right)$ is symmetric matrix

Example 3: $A = \begin{bmatrix} 4 & 1 \\ 5 & 8 \end{bmatrix}$. Show that $\left(A - A^T \right)$ is skewed symmetric matrix.

Solution: $A^T = \begin{bmatrix} 4 & 1 \\ 5 & 8 \end{bmatrix}^T = \begin{bmatrix} 4 & 5 \\ 1 & 8 \end{bmatrix}$

$$\left(A - A^T \right) = \begin{bmatrix} 4 & 1 \\ 5 & 8 \end{bmatrix} - \begin{bmatrix} 4 & 5 \\ 1 & 8 \end{bmatrix} = \begin{bmatrix} 0 & -4 \\ 4 & 0 \end{bmatrix}$$

$$\left(A - A^T \right)^T = \begin{bmatrix} 0 & -4 \\ 4 & 0 \end{bmatrix}^T = \begin{bmatrix} 0 & 4 \\ -4 & 0 \end{bmatrix} = (-) \begin{bmatrix} 0 & -4 \\ 4 & 0 \end{bmatrix} = -\left(A - A^T \right).$$

Since $\left(A-A^T\right)^T =\left(A-A^T\right)$, hence $\left(A-A^T\right)$ is *skewed symmetric* matrix.

Theorem: If $\left(A+A^T\right)$ is symmetric matrix, then $\left(A-A^T\right)$ is *skewed symmetric* matrix.

Example 4: If $A=\begin{bmatrix} 2 & 1 & 3 \\ 4 & 1 & 0 \end{bmatrix}$ and $B=\begin{bmatrix} 1 & -1 \\ 0 & 2 \\ 5 & 0 \end{bmatrix}$ .Show that $\left(AB\right)^T = B^T A^T$

Solution: $AB=\begin{bmatrix} 2 & 1 & 3 \\ 4 & 1 & 0 \end{bmatrix}\times\begin{bmatrix} 1 & -1 \\ 0 & 2 \\ 5 & 0 \end{bmatrix} =\begin{bmatrix} 2\times1+1\times0+3\times5 & 2\times-1+1\times2+3\times0 \\ 4\times1+1\times0+0\times5 & 4\times-1+1\times2+0\times0 \end{bmatrix}$

$$=\begin{bmatrix} 17 & 0 \\ 4 & -2 \end{bmatrix}$$

$$AB=\begin{bmatrix} 17 & 0 \\ 4 & -2 \end{bmatrix} \Rightarrow \left(AB\right)^T =\begin{bmatrix} 17 & 0 \\ 4 & -2 \end{bmatrix}^T =\begin{bmatrix} 17 & 4 \\ 0 & -2 \end{bmatrix}$$

$$B^T =\begin{bmatrix} 1 & -1 \\ 0 & 2 \\ 5 & 0 \end{bmatrix}^T =\begin{bmatrix} 1 & 0 & 5 \\ -1 & 2 & 0 \end{bmatrix},$$

$$A^T =\begin{bmatrix} 2 & 1 & 3 \\ 4 & 1 & 0 \end{bmatrix}^T =\begin{bmatrix} 2 & 4 \\ 1 & 1 \\ 3 & 0 \end{bmatrix}$$

$$B^T A^T =\begin{bmatrix} 1 & 0 & 5 \\ -1 & 2 & 0 \end{bmatrix}\begin{bmatrix} 2 & 4 \\ 1 & 1 \\ 3 & 0 \end{bmatrix}=\begin{bmatrix} 1\times2+0\times1+5\times3 & 1\times4+0\times1+5\times0 \\ -1\times2+2\times1+0\times3 & -1\times4+2\times1+0\times0 \end{bmatrix}.$$

$$=\begin{bmatrix} 2+0+15 & 4+0+0 \\ -2+2+0 & -4+2+0 \end{bmatrix}$$

$$=\begin{bmatrix} 17 & 4 \\ 0 & -2 \end{bmatrix}$$

Hence $\left(AB\right)^T = B^T A^T$

1.2 DETERMINANTS

Let A is a square matrix viz. $A=\begin{bmatrix} a_{11} & a_{12} & ......a_{1n} \\ a_{21} & a_{22} & ......a_{2n} \\ .... \\ a_{n1} & a_{n2} & ......a_{nn} \end{bmatrix}$

Corresponding to A, there exists an expression called **determinants of A** and denoted by 'det.A' or $|A|$ and written as:

$$|A| = \begin{vmatrix} a_{11} & a_{12} & \ldots\ldots a_{1n} \\ a_{21} & a_{22} & \ldots\ldots a_{2n} \\ \ldots\ldots\ldots\ldots\ldots \\ a_{n1} & a_{n2} & \ldots\ldots a_{nn} \end{vmatrix}$$

1.2.1 Value of a determinant

A be a determinant of order 2×2 $|A| = \begin{vmatrix} a_{11} & a_{12} \\ a_{21} & a_{22} \end{vmatrix} = (a_{11}a_{22} - a_{12}a_{21})$

Let B be another determinant $|B| = \begin{vmatrix} a & b & c \\ d & e & f \\ g & h & i \end{vmatrix} = a\begin{vmatrix} e & f \\ h & i \end{vmatrix} - b\begin{vmatrix} d & f \\ g & i \end{vmatrix} + c\begin{vmatrix} d & e \\ g & h \end{vmatrix}$

$$= a(ei - fh) - b(di - fg) + c(dh - eg)$$

Example 1: Find the value of $|A| = \begin{vmatrix} 6 & -3 \\ 7 & -2 \end{vmatrix} = \{6 \times (-2) - 7 \times (-3)\} = -12 + 21 = 9$

Example 2: If $\begin{vmatrix} 4 & m \\ -3 & 5 \end{vmatrix} = 8$, find the value of m.

$$\begin{vmatrix} 4 & m \\ -3 & 5 \end{vmatrix} = (4 \times 5 - m \times -3) = 20 + 3m = 8 \Rightarrow m = \frac{8 - 20}{3} = \frac{-12}{3} = -4$$

Example 3: Evaluate $\Delta = \begin{vmatrix} 3 & 4 & 5 \\ -6 & 2 & -3 \\ 8 & 1 & 7 \end{vmatrix}$

Expanding by 1st row: $\Delta = 3\begin{vmatrix} 2 & -3 \\ 1 & 7 \end{vmatrix} - 4\begin{vmatrix} -6 & -3 \\ 8 & 7 \end{vmatrix} + 5\begin{vmatrix} -6 & 2 \\ 8 & 1 \end{vmatrix}$

$$= 3(14 + 3) - 4(-42 + 24) + 5(-6 - 16)$$

$$= 3(17) - 4(-18) + 5(-22)$$

$$= 51 + 72 - 110 = 13$$

Key Points:

1. $\begin{vmatrix} \lambda a & \lambda b \\ c & d \end{vmatrix} = \lambda\begin{vmatrix} a & b \\ c & d \end{vmatrix} = \begin{vmatrix} a & b \\ \lambda c & \lambda d \end{vmatrix}$ **2** $\begin{vmatrix} a & b \\ c & d \end{vmatrix} = -\begin{vmatrix} c & d \\ a & b \end{vmatrix}$ **3.** det (AB) = det (A).det (B)

4. $\det(A^{-1}) = \dfrac{1}{\det(A)}$

Practice Problems (On Evaluation of Determinants)

Evaluate the following determinants

1. $|A| = \begin{vmatrix} 3 & 0 & -1 \\ 2 & -5 & 4 \\ -3 & 1 & 3 \end{vmatrix}$

2. $|B| = \begin{vmatrix} 2 & -3 & 1 \\ 4 & 2 & -1 \\ -5 & 3 & -2 \end{vmatrix}$

3. $|C| = \begin{vmatrix} 1 & -4 \\ 0 & 3 \end{vmatrix}$

Answers:

1. −44   2. −19   3. 3

1.2.2   Minors and Co-Factors

A "minor" is the determinant of the square matrix formed by deleting one row and one column from some larger square matrix. Since there are lots of rows and columns in the original matrix, we can make lots of minors from it. These minors are labeled according to the row and column deleted. i.e., if we remove the second row and the fourth column of the matrix, the determinant of the smaller matrix is called "the minor $M_{2,4}$".

Minors obtained by removing just one row and one column from square matrices (**first minors**) are required for calculating **cofactors**, which in turn are useful for computing both the determinant and inverse of square matrices.

Minors

The minor of an element of a determinant (say of order n) is a determinant of lesser order $(n-1)$ formed by excluding a row and a column of the element.

Let A is a determinant, $A = \begin{vmatrix} 1 & 3 & 5 \\ 3 & 5 & 7 \\ 5 & 7 & 9 \end{vmatrix}$

Let the minor M_{11} has to be obtained. Then excluding 1st row (1, 3, 5) and 1st column

(1, 3, 5), we get the determinant $M_{11} = \begin{vmatrix} - & - & - \\ - & 5 & 7 \\ - & 7 & 9 \end{vmatrix} = \begin{vmatrix} 5 & 7 \\ 7 & 9 \end{vmatrix} = 5 \times 9 - 7 \times 7 = -4$

M_{11} is known as the minor of element a_{11}. $(a_{11} = 1)$

Similarly for getting minor M_{12}, **row 1** and **column 2** have to be excluded i.e. row and column passing through the element are to be excluded.

$$\therefore M_{12} = \begin{vmatrix} - & - & - \\ 3 & - & 7 \\ 5 & - & 9 \end{vmatrix} = \begin{vmatrix} 3 & 7 \\ 5 & 9 \end{vmatrix} = 27 - 35 = -8$$

M_{13} shall be obtained by excluding 1st row and 3rd column.

$$M_{13} = \begin{vmatrix} - & - & - \\ 3 & 5 & - \\ 5 & 7 & - \end{vmatrix} \begin{vmatrix} 3 & 5 \\ 5 & 7 \end{vmatrix} = 21 - 25 = -4$$

Similarly we can get

$$M_{21} = \begin{vmatrix} 3 & 5 \\ 7 & 9 \end{vmatrix} = 27 - 35 = -8,$$

$$M_{22} = \begin{vmatrix} 1 & 5 \\ 5 & 9 \end{vmatrix} = 9 - 25 = -16,$$

$$M_{23} = \begin{vmatrix} 1 & 3 \\ 5 & 7 \end{vmatrix} = 7 - 15 = -8$$

$$M_{31} = \begin{vmatrix} 3 & 5 \\ 5 & 7 \end{vmatrix} = 21 - 25 = -4$$

$$M_{32} = \begin{vmatrix} 1 & 5 \\ 3 & 7 \end{vmatrix} = 7 - 15 = -8$$

$$M_{33} = \begin{vmatrix} 1 & 3 \\ 3 & 5 \end{vmatrix} = 5 - 9 = -4$$

Cofactors

Cofactor of an element a_{ij} is defined as $C_{ij} = (-1)^{i+j} M_{ij}$

C_{ij} **is the cofactor of the element** a_{ij}. For example C_{11} is cofactor of element a_{11} i.e. element of first row and first column. Similarly C_{23} is cofactor of element a_{23} i.e. element of second row and third column and so on. Cofactor of an element depends on its minor.

We can calculate cofactors from the above minors as follows:

$$C_{11} = (-1)^{1+1} M_{11} = (-1)^2 (-4) = -4$$

$$C_{12} = (-1)^{1+2} M_{12} = (-1)^3 (-8) = +8$$

$$C_{13} = (-1)^{1+3} M_{13} = (-1)^4 (-4) = -4$$

$$C_{21} = (-1)^{2+1} M_{21} = (-1)^3 (-8) = +8$$

$$C_{22} = (-1)^{2+2} M_{22} = (-1)^4 (-16) = -16$$

$$C_{23} = (-1)^{2+3} M_{23} = (-1)^5 (-8) = +8$$

$$C_{31} = (-1)^{3+1} M_{31} = (-1)^4 (-4) = +4$$

$$C_{32} = (-1)^{3+2} M_{32} = (-1)^5(-8) = +8$$

$$C_{33} = (-1)^{3+3} M_{33} = (-1)^6(-4) = -4$$

Cofactor matrix $= \begin{bmatrix} C_{11} & C_{12} & C_{13} \\ C_{21} & C_{22} & C_{23} \\ C_{31} & C_{32} & C_{33} \end{bmatrix}$

From the above cofactors we get-

Cofactor matrix $= \begin{bmatrix} -4 & 8 & -4 \\ 8 & -16 & 8 \\ 4 & 8 & -4 \end{bmatrix}$

Transpose of cofactor matrix is obtained by interchanging the rows and columns

Transpose of cofactor matrix $= \begin{bmatrix} -4 & 8 & -4 \\ 8 & -16 & 8 \\ 4 & 8 & -4 \end{bmatrix}^{T} = \begin{bmatrix} -4 & 8 & 4 \\ 8 & -16 & 8 \\ -4 & 8 & -4 \end{bmatrix} =$ Adjoint (A)

Practice Problems (On Minor-Cofactors)

1. Determine the minors and co-factors of following determinant.

$$A = \begin{vmatrix} 8 & 1 & 9 \\ -2 & -8 & -2 \\ 3 & -9 & -7 \end{vmatrix}$$

Answer:

Minors:

$M_{11} = 38$	$M_{21} = -74$	$M_{31} = 70$
$M_{12} = -20$	$M_{22} = -83$	$M_{32} = -11$
$M_{13} = 42$	$M_{23} = 75$	$M_{33} = -62$

Cofactors:

$C_{11} = -7$	$C_{21} = -9$	$C_{31} = 0$
$C_{12} = -3$	$C_{22} = 8$	$C_{32} = 0$
$C_{13} = 0$	$C_{23} = 0$	$C_{33} = 0$

1.2.3   Inverse of a Matrix

Steps to find Inverse of Matrix- A^{-1}

1. Find determinant of Matrix
2. Find minor
3. Find Cofactor
4. Find Adjoint
5. Replace results in formula below:

Formula for Inverse of $n \times n$ Matrix

The inverse of a general $n \times n$ matrix A can be found by using the following equation

$$A^{-1} = \frac{1}{det\ (A)}\ adj\ (A)$$

Illustration: Inverse of 2x2 Matrix

Let A $= \begin{bmatrix} a & b \\ c & d \end{bmatrix}$

Determinant of A is $|A| = \begin{vmatrix} a & b \\ c & d \end{vmatrix} = (ad - cb)$

Minors:

$M_{11} = d \quad M_{12} = c \quad M_{21} = b \quad M_{22} = a$

Cofactors:

$C_{11} = (-1)^{1+1}\ M_{11} = (-1)^2 \cdot d = d\ C_{12} = (-1)^{1+2}\ M_{12} = (-1)^3 \cdot c = -c$

$C_{21} = (-1)^{2+1}\ M_{21} = (-1)^3 \cdot b = -b\ C_{22} = (-1)^{2+2}\ M_{22} = (-1)^4 \cdot a = a$

Cofactor Matrix of A $= \begin{bmatrix} C_{11} & C_{12} \\ C_{21} & C_{22} \end{bmatrix}$

Transpose of cofactor matrix of A $= \begin{bmatrix} C_{11} & C_{21} \\ C_{12} & C_{22} \end{bmatrix} = \begin{bmatrix} d & -b \\ -c & a \end{bmatrix}$

Hence $adj\ A = \begin{bmatrix} d & -b \\ -c & a \end{bmatrix}$

Inverse of A is given by $A^{-1} = \frac{1}{|A|} adj(A) = \frac{1}{(ad-cb)} \begin{bmatrix} d & -b \\ -c & a \end{bmatrix}$

If $|A| \neq 0$, then inverse of matrix exists, otherwise not.

Solved Examples

1. Find the inverse of matrix A, $A = \begin{bmatrix} 1 & 2 \\ 3 & 4 \end{bmatrix}$

Solution: Find determinant of A.

$$|A| = \begin{vmatrix} 1 & 2 \\ 3 & 4 \end{vmatrix} = (1 \times 4 - 3 \times 2) = -2$$

Since $|A| \neq 0$, hence inverse of matrix (A) exists

Inverse of (A) is given by

$$A^{-1} = \frac{1}{|A|} \begin{bmatrix} d & -b \\ -c & a \end{bmatrix} = \frac{1}{(ad - cb)} \begin{bmatrix} d & -b \\ -c & a \end{bmatrix}$$

Here $a = 1$ $b = 2$ $c = 3$ $d = 4$

For the given matrix

$$A^{-1} = \frac{1}{|A|} \begin{bmatrix} 4 & -2 \\ -3 & 1 \end{bmatrix} = \frac{1}{-2} \begin{bmatrix} 4 & -2 \\ -3 & 1 \end{bmatrix} = \begin{bmatrix} -2 & 1 \\ \frac{3}{2} & \frac{-1}{2} \end{bmatrix}$$

2. Find the inverse of matrix A, $A = \begin{bmatrix} 8 & -5 \\ -3 & 2 \end{bmatrix}$

Solution: Find determinant of A.

$$|A| = \begin{vmatrix} 8 & -5 \\ -3 & 2 \end{vmatrix} = (8 \times 2 - (-3) \times (-5)) = 16 - 15 = 1$$

Since $|A| \neq 0$, hence inverse of matrix A exists.

Apply $A^{-1} = \frac{1}{|A|} \begin{bmatrix} d & -b \\ -c & a \end{bmatrix} = \frac{1}{(ad - cb)} \begin{bmatrix} d & -b \\ -c & a \end{bmatrix}$, as shown in the illustration.

In this problem $a = 8$, $b = -5$, $c = -3$ and $d = 2$

$$\therefore A^{-1} = \frac{1}{1} \begin{bmatrix} 2 & 5 \\ 3 & 8 \end{bmatrix} = \begin{bmatrix} 2 & 5 \\ 3 & 8 \end{bmatrix}$$

3. Find the inverse of matrix A, $A = \begin{bmatrix} 5 & 16 \\ -1 & -3 \end{bmatrix}$

Solution: $|A| = \begin{vmatrix} 5 & 16 \\ -1 & -3 \end{vmatrix} = (5 \times -3 - (16) \times (-1)) = -15 + 16 = 1$

Since $|A| \neq 0$, hence inverse of matrix A exists.

In this problem $a = 5$, $b = 16$, $c = -1$ and $d = -3$

Apply $A^{-1} = \dfrac{1}{|A|}\begin{bmatrix} d & \text{-}b \\ \text{-}c & a \end{bmatrix} = \dfrac{1}{(ad-cb)}\begin{bmatrix} d & \text{-}b \\ \text{-}c & a \end{bmatrix}$, as shown in the illustration

$$A^{-1} = \dfrac{1}{|A|}\begin{bmatrix} d & \text{-}b \\ \text{-}c & a \end{bmatrix} = \dfrac{1}{1}\begin{bmatrix} \text{-}3 & \text{-}16 \\ 1 & 5 \end{bmatrix}$$

Practice Problems (On Inverse of Matrix)

1. Find the inverse of the matrix $A = \begin{bmatrix} 3 & 1 \\ 5 & 2 \end{bmatrix}$

2. Find the inverse of the matrix $A = \begin{bmatrix} 3 & 1 \\ 7 & 4 \end{bmatrix}$

3. If $A = \begin{bmatrix} 2 & 3 \\ 5 & \text{-}2 \end{bmatrix}$, show that $A^{-1} = \dfrac{1}{19}A$

Ans : 1. $\begin{bmatrix} 2 & -1 \\ -5 & 3 \end{bmatrix}$ 2. $\begin{bmatrix} \dfrac{4}{5} & -\dfrac{1}{5} \\ \dfrac{7}{5} & \dfrac{3}{5} \end{bmatrix}$

1.2.4   Singular and Non-singular Matrix

Singular matrix is a square matrix that does not have a matrix inverse, or it is singular if and only if its determinant is zero.

Following matrices are examples of singular matrices:

$$\begin{bmatrix} 0 & 0 \\ 0 & 0 \end{bmatrix}, \begin{bmatrix} 0 & 0 \\ 0 & 1 \end{bmatrix}, \begin{bmatrix} 0 & 0 \\ 1 & 0 \end{bmatrix}, \begin{bmatrix} 0 & 0 \\ 1 & 1 \end{bmatrix}, \begin{bmatrix} 0 & 1 \\ 0 & 0 \end{bmatrix}, \begin{bmatrix} 1 & 0 \\ 0 & 0 \end{bmatrix}, \begin{bmatrix} 1 & 1 \\ 0 & 0 \end{bmatrix}, \begin{bmatrix} 1 & 0 \\ 1 & 0 \end{bmatrix}, \begin{bmatrix} 0 & 1 \\ 0 & 1 \end{bmatrix}$$

The determinants of all the above matrices are zero.

Solved Examples

Example 1: Show that matrix $\begin{bmatrix} 2 & 6 \\ 1 & 3 \end{bmatrix}$ is a singular matrix.

Solution: The determinant $\begin{vmatrix} 2 & 6 \\ 1 & 3 \end{vmatrix} = 2 \times 3 - 6 \times 1 = 6 - 6 = 0$

Since determinant is zero, the given matrix has no inverse.

Hence it is a singular matrix.

Example 2: If $A = \begin{bmatrix} \text{-}3 & \text{-}5 & 1 \\ 9 & 14 & 1 \\ 7+k & 29 & \text{-}2 \end{bmatrix}$ is singular, then find the value of k

Solution: $|A| = \begin{vmatrix} -3 & -5 & 1 \\ 9 & 14 & 1 \\ 7+k & 29 & -2 \end{vmatrix} = (-3)\begin{vmatrix} 14 & 1 \\ 29 & -2 \end{vmatrix} - (9)\begin{vmatrix} -5 & 1 \\ 29 & -2 \end{vmatrix} + (7+k)\begin{vmatrix} -5 & 1 \\ 14 & 1 \end{vmatrix}$

$= (-3)[(-2\times14)-(29\times1)]-(9)[(-5\times-2)-(29\times1)]+(7+k)[(14\times1)-(-5\times-2)]$

$= (-3)[-28-29]-9[10-29]+(7+k)[-5-14]$

$= (-3)[-57]-9[-19]+(7+k)[-19]$

$= -19[(-3)3-9+(7+k)]$

Taking out -19 as common

$= -19[-18+(7+k)]$

Since matrix A is singular, its determinant will be zero i.e. $|A| = 0$

Hence $-19[-18+(7+k)] = 0$

or $[-18+(7+k)] = 0 \Rightarrow k = 13$

Non-singular matrix is a square matrix whose determinant is not zero i.e. its inverse exists.

Examples of non-singular matrix

$$\begin{bmatrix} 0 & 1 \\ 1 & 0 \end{bmatrix}, \begin{bmatrix} 0 & 1 \\ 1 & 1 \end{bmatrix}, \begin{bmatrix} 1 & 0 \\ 0 & 1 \end{bmatrix}, \begin{bmatrix} 1 & 0 \\ 1 & 1 \end{bmatrix}, \begin{bmatrix} 1 & 1 \\ 0 & 1 \end{bmatrix}, \begin{bmatrix} 1 & 1 \\ 1 & 0 \end{bmatrix}$$

The determinants of these matrices are non-zero.

Solved Examples

1. Show that $A = \begin{bmatrix} 2 & 6 \\ 1 & 3 \end{bmatrix}$ is a non-singular matrix

Solution: determinant $|\ \ | = 6\times3-5\times2 = 8 \neq 0$

Hence matrix A is a non-singular matrix.

2. Show Matrix $A = \begin{bmatrix} 1 & 0 \\ 1 & 1 \end{bmatrix}$ is a non sigular matrix

Solution : Determinant $|A| = \begin{vmatrix} 1 & 0 \\ 1 & 1 \end{vmatrix} = |1\times1-0\times1| = 0$

Hence matrix A is a non-singular matrix.

Practice Problems (On Singular and Non-singular Matrices)

Identify the singular and non-singular matrices

(i) $\begin{bmatrix} 4 & 5 \\ 2 & 3 \end{bmatrix}$, (ii) $\begin{bmatrix} 6 & 3 \\ 2 & 1 \end{bmatrix}$, (iii) $\begin{bmatrix} 1 & 2 \\ 4 & 8 \end{bmatrix}$, (iv) $\begin{bmatrix} 9 & 6 \\ 6 & 4 \end{bmatrix}$, (v) $\begin{bmatrix} 1 & 2 \\ 2 & 4 \end{bmatrix}$, (vi) $\begin{bmatrix} 2 & 3 \\ 4 & 2 \end{bmatrix}$, (vii) $\begin{bmatrix} -2 & 6 \\ -1 & 3 \end{bmatrix}$

Answers: Non- singular matrices: (i) and (vi). Rest are singular matrices.

1.3 APPLICATIONS OF MATRICES AND DETERMINANTS

1.3.1 Cramer' Rules: Solution of System of Linear Equations

Consider the system of linear equations:

$$a_1 x + b_1 y = c_1$$
$$a_2 x + b_2 y = c_2$$

The coefficient matrix is $A = \begin{bmatrix} a_1 & b_1 \\ a_2 & b_2 \end{bmatrix}$

The determinant of coefficient matrix $D = \begin{vmatrix} a_1 & b_1 \\ a_2 & b_2 \end{vmatrix}$

The determinant D_x is obtained by replacing the first column of determinant D by right most column of the given equation i.e. $\begin{vmatrix} c_1 \\ c_2 \end{vmatrix}$. Thus $D_x = \begin{vmatrix} c_1 & b_1 \\ c_2 & b_2 \end{vmatrix}$

Similarly the determinant D_y is obtained by replacing the second column of determinant D by right most column of the given equation i.e. $\begin{vmatrix} c_1 \\ c_2 \end{vmatrix}$. Thus $D_y = \begin{vmatrix} a_1 & c_1 \\ a_2 & c_2 \end{vmatrix}$

$$x = \frac{D_x}{D}$$
$$y = \frac{D_y}{D}$$

Solved Examples

1. Solve the following linear equations by Cramer' rule.

$$2x - y = 17$$

$$3x + 5y = 6$$

Solution: First calculate the value of D , the determinant of coefficient matrix.

$$D = \begin{vmatrix} 2 & -1 \\ 3 & 5 \end{vmatrix} = 2 \times 5 - (-1 \times 3) = 10 + 3 = 13$$

Replace first column of D by column of constant terms $\begin{vmatrix} 17 \\ 6 \end{vmatrix}$ to get the determinant D_x.

$$D_x = \begin{vmatrix} 17 & -1 \\ 6 & 5 \end{vmatrix} = 17 \times 5 - (-1 \times 6) = 85 + 6 = 91$$

Replace second column of D by column of constant terms $\begin{vmatrix} c_1 \\ c_2 \end{vmatrix}$ to get the determinant D_y.

$$D_y = \begin{vmatrix} 2 & 17 \\ 3 & 6 \end{vmatrix} = 2 \times 6 - (17 \times 3) = 12 - 51 = -39$$

$$x = \frac{D_1}{D} = \frac{91}{13} = 7$$

$$y = \frac{D2}{D} = \frac{-39}{13} = -3$$

2. Apply Cramer's rule to solve the following linear equations:

$$2x + 3y = 3$$
$$x + 2y = 1$$

Solution: The coefficient matrix is $A = \begin{bmatrix} 2 & 3 \\ 1 & 2 \end{bmatrix}$ and matrix of constant terms is $b = \begin{bmatrix} 3 \\ 1 \end{bmatrix}$

Determinant of coefficient matrix $D = \begin{vmatrix} 2 & 3 \\ 1 & 2 \end{vmatrix} = 2 \times 2 - 3 \times 1 = 4 - 3 = 1$

In order to get D_x, we replace the first column of D with the column of b :

$$D_x = \begin{vmatrix} 3 & 3 \\ 1 & 2 \end{vmatrix} = 3 \times 2 - (1 \times 3) = 6 - 3 = 3$$

And, in order to get D_2, we replace the second column of D with the values of b:

$$D_y = \begin{vmatrix} 2 & 3 \\ 1 & 1 \end{vmatrix} = 2 \times 1 - (1 \times 3) = 2 - 3 = -1$$

Applying Cramer's rule we have solutions:

$$x = \frac{D_x}{D} = \frac{3}{1} = 3$$
$$y = \frac{D_y}{D} = \frac{-1}{1} = -1$$

3. Apply Cramer's rule to solve the following linear equations:

$$2x + y = 3$$
$$4x + 2y = 5$$

The Determinant of Coefficient Matrix is $D = \begin{vmatrix} 2 & 1 \\ 4 & 2 \end{vmatrix} = 2 \times 2 - 4 \times 1 = 0$

Since $D = 0$, the system of equations has no solution. It is inconsistent.

4. Solve the system of equations using Cramer's Rule:

$$x + 2y - z = 0$$
$$2x + 2y - 2z = 2$$
$$3x \qquad -4z = 2$$

Solution: As usual find the values of determinants D, Dx, Dy and Dz

Determinant of coefficient matrix is $D = \begin{vmatrix} 1 & 2 & -1 \\ 2 & 2 & -2 \\ 3 & 0 & -4 \end{vmatrix}$

Expanding row-wise:

$$D = \begin{vmatrix} 1 & 2 & -1 \\ 2 & 2 & -2 \\ 3 & 0 & -4 \end{vmatrix} = 1(2 \times -4 - 0 \times -2) - 2(2 \times -4 - 0 \times -1) + 3(2 \times -2 - 2 \times -1)$$
$$= 1(-8 - 0) - 2(-8 - 0) + 3(-4 + 2)$$
$$= -8 + 16 - 6 = 2$$

$$D_x = \begin{vmatrix} 0 & 2 & -1 \\ 2 & 2 & -2 \\ 2 & 0 & -4 \end{vmatrix} = 0(2 \times -4 - 0) - 2(2 \times -4 - 0) + 2(2 \times -2 - 2 \times -1)$$
$$= -2(-8 - 0) + 2(-4 + 2)$$
$$= 16 - 4 = 12$$

$$D_y = \begin{vmatrix} 1 & 0 & -1 \\ 2 & 2 & -2 \\ 3 & 2 & -4 \end{vmatrix} = 1(2 \times -4 - 2 \times -2) - 0 - 1(2 \times 2 - 2 \times 3)$$
$$= 1(-8 + 4) - 0 - 1(4 - 6)$$
$$= -4 + 2 = -2$$

$$D_z = \begin{vmatrix} 1 & 2 & 0 \\ 2 & 2 & 2 \\ 3 & 0 & 2 \end{vmatrix} = 1(2 \times 2 - 0) - 2(2 \times 2 - 2 \times 3) - 0$$
$$= 1(4 - 0) - 2(4 - 6)$$
$$= 4 + 4 = 8$$

Now Applying Crammer' Rule -

$$x = \frac{D_x}{D} = \frac{12}{2} = 6$$
$$y = \frac{D_y}{D} = \frac{-2}{2} = -1$$
$$z = \frac{D_z}{D} = \frac{8}{2} = 4$$

Practice Problems (On Cramer's Rule)

Solve the following linear equations by Cramer's Rule:

1. $x - 10y = 4$
    $2x + y = 8$

2. $3x - y = 10$
    $-3x + y = -2$

3. $3x + y + z = 3$
    $2x + 2y + 5z = -1$
    $x - 3y - 4z = 2$

Answers:

1. $x = 4, y = 0$

2. $D = 0$, the system of equations has no solution.

3. $x = 1, y = 1, z = -1$

1.3.2 Equation of a Straight Line Passing Through Two Points

Let two points are (x_1, y_1) and (x_2, y_2). Let (x, y) be a point lying on the line joining the given points. The equation of the straight line can be determined by the following determinant:

$$\begin{vmatrix} x & y & 1 \\ x_1 & y_1 & 1 \\ x_2 & y_2 & 1 \end{vmatrix} = 0$$

Expanding the above determinant column-wise we get:

$$x(y_1 - y_2) - y(x_1 - x_2) + 1(x_1 y_2 - x_2 y_1) = 0$$

Solved Examples

1. Obtain the equation of the straight line passing through the points:

(i) $(2, 3)$ and $(-4, 9)$ (ii) $(a, 0)$ and $(0, b)$

Solution: (i) $x_1 = 2$, $y_1 = -3$ and $x_2 = -4$, $y_2 = 9$

Applying $\begin{vmatrix} x & y & 1 \\ x_1 & y_1 & 1 \\ x_2 & y_2 & 1 \end{vmatrix} = 0$

We get $\begin{vmatrix} x & y & 1 \\ 2 & -3 & 1 \\ -4 & 9 & 1 \end{vmatrix} = 0$

Expanding the above determinant column-wise we get:

$$x(-3-9)-y(2-(-4))+1(2\times9-(-4)(-3))=0$$

$$x(-12)-y(6)+1(6)=0$$

$$-12x-6y+6=0$$

$$or\ 2x+y=6$$

(ii) $x_1=a,\ y_1=0$ and $x_2=0,\ y_2=b$

Applying $\begin{vmatrix} x & y & 1 \\ x_1 & y_1 & 1 \\ x_2 & y_2 & 1 \end{vmatrix}=0$

We get $\begin{vmatrix} x & y & 1 \\ a & 0 & 1 \\ 0 & b & 1 \end{vmatrix}=0$

Expanding the above determinant column-wise we get:

$$x(0-b)-y(a-0)+1(ab-0\times0)=0$$

$$-bx-ay+ab=0$$

$$ay+bx=ab$$

Practice Problems (Equation of straight line)

1. Obtain the equation of the straight line passing through the following points:

(i) $(3,-4)$ and $(-2,6)$ (ii) $(-3,4)$ and $(1,7)$

Answers:

(i) $2x+y=2$ (ii) $4y-3x=25$

1.3.3 The Area of Triangle

Example 1: Using determinant find the area of the triangle with vertices: $(-5,2),(4,3),(5,-2)$

Method I

Solution: Construct the determinant A $= \pm\dfrac{1}{2}\begin{vmatrix} -5 & 2 & 1 & -5 & 2 \\ 4 & 3 & 1 & 4 & 3 \\ 5 & -2 & 1 & 5 & -2 \end{vmatrix}$

In the above determinant the first two columns are repeated as last two columns

Multiply numbers diagonally from first row to third row and keep it in first bracket

Multiply numbers diagonally from third row to first row and keep it in second bracket

Subtract the 2nd bracket from the 1st bracket.

$$= \pm\frac{1}{2}\{(-5.3.1+2.1.5+1.4.-2)-(5.3.1+-2.1.-5+1.4.2)\}$$

$$= \pm\frac{1}{2}\{(-15+10-8)-(15+10+8)\}$$

$$= \pm\frac{1}{2}\{(-13)-(33)\} = 23 \; sq.units$$

Method II

The area of triangle in determinant form is $\Delta = \pm\frac{1}{2}\begin{vmatrix} -5 & 2 & 1 \\ 4 & 3 & 1 \\ 5 & -2 & 1 \end{vmatrix}$

$$= \pm\frac{1}{2}\{-5(3+2)-4(2+2)+5(2-3)\}$$

$$= \pm\frac{1}{2}\{-25-16-5\} \qquad \text{\{Expanding by 1}^{st}\text{ Column\}}$$

$$= \pm\frac{1}{2}\{-46\} = 23 \; sq.units$$

Method III

Points of Triangle are:

$$x_1 = -5, \; y_1 = 2$$

$$x_2 = 4, \; y_2 = 3$$

$$x_1 = 5, \; y_1 = -2$$

Area of the triangle may be determined as shown below:

$$\Delta = \pm\frac{1}{2}\begin{vmatrix} (x_2-x_1) & (y_2-y_1) \\ (x_3-x_1) & (y_3-y_1) \end{vmatrix}$$

$$= \pm\frac{1}{2}\begin{vmatrix} (4+5) & (3-2) \\ (5+5) & (-2-2) \end{vmatrix}$$

$$= \pm\frac{1}{2}\begin{vmatrix} (9) & (1) \\ (10) & (-4) \end{vmatrix} = \pm\frac{1}{2}\{-36-10\} = 23 \text{ sq. units}$$

Example 2. Determine whether the points (4, 2), (3, 3) and (2, 4) are collinear?

Solution: Let us calculate the area of the triangle formed by the given points.

$$\Delta = \pm\frac{1}{2}\begin{vmatrix} (x_2-x_1) & (y_2-y_1) \\ (x_3-x_1) & (y_3-y_1) \end{vmatrix}$$

$$= \pm\frac{1}{2}\begin{vmatrix} (3-4) & (3-2) \\ (2-4) & (4-2) \end{vmatrix} = \pm\frac{1}{2}\begin{vmatrix} -1 & 1 \\ -2 & 2 \end{vmatrix} = \pm\frac{1}{2}[-2+2] = 0$$

Since area of triangle is zero, hence the points are collinear.

Example 3: If the points (2,-3), (k, -1) and (0, 4) are collinear, then find the value of k.

Solution: The given points are collinear if area of triangle formed by them is zero.

The area of triangle in determinant form is $\Delta = \pm\dfrac{1}{2}\begin{vmatrix} 2 & -3 & 1 \\ k & -1 & 1 \\ 0 & 4 & 1 \end{vmatrix} = 0$

$2[-1-4] - k[-3-4] = 0$

$-10 + 7k \Rightarrow k = \dfrac{1}{7}$

Practice Problems (Area of Δ by determinant)

1. Using determinant, find the area of the triangle whose vertices are: (-2, 1), (0, 3) and (7, 12)

2. Using determinant, find the area of the triangle whose vertices are: (a, b), (b, c) and (c, a)

3. Show using determinant that the points $(-1, 8)$, $(1, -2)$ and $(2,1)$ are collinear.(Hint: $\Delta = 0$)

Answers:

1. 2 sq. units

2. $(ab + bc + ca) - (a^2 + b^2 + c^2)$

1.3.4   Balancing of Chemical Equation

We can use matrices to solve and balance chemistry equations.

Example 1. Let us take the following reaction

$$Cr + O_2 \rightarrow Cr_2 O_3$$

For balancing it , let us write it as follows-

$$xCr + yO_2 \rightarrow zCr_2 O_3$$

We have to find the value of x, y, z .Since there are two different elements Cr and O, so we will need two different equations:

$$Cr: \quad 1x \quad + 0.y = 2z$$

(1 × indicates 1 Cr, 0y indicates zero Cr and 2z indicates 2 Cr)

$$O: \quad 0x + 2y = 3z$$

(0 × indicates zero Oxygen; 2y indicates 2 Oxygen and 3z 3 Oxygen)

We have two matrices: $A = \begin{bmatrix} 1 & 0 \\ 0 & 2 \end{bmatrix}$ and $B = \begin{bmatrix} 2 \\ 3 \end{bmatrix}$

To find the value of z, calculate determinant (D) of A:

$$D = \begin{vmatrix} 1 & 0 \\ 0 & 2 \end{vmatrix} = 1 \times 2 - 0 \times 0 = 2$$

To find the value of x, replace column 1 of determinant (D) of A by column of B

$$D = \begin{vmatrix} 2 & 0 \\ 3 & 2 \end{vmatrix} = 2 \times 2 - 0 \times 3 = 4$$

To find the value of y, replace column 2 of determinant (D) of A by column of B

$$D = \begin{vmatrix} 1 & 2 \\ 0 & 3 \end{vmatrix} = 1 \times 3 - 2 \times 0 = 3$$

Thus x = 4, y = 3, z = 2.

Hence the balanced equation is: $4Cr + 3O_2 \rightarrow 2Cr_2O_3$

Example 2: Balance the reaction - $MgO + Fe \rightarrow Fe_2O_3 + Mg$

Let us write it as:

$$a.MgO + b.Fe \rightarrow c.Fe_2O_3 + d.Mg \qquad \qquad ...(1)$$

Writing the equations in terms of respective elements-Mg, Fe, and O

Mg: 1a + 0b – 0c = 1d		(in terms of Mg)
Fe: 0a + 1b – 2c = 0d		(in terms of Fe)
O: 1a + 1b – 3c = 0d		(in terms of O)

$$D = \begin{vmatrix} 1 & 0 & 0 \\ 0 & 1 & -2 \\ 1 & 0 & -3 \end{vmatrix} = \left| 1\{(1 \times -3) - (0 \times -2)\} - 0\{(0 \times -3) - (1 \times -3)\} + 0\{0 - 1 \times 1\} \right| = |-3| = 3$$

This gives the vale of D i.e. D = 3

To get the value of **a** , replace column 1 of D by coefficients of d, we get determinant D_1

$$D_1 = \begin{vmatrix} 1 & 0 & 0 \\ 0 & 1 & -2 \\ 0 & 0 & -3 \end{vmatrix}$$

Expanding the above determinant by first row.

$$D_1 = \begin{vmatrix} 1 & 0 & 0 \\ 0 & 1 & -2 \\ 0 & 0 & -3 \end{vmatrix} = \left| \{1(1 \times -3) - (-2 \times 0)\} \right| = |-3| = 3$$

This gives the vale of a i.e, a = 3

Replacing column 2 of determinant D by coefficients of d, we get determinant D_2

Now expanding the above determinant by second column

$$D_2 = \begin{vmatrix} 1 & 1 & 0 \\ 0 & 0 & -2 \\ 1 & 0 & -3 \end{vmatrix} = \left|1\{(0 \times -3)-(-2 \times 1)\}\right| = |0+2| = 2$$

Hence value of $b = 2$

To get the value of c, replace column 3 of determinant D by coefficients of d, then we get the determinant D_2

$$D_3 = \begin{vmatrix} 1 & 0 & 1 \\ 0 & 1 & 0 \\ 1 & 0 & 0 \end{vmatrix}$$

Now expanding the above determinant by third column

$$D_3 = \begin{vmatrix} 1 & 0 & 1 \\ 0 & 1 & 0 \\ 1 & 0 & 0 \end{vmatrix} = |1(0 \times 0 - 1 \times 1)| = |-1| = 1$$

Thus $c = 1$

Hence substituting the values of a, b, c, and d in the equation (1), the balanced equation is obtained as below-

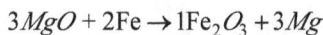

$$3MgO + 2Fe \rightarrow 1Fe_2O_3 + 3Mg$$

Practice Problems (Balancing of Chemical Equation)

1. Balance the chemical equation: $\underline{a}FeCl_2 + \underline{b}Na_3(PO_4) \rightarrow \underline{c}Fe_3(PO_4)_2 + \underline{d}NaCl$

   (**Answer** : $\underline{3}FeCl_2 + \underline{2}Na_3(PO_4) \rightarrow \underline{1}Fe_3(PO_4)_2 + \underline{6}NaCl$)

2. Balance the chemical equation: $[A]Fe_2O_3 + [B]Al \rightarrow [C]Al_2O_3 + [D]Fe$

   (Hint: In this reaction there are 3 elements involved and 4 unknown coefficients. This can be shown as:)

	Fe_2O_3	+	Al	=	Al_2O_3	+	Fe
Fe	2		0		0		1
O	3		0		3		0
Al	0		1		2		0

The linear equations will be :

$$2A + 0B = 0\,C + 1D \qquad \text{(In terms of Fe)}$$
$$3A + 0B = 3C + 0\,D \qquad \text{(In terms of O)}$$
$$0A + 1B = 2C + 0\,D \qquad \text{(In terms of Al)}$$

Answer:

$A = 2, B = 2, C = 1, D = 4$

2. Evaluate the followings:

(i) $\begin{bmatrix} 8 & 4 \\ -1 & -5 \end{bmatrix} - \begin{bmatrix} 3 & -1 \\ -2 & 6 \end{bmatrix}$

(ii) $\begin{bmatrix} 7 & 4 \\ 1 & 1 \end{bmatrix} + \begin{bmatrix} 0 & -2 \\ 3 & 1 \end{bmatrix}$

(iii) $\begin{bmatrix} -1 & 4 \\ 0 & 2 \end{bmatrix} \begin{bmatrix} 1 & 1 & -1 \\ 2 & -1 & 3 \end{bmatrix}$

(iv) Transpose of $A = \begin{bmatrix} 2 & -3 & 4 \\ -1 & 1 & 2 \end{bmatrix}$

(v) Solve: $\begin{array}{l} 3x - 5y = -1 \\ 2x + y = 8 \end{array}$

(vi) Inverse of A, $A = \begin{bmatrix} 10 & -6 \\ 4 & -2 \end{bmatrix}$

(vii) If, $A = \begin{bmatrix} 2 & -4 \\ 1 & 3 \end{bmatrix}$, $B = \begin{bmatrix} 4 & -1 \\ 2 & 0 \end{bmatrix}$, $C = \begin{bmatrix} 4 \\ 3 \end{bmatrix}$, $D = \begin{bmatrix} 3 & 2 \\ 4 & 1 \end{bmatrix}$, $E = \begin{bmatrix} -3 & 2 & 0 \\ 1 & -1 & -2 \end{bmatrix}$

Find (a) $A + B$ (b) $3B$, (c) AC, (e) $7B + D$, (f) $8B - 2A$

Answers:

(i) $\begin{bmatrix} 5 & 5 \\ 1 & -11 \end{bmatrix}$,

(ii) $\begin{bmatrix} 7 & 2 \\ 4 & 2 \end{bmatrix}$,

(iii) $\begin{bmatrix} 7 & 3 & 13 \\ 4 & -2 & 6 \end{bmatrix}$

(iv) $\begin{bmatrix} 2 & -1 \\ -3 & 1 \\ 4 & 2 \end{bmatrix}$, (v) x = 3, y = −2 (vi) $\begin{bmatrix} \frac{1}{2} & \frac{3}{2} \\ -1 & \frac{5}{2} \end{bmatrix}$,

(vii) a. $\begin{bmatrix} 6 & -5 \\ 3 & 3 \end{bmatrix}$, b. $\begin{bmatrix} 12 & -3 \\ 6 & 0 \end{bmatrix}$, c. $\begin{bmatrix} -4 \\ 13 \end{bmatrix}$, e. $\begin{bmatrix} 31 & -5 \\ 18 & 1 \end{bmatrix}$ f. . $\begin{bmatrix} 28 & 0 \\ 14 & -6 \end{bmatrix}$

Objective Questions: (Matrices & Determinants)

1. If the order of matrix A is $m \times n$ and the order of matrix B is $n \times p$, then the order of AB is:

   (a) $m \times p$l

   (b) $p \times n$

   (c) $n \times p$

   (d) $p \times m$

2. If A and B are two matrices of same order, then which of the following is true?

   (a) $A + B = B + A$

   (b) $AB = BA$

   (c) $(A + B)^T = (B + A)^T$

   (d) all are true

3. If What is the value of x if $A = \begin{bmatrix} 1 & 3 \\ 3 & x \end{bmatrix}$

(a) 6   (b) 8

(c) 9   (d) None of these

4. If $A = \begin{bmatrix} 2i & i \\ i & -i \end{bmatrix}$, then value of $|A|$ is:

(a) 2   (b) 3

(c) 4   (d) 5

5. The matrix $A = \begin{bmatrix} a & b & c \\ b & x & y \\ c & y & z \end{bmatrix}$ is a:

(a) Symmetric   (b) Square

(c) Skewed   (d) Singular

6. The matrix $A = \begin{bmatrix} a & 0 & 0 \\ 0 & a & 0 \\ 0 & 0 & a \end{bmatrix}$ is a:

(a) diagonal matrix   (b) scalar matrix

(c) even matrix   (d) identity matrix

7. If matrix $A = \begin{bmatrix} 4 & 3 & 2 \\ 0 & 4 & 2 \\ 0 & 0 & 1 \end{bmatrix}$, the $|A|$ is:

(a) 4   (b) 3

(c) 2   (d) 0

8. If matrices $A = \begin{bmatrix} 2 & 3 \\ 3 & 2 \end{bmatrix}$, $B = \begin{bmatrix} 1 & 4 \\ -2 & 5 \end{bmatrix}$ then A + B is:

(a) $\begin{bmatrix} 3 & 6 \\ 2 & 7 \end{bmatrix}$ (b) $\begin{bmatrix} 3 & 7 \\ 2 & 7 \end{bmatrix}$

(c) $\begin{bmatrix} 3 & 7 \\ 1 & 7 \end{bmatrix}$ (d) $\begin{bmatrix} 3 & 6 \\ 5 & 7 \end{bmatrix}$

9. If $A = \begin{bmatrix} -10 & 5 \\ 4 & -8 \end{bmatrix}$, $B = \begin{bmatrix} 0 & -10 \\ 6 & -9 \end{bmatrix}$ and $c = \begin{bmatrix} 1 & -4 \\ 2 & 7 \end{bmatrix}$, then A + B – C is equal to:

(a) $\begin{bmatrix} -11 & 10 \\ 8 & -24 \end{bmatrix}$ (b) $\begin{bmatrix} -11 & -1 \\ 8 & -24 \end{bmatrix}$

(c) $\begin{bmatrix} -11 & 19 \\ -4 & -6 \end{bmatrix}$ (d) $\begin{bmatrix} -9 & 11 \\ 0 & 8 \end{bmatrix}$

10. If $A = \begin{bmatrix} 12 & 5 \\ 7 & -1 \end{bmatrix}$, $B = \begin{bmatrix} 5 & 7 \\ 6 & -5 \end{bmatrix}$ and $c = \begin{bmatrix} 3 & 5 \\ 2 & 1 \end{bmatrix}$, then A – B – C is equal to:

(a) $\begin{bmatrix} 20 & 17 \\ 15 & 7 \end{bmatrix}$

(b) $\begin{bmatrix} 14 & 7 \\ 11 & 7 \end{bmatrix}$

(c) $\begin{bmatrix} 4 & -7 \\ -1 & 3 \end{bmatrix}$

(d) $\begin{bmatrix} 4 & 7 \\ 1 & 3 \end{bmatrix}$

11. If $A = \begin{bmatrix} 5 & 7 & -7 \end{bmatrix}$ and $B = \begin{bmatrix} 5 \\ -1 \\ 3 \end{bmatrix}$, then product AB is equal to:

(a) $\begin{bmatrix} 25 \\ -7 \\ -21 \end{bmatrix}$

(b) $\begin{bmatrix} -3 \end{bmatrix}$

(c) $\begin{bmatrix} -23 \end{bmatrix}$

(d) $\begin{bmatrix} 25 & -7 & -21 \end{bmatrix}$

12. If $A = \begin{bmatrix} -9 & -8 & 1 \\ -7 & 3 & 9 \end{bmatrix}$ and $B = \begin{bmatrix} 5 \\ -1 \\ 3 \end{bmatrix}$, then the product AB is equal to:

(a) $\begin{bmatrix} -45 & 8 & 3 \\ -35 & -3 & 27 \end{bmatrix}$

(b) $\begin{bmatrix} -34 \\ -14 \end{bmatrix}$

(c) $\begin{bmatrix} -34 & -14 \end{bmatrix}$

(d) $\begin{bmatrix} -45 & 11 \\ -35 & 21 \end{bmatrix}$

13. If $A = \begin{bmatrix} 1 & 6 & 5 \end{bmatrix}$ and $B = \begin{bmatrix} 8 \\ -9 \\ 5 \end{bmatrix}$, then the product AB is equal to:

(a) $\begin{bmatrix} 8 & -54 & 25 \end{bmatrix}$

(b) $\begin{bmatrix} 8 \\ -54 \\ 25 \end{bmatrix}$

(c) $\begin{bmatrix} -21 \end{bmatrix}$

(d) $\begin{bmatrix} 1 & 6 & 8 \\ 8 & -9 & 5 \end{bmatrix}$

14. If $A = \begin{bmatrix} 1 & 1 \\ 0 & 1 \end{bmatrix}$ and $B = \begin{bmatrix} 0 & 1 \\ 1 & 0 \end{bmatrix}$, then AB is equal to:

(a) $\begin{bmatrix} 0 & 0 \\ 0 & 0 \end{bmatrix}$

(b) $\begin{bmatrix} 1 & 1 \\ 1 & 0 \end{bmatrix}$

(c) $\begin{bmatrix} 1 & 0 \\ 0 & 1 \end{bmatrix}$

(d) $\begin{bmatrix} -1 & 0 \\ 0 & 1 \end{bmatrix}$

15. The Value of $\begin{vmatrix} 12 & -3 \\ 10 & 5 \end{vmatrix}$

(a) 105

(b) -85

(c) 24

(d) 14

16. $A = \begin{bmatrix} 6 & -3 & 6 \\ 0 & -3 & 7 \\ 0 & -2 & 5 \end{bmatrix}$, then the value of $|A|$ is:

(a) -6

(b) 6

(c) 7

(d) 5

17. If $A = \begin{bmatrix} 5 & x \\ y & 0 \end{bmatrix}$ and A A , then

(a) $x + y = 5$

(b) $x = 0, y = 5$

(c) $x + y = 5$

(d) $x = y$

18. If $A = \begin{bmatrix} 1 & 2 & 3 \\ -3 & 2 & -1 \\ 2 & -4 & 3 \end{bmatrix}$, then minor M_{11} is:

(a) 2

(b) -3

(c) 4

(d) 3

19. If $A = \begin{bmatrix} 1 & 2 & 3 \\ -3 & 2 & -1 \\ 2 & -4 & 3 \end{bmatrix}$, then cofactor C_{11} is:

(a) 2

(b) -3

(c) 4

(d) 3

20. Whether the system of equations: $2x + y = 3$ and $4x + y = 5$ is

(a) Consistent

(b) Inconsistent

(c) Independent

(d) None of these

21. If $A = \begin{bmatrix} 1 & -2 \\ 5 & 3 \end{bmatrix}$, then $A + A^T$ is equal to:

(a) $\begin{bmatrix} 2 & 3 \\ 3 & 6 \end{bmatrix}$

(b) $\begin{bmatrix} 2 & -4 \\ 10 & 6 \end{bmatrix}$

(c) $\begin{bmatrix} 2 & 4 \\ -10 & 6 \end{bmatrix}$

(d) None of these

22. If A and B are square matrices of same order, then $(AB)^T$ is equal to:

    (a) $A^T B^T$ (b) $B^T A^T$

    (c) AB^T (d) BA^T

Answers

1a	2d	3c	4b	5a	6b	7a
8c	9b	10c	11b	12b	13c	14b
15c	16a	17d	18a	19a	20b	21a
22b						

CHAPTER

8

Permutation and Combination

1.1 FACTORIALS

1.1.1 Brief History

Multiple scientists worked on factorials , but the principal inventors are J. Sterling in 1730 who gives the asymptotic formula after some work in collaboration with De Moivre, then Euler in 1751 and finally *Christian Kramp* who introduced between 1808 and 1816 the actual notation: $n!$ or $\lfloor n$

The factorial is the continuous product of n integer and its subsequent values less than n. It is expressed with a sign of exclamation (!).

$$n! = n(n-1)(n-2)\ldots\ldots\ldots3.2.1$$

$n!$ is read as "n factorial" .

Factorial value of zero is one i.e. $0! = 1$. The factorial values for negative integers are not defined.

Example: $6! = 6 \times 5 \times 4 \times 3 \times 2 \times 1$

Also 6! Can be written as $6! = 6 \times 5!$ or $6! = 6 \times 5 \times 4!$

From $n! = n(n-1)!$

If $n = 1$ then $n! = n(1-1)! = n.0! = n$

1.1.2 Multiplication of Two Factorials

Example 1: $3! \times 4! = (3 \times 2 \times 1) \times (4 \times 3 \times 2 \times 1)$

$$= (6) \times (24)$$

$$= 144$$

Example 2: $5! \times 10 = (5 \times 4 \times 3 \times 2 \times 1) \times 10$

$$= 120 \times 10 = 1200$$

1.1.3 Division of Two Factorials

Example 1: $4! \div 3! = \dfrac{4!}{3!} = \dfrac{4 \times 3 \times 2 \times 1}{3 \times 2 \times 1} = 4$

Also we know $n! = n(n-1)!$

$\therefore 4! = 4(4-1)! = 4(3)!$

$$4! \div 3! = \frac{4!}{3!} = \frac{4 \times 3!}{3!} = 4$$

1.1.4 Addition of Two Factorials

Example 1: $6! + 5! = (6 \times 5 \times 4 \times 3 \times 2 \times 1) + (5 \times 4 \times 3 \times 2 \times 1)$

$$= (720) + (120) = 840$$

1.1.5 Subtraction of Two Factorials

Example 1: $8! - 5! = (8 \times 7 \times 6 \times 5!) - 5!$

$$= (336 \times 5!) - 5!$$

$$= 5!(336 - 1) = 5! \times 335 = 120 \times 335 = 40200$$

Points to remember

$$n! = n(n-1)!$$

$$(n \times m)! \neq n! \times m!$$

$$(n + m)! \neq n! + m!$$

$$(n - m)! \neq n! - m!$$

$$0! = 1$$

0! is an special case of factorial

In $n!$, n must always be positive.

Solved Examples

1. Evaluate: (i) $\dfrac{15!}{14!}$, (ii) $\dfrac{11!-10!}{9!}$ (iii) $\dfrac{4!}{2!2!}$

Solution: (i) $\dfrac{15!}{14!} = \dfrac{15.14!}{14!} = 15$ From $n! = n(n-1)!$

(ii) $\dfrac{11!-10!}{9!} = \dfrac{11.10!-10!}{9!} = \dfrac{10!(11-1)}{9!} = \dfrac{10!.10}{9!} = \dfrac{10.9!.10}{9!} = 100$

(iii) $\dfrac{4!}{2!2!} = \dfrac{4.3.2.1}{2.1.2.1} = 6$

2. Write 5.6.7.8.9.10 into factorial form

Solution: $5.6.7.8.9.10 = \dfrac{1.2.3.4.5.6.7.8.9.10}{1.2.3.4} = \dfrac{10!}{4!}$

3. Fin the value of n if (i) $(n+2)! = 56.n!$ (ii) $(n+1)! = 30(n-1)!$

Solution: (i) $(n+2)! = 56.n!$

$\qquad (n+2)! = 56.n!$

$\qquad (n+2)(n+1)n! = 56.n!$

$\qquad (n+2)(n+1) = 56$

$\qquad n^2 + 3n + 2 = 56$

$\qquad n^2 + 3n - 54 = 0$

$\qquad n^2 + 9n - 6n - 9 \times 6 = 0$

$\qquad n(n+9) - 6(n+9) = 0$

$\qquad (n+9)(n-6) = 0 \Rightarrow n = 6$ or -9

$\quad n = 6$ since n can not be negative

4. Find n if $\dfrac{n!}{3!(n-3)!}$ and $\dfrac{n!}{4!(n-4)!}$ are in the ratio 2 : 1

Solution: $\dfrac{n!}{3!(n-3)!} = \dfrac{n(n-1)(n-2)(n-3)!}{3!(n-3)!} = \dfrac{n(n-1)(n-2)}{3.2.1}$

$\qquad \dfrac{n!}{4!(n-4)!} = \dfrac{n(n-1)(n-2)(n-3)(n-4)!}{4!(n-4)!} = \dfrac{n(n-1)(n-2)(n-3)}{24}$

$\qquad\qquad \dfrac{n(n-1)(n-2)/6}{n(n-1)(n-2)(n-3)/24} = \dfrac{2}{1}$

$\qquad\qquad\qquad \dfrac{24}{6(n-3)} = 2$

$\qquad\qquad 4 = 2(n-3) \Rightarrow (n-3) = 2 \Rightarrow n = 5$

5. Given: $\dfrac{1}{5!}+\dfrac{1}{6!}=\dfrac{x}{7!}$. Find the value of x.

Solution: $\because \dfrac{1}{5!}+\dfrac{1}{6!}=\dfrac{x}{7!}$

$\Rightarrow \quad \Rightarrow \dfrac{1}{5!}+\dfrac{1}{6.5!}=\dfrac{x}{7.6.5!}$

$\Rightarrow 1+\dfrac{1}{6}=\dfrac{x}{7.6}$ Taking $\dfrac{1}{5!}$ common from both sides.

$\Rightarrow \dfrac{7}{6}=\dfrac{x}{7.6}$

$\Rightarrow 7=\dfrac{x}{7}$ Taking $\dfrac{1}{6}$ common from both sides.

$\Rightarrow x=49$

Practice Problems (Factorials)

1. Evaluate:

(*i*) $\dfrac{10!}{8!}$,

(*ii*) $\dfrac{10!}{8!2!}$,

(*iii*) $\dfrac{8!}{5!4!}$

(*iv*) $4!.0!$

(*v*) $\dfrac{4!}{0!}$

(*vi*) $\dfrac{(n+2)!}{n!}$

(*vii*) $\dfrac{(n-1)!}{(n+1)!}$

(*viii*) $\dfrac{5!+4!}{4!}$

(*ix*) $\dfrac{5!}{2!3!}$

(*x*) $\dfrac{x}{(x-1)!}$

2. Show that $(n+2)!-(n+1)!=(n+1)(n+1)!$

3. Show that $(k+3)!(k+4)=(k+4)!$

4. Simplify: $\dfrac{8!(n-1)!}{6!n!}$

5. Show that $\dfrac{1}{3!}=\dfrac{4}{4!}$

6. Show that $\dfrac{(n-k)}{(n-k)!}=\dfrac{1}{(n-k-1)}$

7. Show that $\dfrac{(2n+2)!}{(2n)!} = (2n+2)(2n+1)$

8. If $\dfrac{1}{6!} + \dfrac{1}{7!} = \dfrac{x}{8!}$. Find the value of x.

Answers:

1. (*i*) 90   (*ii*) 45,   (*iii*) 14

(*iv*) 24   (*v*) 24   (*vi*) $(n+1)(n+2)$

(*vii*) $1/n(n+1)$ (*viii*) 6   (*ix*) 10

(*x*) x

4. $56/n$

8. $x = 64$

1.2   Permutations and Combinations

Permutations and combinations, both are different ways of arranging a given set of objects.

Permutations are different arrangements of the objects in any order. If the order is changed a new permutation is obtained.

The rule to determine the number of permutations of n objects was known in Indian culture at least as early as around 1150: it is mentioned in the Lilavati by the Indian mathematician Bhaskara II .

The concepts of permutations and combinations can be traced back to the advent of Jainism in India and perhaps even earlier. The credit, however, goes to the Jains who treated its subject matter as a self-contained topic in mathematics, under the name Vikalpa.

Blaise Pascal and Pierre de Fermat are given credit for first investigating about it.

Permutations are when the order matters, combinations are when it doesn't.

If A-B-C is different from C-B-A, it's a permutation; if A-B-C and C-B-A are the same, it's a combination.

Let there are 3 letters: a, b, c.

These letters can be arranged as: abc, acb, bac, bca, cab, cba.

We say that 3 letters can be permuted in 6 ways.

The permutation can also be expressed in terms of factorials.

$^{n}P_{r} = \dfrac{n!}{(n-r)!}$ i.e. r things taken at a time out of given n things

$P(n,r)$ is another way of writing $^{n}P_{r}$

Now 3 letters taken at a time can be permutated as:

$$^3P_3 = \frac{3!}{(3-3)!} = \frac{3!}{0!} = \frac{3 \times 2 \times 1}{1} = 6$$

Thus there can be 6 ways of writing 3 letters under permutation

In combinations, abc can be written in 1 way only. The 6 ways in permutation are same for combination i.e. **abc** is same as **acb or bac or bca......**

The combination of r objects taken out of n objects can be expressed as:

$$^nC_r = \frac{n!}{(n-r)!\,r!}$$

^nC_r and $C(n,r)$ are same.

The combination of 3 letters taken 3 at a time will be:

$$^3C_3 = \frac{3!}{(3-3)!3!} = 1$$

This shows that there can be only 1 way of writing the three letters taken all together.

The combinations abc, acb, bac, bca, cab, cba, are same but in permutation they are different.

Thus we can say that there can be only 1cambination of 3 letters taken 3 at a time.

But we see that in permutation there can be 6 ways of writing 3 letters taken 3 at a time.

1.2.1 Difference between Permutation and Combinations

1. Permutations are when the order matters, combinations are when it doesn't.

If A-B-C is different from C-B-A, it's a permutation; if A-B-C and C-B-A are the same, it's a combination.

2. Permutation corresponds to arrangements and combination corresponds to selection

Again let us have 4 letters: abcd.

The combination of 3 letters out of 4 can be:

$$^4C_3 = \frac{4!}{(4-3)!3!} = \frac{4 \times 3!}{3!} = 4 \text{ viz. abc, bcd, adb, and acd.}$$

Above in each combination the letters are different.

The permutation of 3 letters out of 4 letters will be:

$$^4P_3 = \frac{4!}{(4-3)!} = \frac{4 \times 3 \times 2 \times 1}{1!} = 24$$

abc can be written as abc, acb, bca, bac, cab, cba = 6 ways

bcd can be written as bcd, bdc, dbc, dcb, cbd, cdb = 6 "

adb can be written as adb, abd, bad, bda, dab, dba = 6 "

acd can be written as acd, adc, cda, cad, dca, dac. = 6 "

1.2.2 Relation between Permutation and Combination

$$^{n}P_{r} = \frac{n!}{(n-r)!}$$

$$^{n}C_{r} = \frac{n!}{(n-r)!r!}$$

$$^{n}C_{r} = \frac{^{n}P_{r}}{r!}$$

1.3 PERMUTATIONS

The study of permutations began in ancient times. The number of ways of arranging an n number of items was known to be $n!$ for at least 2500 years.

The number of possible arrangements of r objects from the set of n objects is known as permutation and it is denoted by $^{n}P_{r}$ or P(n, r). n and r are positive integers such that $1 \le r \le n$

$$^{n}P_{r} = \frac{n!}{(n-r)!}$$

$$^{n}P_{n} = \frac{n!}{(n-n)!} = \frac{n!}{0!} = n! \qquad\qquad \therefore 0! = 1$$

i.e. the number of all permutations of n distinct objects taken all at a time is n!

1.3.1 Permutation when some objects are alike.

If there are n objects of which p_1 are alike of one kind; p_2 are alike of another kind; p_3 are alike of third kind and so on and p_r are alike of rth kind, such that $(p_1 + p_2 + ... p_r) = n$.

Then, number of permutations (arrangements) of these n objects is $\dfrac{n!}{p_1! p_2! .....p_r!}$

Example 1: In how many ways the letters of the word "DOON" can be arranged?

Solution: The letter O repeats two times . Total letters are 4.

Hence the letters of given word can be arranged in $\dfrac{4!}{2!} = \dfrac{4 \times 3 \times 2!}{2!} = 12$ ways

Example 2: In how many ways the letters of the word "Dum dum" can be arranged?

Solution: Total letters are 6. They can be arranged in 6! Ways

The letters D, U, M repeat two times each.

Hence the letters of the given word can be arranged in $\dfrac{6!}{2!2!2!} = \dfrac{6 \times 5 \times 4 \times 3 \times 2 \times 1}{2 \times 1 \times 2 \times 1 \times 2 \times 1} = 30$

1.3.2 Circular Permutations

The number of circular permutation of n distinct objects is given by $(n-1)!$

A circular permutation has no starting point and no ending point. It has a reference but no reference point.

Example 1. In how many ways can 6 people be seated around a circular table?

Solution: The number of ways $= (n-1)! = (6-1)! = 5! = 5 \times 4 \times 3 \times 2 \times 1 = 120$

Example 2: In how many ways 5 people A, B, C, D and E can be seated such that B and C always sit together.

Solution: Since BC always sit together, so they may be treated as one unit. We have to arrange the remaining 4 people.

Hence number of ways $= (n-1)! = (4-1)! = 3! = 6$

Solved Examples

1. Evaluate: (i) $^{10}P_2$ **(ii)** $9P_7$ **(iii)** $^4P_4$ **(iv)** $^7P_0$

Solution: (i) $^{10}P_2 = \dfrac{10!}{(10-2)!} = \dfrac{10.9.8!}{8!} = 90 \quad \therefore {}^nP_r = \dfrac{n!}{(n-r)!}$

(ii) $9P_3 = \dfrac{9!}{(9-3)!} = \dfrac{9!}{6!} = \dfrac{9.8.7.6!}{6!} = 9.8.7 = 504$

(iii) $^4P_4 = \dfrac{4!}{(4-4)!} = \dfrac{4!}{0!} = \dfrac{4.3.2.1}{1} = 24 \quad \because 0! = 1$

(iv) $^7P_0 = \dfrac{7!}{(7-0)!} = \dfrac{7!}{7!} = 1$

2. Express in factorial notations: (i) ^nP_k **(ii)** $^{12}P_8$ **(iii)** $^mP_0$

Solution: (i) $^nP_k = \dfrac{n!}{(n-k)!} \quad \therefore {}^nP_r = \dfrac{n!}{(n-r)!}$

(ii) $^{12}P_8 = \dfrac{12!}{(12-8)!} = \dfrac{12!}{4!}$

(iii) $^{m}P_{0} = \dfrac{m!}{(m-0)!} = \dfrac{m!}{m!}$

3. Given: $2\left(^{n}P_{2}\right) = 60$. **Find the value of n**

Solution: $2\left(^{n}P_{2}\right) = 60$

$$\left(^{n}P_{2}\right) = \frac{60}{2} = 30$$

$$\frac{n!}{(n-2)!} = 30$$

$$\frac{n(n-1)(n-2)!}{(n-2)!} = 30$$

$$n^2 - n - 30 = 0$$

$$n^2 - 6n + 5n - 30 = 0$$

$$n(n-6) + 5(n-6) = 0$$

$$(n-6)(n+5) = 0$$
$$n = 6 \text{ or } -5$$

4. Find m and n if $P(m,n) = \dfrac{9!}{6!}$

Solution: $P(m,n) = \dfrac{9!}{6!}$

$$\frac{m!}{(m-n)!} = \frac{9!}{6!}$$

$\Rightarrow m! = 9! \Rightarrow m = 9$ Comparing numerators of both sides

Also $(m-n)! = 6!$ Comparing denominators of both sides

$\Rightarrow m - n = 6$
i.e. 9-n =6
n = 9- 6 = 3
m = 9 and n = 3

5. Show: $P(n,n) = P(n, n-1)$.

Solution: LHS = $P(n,n) = \dfrac{n!}{(n-n)!} = \dfrac{n!}{0!} = n!$ $\because 0! = 1$

RHS $= P(n, n-1) = \dfrac{n!}{\left[n-(n-1)\right]!} = \dfrac{n!}{1!} = n!$

Hence LHS = RHS

6. If show that $P(n, n) = P(n, n-2)$

Solution: LHS $= P(n, n) = \dfrac{n!}{(n-n)!} = \dfrac{n!}{0!} = n! \quad \because 0! = 1$

RHS $= = 2.P(n, n-2) = \dfrac{2.n!}{\left[n-(n-2)\right]!} = \dfrac{2.n!}{2!} = \dfrac{2.n!}{2} = n!$

Hence LHS = RHS

Practice Problems (On Permutation)

1. Evaluate: (*i*) $^5P_2$ (*ii*) $^{10}P_1$ (*iii*) $^3P_3$

2. $^nP_2 = 72$. **Solve for n**

3. $^nP_3 = 42n$. **Find the value of n.**

4. $P(n, 5) = 12P(n, 4)$. **Find the value of n.**

5. Show that $(n+1)\,^nP_r = {}^{n+1}P_{r+1}$

Answers:

1. (*i*) 20 (*ii*) 10 (*iii*) 6	**2.** 8		**3.** 8
4. 16	**5.**		

1.3.3 Word Problems on Permutation

Solved Examples

1. In how many ways 3 different vases can be arranged?

Solution: Using permutation formula $^nP_r = \dfrac{n!}{(n-r)!}$

$\therefore\ ^3P_3 = \dfrac{3!}{(3-3)!} = \dfrac{3!}{0!} = \dfrac{3!}{1} = 3.2.1 = 6$

Thus 3 vases can be arranged in 3! ways

If there are n vase, the can be arranged in $n!$ ways.

2. How many permutations of the letters o, p, q, r, s and t we can have such that each permutation begin with letters o, p .

Solution: There are 6 letters. OP is in the beginning, so 4 letters are to be permuted.

Hence 4! = 4.3.2.1 = 24 ways

3. In how many different ways can the letters of the word CALCULUS be arranged?

Solution: The number of permutations of n objects of which p are same and q are same

is given by $\dfrac{n!}{p!q!}$

There are 8 letters i.e. n = 8

There are two Cs i.e. p =2

There are two U i.e. q =2

The number of ways $=\dfrac{n!}{p!q!}=\dfrac{8!}{2!2!}=\dfrac{8.7.6.5.4.3.2.1}{2.1\times2.1}=10080$

4. In how many different ways can the letters of the word CLASS be arranged?

Solution: There are 5 letters i.e. n = 5

There are two S i.e. p =2

The number of ways $=\dfrac{n!}{p!}=\dfrac{5!}{2!}=\dfrac{5.4.3.2.1}{2.1}=5.4.3=60$

5. In how many different ways can the letters of the word TIHAR be arranged?

Solution: There are 5 letters. The 5 letters can be permuted in 5! Ways.

The number of ways 5 letters can be written = 5! =5.4.3.2.1=120

6. In how many different ways can the letters of the word APPLE be arranged?

Solution: There are 5 letters.

There are two Ps i.e. p =2

The number of ways $=\dfrac{n!}{p!}=\dfrac{5!}{2!}=\dfrac{5.4.3.2.1}{2.1}=5.4.3=60$

7. In how many ways can 4 girls and 5 boys be arranged in a row so that all the four girls remain together?

Solution: 4 girls remain together, so they are as one unit.

Thus 5 boys and girls form as 6 units. 6 units can be arranged in 6! Ways.

The 4 girls within the unit can be arranged in 4! Ways.

Hence total number of ways = 6!. 4! = 6.5.4.3.2.1 x 4.3.2.1 =720 x 24 = 17280

8. In how many ways can the 5 lions and 4 tigers be arranged in a row such that no two lions are together?

Solution: 5 lions can be arranged in 5 placed in 5! Ways.

And 4 tigers can be arranged each one between two lions in 4! Ways.

Hence total number of ways = 5!. 4! = 5.4.3.2.1 x 4.3.2.1 =120 x 24 = 2880

9. There are 5 books on mathematics, 4 books on science and 3 books on computer. I how many ways the books can be arranged such that books on one subject are together.

Solution: 5 books of mathematics can be arranged in 5! =5.4.3.2.1 = 120 ways.

Similarly 4 books of science can be arranged in 4! = 4.3.2.1 = 24 ways.

And 3 books of computer can be arranged in 3! = 3.2.1 = 6 ways.

Hence total number of ways of arrangement will be = 5! 4! 3! =120 x 24 x 6 =17280

Above the books are in the given order of subject.

But if the order of the subject is not strict then they can be arranged in 3! = 6 ways

Now the total number of ways of arrangement will be = 3! (17280) = 6(17280) = 1,03, 680

10. In how many ways the letters of the word GARHWAL can be arranged.

Solution: There are 7 letters but letter A occurs two times.

Hence the number of ways letter can be arranged is $= \dfrac{7!}{2!} = \dfrac{7.6.5.4.3.2.1}{2.1} = 7.6.5.4.3 = 2520$

11. In how many ways can 5different colored beads be arranged in (i) a line, (ii) circle.

Solution: (i) The number of ways of linear arrangement is $^5P_5 = 5! = 5.4.3.2.1 = 120$

(ii) The number of ways of circular arrangement is $(n-1)! = 4! = 4.3.2.1 = 24$

12. In how many ways can 8 persons be seated in a (i) line (ii) circle.

Solution: (i) the linear permutation is $^8P_8 = 8! = 8.7.6.5.4.3.2.1 = 40320$

(ii) The number of ways of circular arrangement is $(n-1)! = 7! = 7.6.5.4.3.2.1 = 5040$

1.4   COMBINATIONS

Each of the different groups or selections which can be formed by taking some or all of a number of objects is called a *combination*.

Selection of r items taken at a time out of n items is given by: $C_r^n = \dfrac{n!}{(r!)(n-r)!}$

The combination C_r^n can be expressed as $\begin{pmatrix} n \\ r \end{pmatrix}$ or $C(n,r)$ or nC_r

From the formula of combination: $C_r^n = \dfrac{n!}{(r!)(n-r)!}$

(*i*) If r is replaced by n then: $C_n^n = \dfrac{n!}{(n!)(n-n)!} = \dfrac{n!}{(n!)(0)!} = 1 \quad 0! = 1$

(*ii*) If r is replaced by 0 then: $C_0^n = \dfrac{n!}{(0!)(n-0)!} = \dfrac{n!}{(0!)(n)!} = 1$

(*iii*) If r is replaced by 1 then: $C_1^n = \dfrac{n!}{(1!)(n-1)!} = \dfrac{n(n-1)!}{(1!)(n-1)!} = n$

(*iv*) If r is replaced by (n-r) then: $C_{n-r}^n = \dfrac{n!}{(n-r!)(n-n+r)!} = \dfrac{n!}{(n-r)!r!} = C_r^n$

(*v*) If $^nC_x = {^nC_y} \Rightarrow x+y = n$ **i.e.** $C_r^n = C_{n-r}^n$

Example: $C_5^7 = C_{7-5}^7 = C_2^7$ .This simplifies the calculation.

Solved Examples

1. Evaluate : (i) C_7^{10} , (ii) C_4^6 , (iii) C_2^{100} , (iv) C_n^{n+1}

Solution: (*i*) $C_7^{10} = C_{10-7}^{10} = C_3^{10} = \dfrac{10 \times 9 \times 8}{3 \times 2 \times 1} = 120$

(*ii*) $C_4^6 = C_{6-4}^6 = C_2^6 = \dfrac{6 \times 5}{2 \times 1} = 15$

(*iii*) $C_2^{100} = \dfrac{100!}{2!(100-2)} = \dfrac{100 \times 99 \times 98!}{2!98!} = \dfrac{100 \times 99}{2} = 4950$

(*iv*) $C_n^{n+1} = C_{n+1-n}^{n+1} = C_1^{n+1} = (n+1) \quad \because C_1^n = n$

2. Prove that $C_4^8 + C_3^8 = C_4^9$

Solution: LHS $= C_4^8 + C_3^8 = \dfrac{8.7.6.5}{4.3.2.1} + \dfrac{8.7.6}{3.2.1} = 7.2.5 + 8.7 = 70 + 56 = 126$

RHS $= C_4^9 = \dfrac{9.8.7.6}{4.3.2.1} = 9.7.2 = 126$

LHS = RHS

3. If $C_7^n = C_5^n$ **, find the value of n**

Solution: $C_7^n = \dfrac{n(n-1)(n-2)(n-3)(n-4)(n-5)(n-6)}{7.6.5.4.3.2.1}$

$$C_5^n = \dfrac{n(n-1)(n-2)(n-3)(n-4)}{5.4.3.2.1}$$

$\because C_7^n = C_5^n$

$$\dfrac{n(n-1)(n-2)(n-3)(n-4)(n-5)(n-6)}{7.6.5.4.3.2.1} = \dfrac{n(n-1)(n-2)(n-3)(n-4)}{5.4.3.2.1}$$

$\dfrac{n(n-1)(n-2)(n-3)(n-4)(n-5)(n-6)}{n(n-1)(n-2)(n-3)(n-4)} = \dfrac{7.6.5.4.3.2.1}{5.4.3.2.1}$ **(by cross-multiplication)**

$$(n-5)(n-6) = 7.6$$

$$n^2 - 11n + 30 = 42$$

$$n^2 - 11n - 12 = 0$$

$$n^2 - 12n + n - 12 = 0$$

$$(n+1)(n-12) = 0$$

$\Rightarrow n = -1$ or 12

$n = 12$ **(since n cannot be negative)**

4. $^{24}C_x = {}^{24}C_{2x+3}$. **Find the value of x.**

Solution: We know if $^nC_x = {}^nC_y \Rightarrow x + y = n$

$$x + 2x + 3 = 24$$

$$3(x+1) = 24 \Rightarrow x + 1 = 8 \Rightarrow x = 7$$

5. If $^nC_2 = 36,\, {}^nC_3 = 84,\, then\ show\ {}^nC_4 = 126$.

Solution: $^nC_r = \dfrac{n!}{(n-r)!\,r!}$

When r = 2

$$^nC_2 = \dfrac{n!}{(n-2)!\,2!} = 36$$

When r = 3

$$^nC_3 = \frac{n!}{(n-3)!3!} = 84$$

$$\therefore \frac{^nC_3}{^nC_2} = \frac{(n-2)!2!}{(n-3)!3!} = \frac{84}{36}$$

$$\frac{(n-2)(n-3)!2!}{(n-3)!3.2!} = \frac{84}{36}$$

$$\because (n-2)! = (n-2)(n-3)!$$

$$\frac{(n-2)}{3} = \frac{7}{3} \Rightarrow n-2 = 7 \Rightarrow n = 9$$

$$^nC_4 = {}^9C_4 = \frac{9 \times 8 \times 7 \times 6}{4 \times 3 \times 2 \times 1} = 126$$

6. $^{15}C_r : {}^{15}C_{r-1} = 11:5$ **. Find r**

Solution: $^{15}C_r = \frac{15!}{(15-r)!r!}$

$$^{15}C_{r-1} = \frac{15!}{(15-r+1)!(r-1)!} = \frac{15!}{(16-r)!(r-1)!}$$

$$\frac{^{15}C_r}{^{15}C_{r-1}} = \frac{(15-r)!r!}{(16-r)!(r-1)!} = \frac{(15-r)!r.(r-1)!}{(16-r)(15-r)!(r-1)!} = \frac{r}{(16-r)}$$

$$\frac{^{15}C_r}{^{15}C_{r-1}} = 5 \text{ (given)}$$

$$\therefore \frac{r}{(16-r)} = \frac{11}{5}$$

$$5r = 176 - 11r \Rightarrow 16r = 176 \Rightarrow r = 11$$

Practice Problems (On Combinations)

1. Evaluate the followings: (i) $^6C_4$ (ii) $^{10}C_4$ (iii) $^{10}C_{10}$ (iv) $^{12}C_0$

2. If $C_4^n = C_6^n$, find the value of n.

3. If $C_{3r}^{18} = C_{2r-7}^{18}$, find the value of r

4. If $C_8^n = C_6^n$, find the value of $^nC_2$

5. $^{10}C_x = {}^{10}C_{x+4}$. **Find x**

Answers:

1 (*i*) 6 (*ii*) 210 (*iii*) 1 (*iv*) 1   **2.** 10   **3.** 5

4. 91   **5.** 3

1.4.1 Word Problems on Combination

1. In a competitive exam, each question has 5 different options with 2 correct options. How many ways an applicant can select correct options for each question?
Solution: Given total number of options in the question is 5 = n. For a applicant to select a correct option for each question is 2 = r

$$\text{The ways of selection} = {}^5C_2 = \frac{5!}{(5-2)!2!} = \frac{5.4.3.2.1}{3.2.1.2.1} = 10$$

2. In how many ways 4 cards cads can be drawn from the pack of 52 cards?
Solution: Total cards n = 52 and 4 cards have to be drawn, so r = 4

$$\text{The ways of selection of cards may be } {}^{52}C_4 = \frac{52!}{(52-4)!4!} = \frac{52!}{48!4!} = \frac{52.51.50.49.48!}{48!.4!}$$

$$= \frac{52.51.50.49}{4!} = \frac{52.51.50.49}{4.3.2.1} = 13.17.25.49 = \mathbf{270725}$$

3. A college has 10 basket ball players. A 5 member team and a captain have to be selected out of 10 players. How many different selections can be made?
Solution: A team of 6 members have to be selected out of 10 players.

$$\text{This can be done in } {}^{10}C_6 = \frac{10!}{(10-6)!6!} = \frac{10!}{4!6!} = \frac{10.9.8.7.6!}{4.3.2.1.6!} = 10.3.7 = 210 \text{ ways.}$$

Out of 6 players, one captain can be selected in 6 ways

Hence total number of ways of selection = 210 x 6 =1260

4. How many triangles can be formed by joining vertices of a hexagonal?
Solution: There are 6 vertices. n = 6

A Triangle is formed by 3 vertices. Again each vertex is connected with 3 triangles.

$$\text{Hence total number of triangles} = \frac{6!}{3!3!} = \frac{6.5.4.3.2.1}{3.2.1.3.2.1} = 20$$

5. How many 7 -card hands are possible, if all kings must be in hand.

Solution: There are 4 kings. All the kings can be selected in $^4C_4$ ways

Remaining 3 cards can be selected out of 48 cards in $^{48}C_3$

The possible hand of 7-cards can be selected in

$$^4C_4 \times {}^{48}C_3 = 1 \times \frac{48.47.46}{3.2.1} = 8.47.46 = 17296$$

6. In how many ways a boy can picks up 5 fruits from given varieties of fruits – apple, banana, mango, orange, peach and guava?

Solution: there are 6 fruits of different varieties.

5 fruits can be drawn in $^6C_5 = \dfrac{6.5!}{5!} = 6$ ways

7. In how many ways can a student choose 5 courses out of 9 courses if 2 courses are compulsory?

Solution: 2 courses are compulsory. Hence remaining 3 courses are to be selected from remaining (9-2) = 7 courses.

This can be done in $^7C_3 = \dfrac{7!}{(7-3)!3!} = \dfrac{7!}{4!3!} = \dfrac{7.6.5.4!}{4!3!} = \dfrac{7.6.5}{3.2.1} = 35$

Thus the student can choose the 3 courses in 35 ways.

8. There are 15 football players in a college. A team of 11 players are to be selected. In how many ways it can be done?

Solution: 11 players can be selected in $^{15}C_{11} = \dfrac{15!}{(15-11)!11!} = \dfrac{15!}{4!11!}$

$$= \frac{15.14.13.12.11!}{4!11!} = 1365$$

9. In how many ways can a cricket team of 11 players be selected, such that:

There is no condition

A particular player is always included in the team.

A particular players is always dropped.

Solution: (*i*) when there is no condition imposed:

The team can be selected in $^{15}C_{11}$ ways $= \dfrac{15!}{(15-11)!11!} = \dfrac{15!}{4!11!} = 1365$ ways

(*ii*) when a particular player is always selected.

So 10 players has to be selected from 14 players.

This can be done in $^{14}C_{10}$ ways $= \dfrac{14!}{10!} = \dfrac{14!}{4!10!} = 1001$ ways

(*iii*) when a particular player is never included.

When a player is not included, the total players are 14.

Hence 11 players have to be selected fro 14 players.

This can be done in $^{14}C_{11}$ ways $= \dfrac{14!}{(14-11)!11!} = \dfrac{14!}{3!11!} = \dfrac{14.13.12.11!}{3!11!} = 364$ ways.

10. From a group of 7 men and 6 women, 5 persons are to be select to form a committee, with the condition that at least 3 men should be in the committee. In how many ways this can be done?

Solution: At least 3 men means number of men ≥ 3, i.e. 3, 4 or 5

If men are 3, the women will be 2 (because committee consists of 5 persons).

This can be done in $^{7}C_{3} \times {}^{6}C_{2} = \dfrac{7!}{4!3!} \times \dfrac{6!}{4!2!} = \dfrac{7.6.5.4!}{4!3.2.1} \times \dfrac{6.5.}{4!2.1}4! = 35 \times 15 = 525$ ways

If men are 4, the women will be 1 (because committee consists of 5 persons).

This can be done in $^{7}C_{4} \times {}^{6}C_{1} = \dfrac{7!}{3!4!} \times \dfrac{6!}{4!2!} = \dfrac{7.6.5.4!}{3.2.1.4!} \times 6 = 35 \times 6 = 210$ ways

If men are 5, the women will be 0 (because committee consists of 5 persons).

This can be done in $^{7}C_{5} \times {}^{6}C_{0} = \dfrac{7!}{2!5!} \times 1 = \dfrac{7.6.5!}{2.1.5!} \times 1 = 21 \times 1 = 21$ ways

Hence total number of ways $= 525 + 210 + 21 = 756$

1.4.2 Permutation Vs Combination

Permutation	Combination
Number of ways of selecting r items out of n items	
Order is important:	Order is not important.
As the order changes new permutation is obtained	As the order changes combination remain same
Arrangements of n items taking r at a time	Selection of r items out of n items
$P(n,r) = \dfrac{n!}{(n-r)!}$	$C(n,r) = \dfrac{P(n,r)}{r!}$
Clue words: arrangement, schedule	Clue words: selection, formation

Objective Questions: Factorial

1. n! is equal to:

(a) n

(b) $n(n-1)$

(c) $n(n-1)!$

(d) $n(n+1)!$

2. $C(n,r)$ is equal to:

(a) $\dfrac{n!}{r!}$

(b) $\dfrac{n!}{(n-r)!}$

(c) $\dfrac{n!}{r!(n-r)!}$

(d) $n!r!$

3. 0! is equal to:

(a) 0

(b) 1

(c) -1

(d) None of these

4. $3! \times 4!$ is equal to:

(a) 12

(b) 7

(c) 24

(d) 144

5. $\dfrac{6!}{3!}$ is equal to:

(a) 2

(b) 9

(c) 120

(d) 18

6. $4! + 3!$ is equal to:

(a) 7

(b) 12

(c) 24

(d) 30

7. $4! - 3!$ is equal to:

(a) 1

(b) 0

(c) 12

(d) 18

8. $3! \times 0!$ is equal to:

(a) 0

(b) 3

(c) 6

(d) None of these

9. $\dfrac{n!}{(n-1)!}$ is equal to:

(a) n

(b) $(n+1)$

(c) $n!$

(d) $(n-1)!$

10. $\dfrac{(n-k-1)!}{(n-k)!} = \ldots$

    (a) $(n-k)$ (b) $(n-k-1)$

    (c) $(n-k+1)$ (d) $n!$

11. $\dfrac{4!}{0!} = \ldots.$

    (a) ∞ (b) 4

    (c) 24   (d) 12

12. $\dfrac{(n+2)!}{n!} = \ldots$

    (a) $\left(n^2 + 3n\right)$ (b) $\left(n^2 + n + 1\right)$

    (c) $\left(n^2 + 3n + 1\right)$ (d) $(n+2)$

13. If $1! = 1$ and $2! = 2$, then $3! = \ldots$

    (a) 3   (b) 4

    (c) 5   (d) 6

14. $(n \times m)! = \ldots$

    (a) $n! \times m!$ (b) $n(m)!$

    (c) $(n!)m$ (d) None of these.

15. $2! 3! 4! = \ldots$

    (a) 24   (b) 144

    (c) 288   (d) 576

16. $2!(1! + 0!) = \ldots$

    (a) 9   (b) 4

    (c) 6   (d) 0

17. The value of $(-5)! = \ldots$

    (a) 1   (b) 5

    (c) 20   (d) Undefined.

18. Which one is true?

(a) $(n-1)! = n! - 1!$

(b) $n! = n(n-1)$

(c) $n! = n(n-1)!$

(d) None of these

19. $\dfrac{5!}{3!(5-3)!} = ...$

(a) 12

(b) 10

(c) 20

(d) 24

20. $\dfrac{1}{9!} + \dfrac{1}{10!} = \dfrac{n}{11!}$, then value of n is:

(a) 12

(b) 10

(c) 120

(d) 121

21. Value of $\dfrac{5! + 4!}{5! - 4!}$ is:

(a) $\dfrac{3}{2}$

(b) $\dfrac{5}{4}$

(c) $\dfrac{1}{2}$

(d) $\dfrac{4}{3}$

Answers:

1. c	2. c	3. b	4. d	5. c	6. d	7. d
8. c	9. a	10. c	11. c	12. c	13. d	14. d
15. c	16. b	17. d	18. c	19. b	20. b	21. a

Objective Questions: Permutation

1. Value of $^{5}P_{2}$ is:

(a) 5

(b) 10

(c) 20

(d) 15

2. If $^{n}P_{2} = 30$, then value of n is:

(a) 5

(b) 6

(c) 7

(d) 8

3. If $^{n}P_{6} = 3 \, ^{n}P_{5}$, then value of n is:

(a) 5

(b) 6

(c) 7

(d) 8

4. If $P(m,n) = \dfrac{9!}{6!}$, then m, n are:

    (a) 4,5   (b) 8,2

    (c) 9,3   (d) 9,6

5. Value of $\left({}^4P_2\right)^2$ is:

    (a) 12   (b) 144

    (c) 156   (d) 64

6. $\dfrac{60}{{}^5P_3} = \ldots$

    (a) 1   (b) 2

    (c) 3   (d) 4

7. $\dfrac{6!}{{}^6P_2}$ is equal to:

    (a) 12   (b) 24

    (c) 8   (d) 6

8. $\dfrac{P(4,3)}{P(3,2)}$ is equal to:

    (a) 3   (b) 4

    (c) 5   (d) 6

9. $5 \times {}^nP_3 = 4 \times {}^{n+1}P_3$, then value of n is:

    (a) 10   (b) 12

    (c) 14   (d) 16

10. $P(8,8) = \ldots$

    (a) 6!   (b) 7

    (c) 8!   (d) 9!

11. In how many ways 6 books can be arranged in the self? $\left({}^nP_n = n! \ ways\right)$

    (a) 4!   (b) 5!

    (c) 6!   (d) 7!

12. In how many ways, 6 New Year cards can be sent to 4 friends? $\left({}^nP_r \ ways\right)$

    (a) 120   (b) 180

    (c) 240   (d) 360

13. How many arrangements of letters in the word BANK can be made? $(n!$ ways$)$

    (a) 12   (b) 180

    (c) 24   (d) 36

14. How many words can be formed with the letters of the word ADITI? $\left(\dfrac{n!}{r!}\right)$

    (a) 40   (b) 50

    (c) 60   (d) 90

15. How many 4 digits number can be formed using the digits 3, 6, 7 and 8? $\left(^{n}P_{r}\right)$

    (a) 12   (b) 24

    (c) 36   (d) 4

16. How many 3 letters word can be formed with the letters of SHEEL? $\left(^{n}P_{r}\right)$

    (a) 24   (b) 60

    (c) 36   (d) 40

17. How many permutations can be made of the letters RITA? $(n!$ $)$

    (a) 24   (b) 60

    (c) 36   (d) 40

Answers

1. c	2. b	3. d	4. c	5. b	6. a	7. b
8. b	9.	10. c	11. c	12. d	13. c	14. c
15. b	16. b	17.a				

Objective Questions: Combination

1. $^{100}C_{98}$ is equal to:

    (a) 9900   (b) 100

    (c) 98   (d) 9800

2. $^{n}C_{3} = {}^{n}C_{5}$, then n is equal to:

    (a) 3   (b) 4

    (c) 5   (d) 8

3. $^{18}C_{r} = {}^{18}C_{r+2}$, then r is equal to:

    (a) 8   (b) 10

    (c) 12   (d) 6

4. $C(2n, r) = C(2n, r+2)$, then r is equal to:

(a) n

(b) $(n-1)$

(c) $(n+1)$

(d) $(n-2)$

5. $^{n+1}C_3 = 2\,^nC_2$, then n is equal to:

(a) 3

(b) 4

(c) 5

(d) 6

6. $C(19,17) + C(19,18) = \ldots$

(a) 180

(b) 190

(c) 200

(d) 210

7. $C(12,3).C(10.2) = \ldots$

(a) 900

(b) 9000

(c) 9900

(d) 1000

8. $^mC_n = 13$, them value of m and n are:

(a) 13,12

(b) 13,1

(c) 13,3

(d) 13,4

9. The value of $^4C_5$ is:

(a) 5

(b) 4

(c) 0

(d) Not defined

10. $C(a,3)$ is equal to:

(a) $C(a, 3-a)$

(b) $C(a-3, 3)$

(c) $C(a, a-3)$

(d) $C(3, a)$

11. If $^6P_r = 360$, $^6C_r = 15$, then r is:

(a) 4

(b) 5

(c) 6

(d) 8

12. If $^7P_4 + r^7C_4 = 0$, then value of r is:

(a) -24

(b) 24

(c) -27

(d) 27

13. $^9P_2 + ^9C_2 = \ldots$

(a) 108

(b) 36

(c) 118

(d) 72

14. $^nP_r = 6720$ and $^nC_r = 56$, then value of r is:

(a) 3

(b) 4

(c) 5

(d) 6

15. If $^{18}C_r = {}^{18}C_{r+2}$, the r is equal to:

(a) 8

(b) 10

(c) 12

(d) 16

16. If $^nC_4 = {}^nC_6$, the n is equal to:

(a) 6

(b) 8

(c) 10

(d) 12

17. How many diagonals can be drawn in a heptagon? $\left({}^nC_2 - n \right)$

(a) 12

(b) 14

(c) 21

(d) 8

18. There are 15 dots on a circle. How many triangles can be formed?

(a) 45

(b) 135

(c) 455

(d) 555

Answers

1. a	2. d	3. a	4. b	5. c	6. b	7. c
8. a	9. d	10. c	11. a	12. a	13. a	14. c
15. a	16. c	17. b	18. c			

9

Binomial Theorem

1.0 BRIEF HISTORY

A binomial is a polynomial with two terms. The most general case of the binomial theorem is the binomial series identity. In elementary algebra, the binomial theorem describes the algebraic expansion of powers of a binomial.

A more general binomial theorem and the so-called ``**Pascal's triangle**'' were known in the 10th-century AD to Indian mathematician Halayudha and Persian mathematician Al-Karaji, in the 11th century to Persian poet and mathematician Omar Khayyam, and in the 13th century to Chinese mathematician Yang Hui, who all derived similar results.

Sir Isaac Newton is generally credited with the generalised binomial theorem, valid for any exponent.

There are several closely related results that are alternatively known as the binomial theorem: as the binomial formula, binomial expansion, and binomial identity, or sometimes simply called the "binomial series" rather than "binomial theorem."

1.1 BINOMIAL THEOREM FOR POSITIVE INTEGRAL INDEX

If x, y are two real numbers, then their expansion for different powers is given by:

$$(x+y)^0 = 1$$

$$(x+y)^1 = x+y$$

$$(x+y)^2 = x^2 + 2xy + y^2$$

$$(x+y)^3 = x^3 + 3x^2y + 3xy^2 + y^3$$

$(x+y)^4 = x^4 + 4x^3y + 6x^2y^2 + 4xy^3 + y^4$ and so on.

It may be noted that if the power is n, then *number of terms in the expansion is $(n+1)$*

We notice that the coefficients of (the numbers before) x and y are:

Binomial Expansion	Exponent	Coefficients in the expansion
$(x+y)^0$	0	1
$(x+y)^1$	1	1 1
$(x+y)^2$	2	1 2 1
$(x+y)^3$	3	1 3 3 1
$(x+y)^4$	4	1 4 6 4 1

The coefficients form a triangle called as **Pascal's triangle,** which is an easy way to find the coefficients of a binomial expansion.

In the form of combinations, the coefficients may be written as

$$^0c_0$$

$$^1c_1 \; ^1c_1$$

$$^2c_0 \; ^2c_1 \; ^2c_2$$

$$^3c_0 \; ^3c_1 \; ^3c_2 \; ^3c_3$$

Points to note:

There are $n+1$ terms in the expansion of $(x+y)^n$

The degree of each term is n

The powers on x begin with n and decrease to 0

The powers on y begin with 0 and increase to n

The coefficients are symmetric

The generalized form of Binomial Expansion (i.e. power = n):

The Binomial Expansion or Theorem can be written in summation notation, which is very compact and manageable.

$$(x+y)^n = \sum_{k=0}^{n} \binom{n}{k} x^{n-k} y^k$$

or

$$(x+y)^n = \sum_{k=0}^{n} C_k^n x^{n-k} y^k$$

or

$$(x+y)^n = C_0^n x^n + C_1^n x^{n-1} y + C_2^n x^{n-2} y^2 + \ldots + C_r^n x^{n-r} y^r + \ldots + C_n^n y^n$$

Solved Examples

1. Write the expansion of $(x+6)^6$

Solution: $(x+6)^6 = C_0^6 x^6 + C_1^6 x^5 y + C_2^6 x^4 y^2 + C_3^6 x^3 y^3 + C_4^6 x^2 y^4 + C_5^6 xy^5 + C_6^6 y^6$

$$= 1x^6 + 6x^5 y + 15x^4 y^2 + 20x^3 y^3 + 15x^2 y^4 + 6xy^5 + y^6$$

2. Write the expansion of $(x-5)^4$

Solution: $(x-5)^4 = C_0^4 x^4 + C_1^4 x^3 y + C_2^4 x^2 y^2 + C_3^4 xy^3 + C_4^4 y^4$

$$= x^4 + 4x^3(-5) + 6x^2(-5)^2 + 4x(-5)^3 + (-5)^4$$

$$= x^4 - 20x^3 + 150x^2 - 500x + 625$$

3. Write the expansion of $(2x+3y)^5$

Solution: $(2x+3y)^5 = C_0^5(2x)^5 + C_1^5(2x)^4(3y) + C_2^5(2x)^3(3y)^2$

$$+C_3^5(2x)^2(3y)^3 + C_4^5(2x)(3y)^4 + C_5^5(3y)^5$$

$$= (2x)^5 + 5(2x)^4(3y) + 10(2x)^3(3y)^2 + 10(2x)^2(3y)^3 + 5(2x)(3y)^4 + (3y)^5$$

$$= 32x^5 + 5.16.3x^4 y + 10.8.9x^3 y^2 + 10.4.27x^2 y^3 + 5.2.81xy^4 + 243 y^5$$

$$= 32x^5 + 240x^4 y + 720x^3 y^2 + 1080x^2 y^3 + 810xy^4 + 243 y^5$$

1.1.2 General term of a Binomial Expansion

It is given by- $T_{r+1} = C_r^n x^{n-r} y^r$

Binomial Expansion	General Term i.e. (r + 1)th term
$(x+y)^n$	$^n c_r x^{n-r} y^r$
$(x-y)^n$	$(-1)^r \, ^n c_r x^{n-r} y^r$
$(1+x)^n$	$^n c_r x^r$
$(1-x)^n$	$(-1)^r \, ^n c_r x^r$

Solved Examples

1. Find the 4th term in the expansion of $(3x-2)^4$

Solution: $T_{r+1} = C_r^n x^{n-r} y^r$

$$r+1 = 4 \Rightarrow r = 3$$

$$\therefore T_{3+1} = C_3^4 (3x)^{4-3} (-2)^3 = 4.3x(-8) = -96x$$

2. Find the 3rd term in the expansion $\left(\dfrac{4x}{5} + \dfrac{5}{2x}\right)^5$

Solution: n = 5 and $r+1 = 3 \Rightarrow r = 2$

$$T_3 = C_2^5 \left(\frac{4x}{5}\right)^{5-2} \left(\frac{5}{2x}\right)^2 = \frac{5 \times 4}{2 \times 1} \left(\frac{4}{5}\right)^3 (x)^3 \left(\frac{5}{2}\right)^2 \left(\frac{1}{x}\right)^2$$

$$= 10. \left(\frac{4}{5}\right)^3 \left(\frac{5}{2}\right)^2 (x)^3 \left(\frac{1}{x}\right)^2$$

$$= 10. \frac{4^2}{5} x = 32x$$

1.1.3   rth term from the end of Binomial Expansion of $(x+y)^n$

The number of terms in the expansion of $(x+y)^n$ is $(n+1)$

The rth term from the end = $\{n+1-(r-1)\} = \{n-r+2\}$ th term from the beginning

Example: Find the 3 rd term from the end of the expansion of $(x+y)^6$.

Solution: 3rd term from the end = $\{6-3+2\} = 5^{th}$ term from the beginning.

$$\therefore T_5 = T_{4+1} = {}^6 c_4 x^{6-4} y^4 = {}^6 c_2 x^2 y^4 \quad \because {}^n c_r = {}^n c_{n-r}$$

$$= \frac{6 \times 5}{1 \times 2} x^2 y^4 \quad 15 x^2 y^4$$

1.1.4   Middle Term in Binomial Expansion

Let us find the middle term of the binomial expansion of $(x+y)^n$

Now two cases arises regarding the index (power) *n*

Case 1. If n is odd , then there shall be two middle terms.

One is $T_{(n+1)/2}$ and the other is $T_{(n+3)/2}$

Example: Find the middle term in the expansion $(3x-2)^5$

Solution: $n = 5$ i.e. an odd number. Hence there will be two middle terms.

(i) $T_{(n+1)/2} = T_{(5+1)/2} = T_3 = \mathbf{3^{rd}}$ **term**

Applying $T_{r+1} = {}^n c_r x^{n-r} y^r$

$$\therefore T_3 = T_{2+1} = {}^5 c_2 (3x)^{5-2} (-2)^2 = \frac{5 \times 4}{1 \times 2} \times 4 \times 27 x^3 = 1080 x^3$$

(ii) $T_{(n+3)/2} = T_{(5+3)/2} = T_4 = \mathbf{4^{th}}$ **term**

$$\therefore T_4 = T_{3+1} = {}^5 c_3 (3x)^{5-3} (-2)^3 = -\frac{5 \times 4 \times 3}{1 \times 2 \times 3} \times 8 \times 9 x^2 = -720 x^2$$

Case I1. If n is eve, then there shall be only one middle term viz. $_{/2}$ **th term**

Example: Find the middle term in the expansion $(x-2)^6$

Solution: n = 6 i.e. an even number. Hence there will be one middle term

The middle term is $T_{n/2} = T_{6/2} = T_3$. i.e. 3rd term.

Applying $T_{r+1} = {}^n c_r x^{n-r} y^r$

$$\therefore T_3 = T_{2+1} = {}^6 c_2 (x)^{6-2} (-2)^2 = \frac{6 \times 5}{1 \times 2} \times 4 \times x^4 = 60 x^4$$

1.1.5   Term Independent of x

1. Find the term independent of x in the expansion of $\left(x + \dfrac{1}{x}\right)^{10}$

Solution: $T_{r+1} = C_r^n x^{n-r} y^r$

$$T_{r+1} = C_r^{10} x^{10-r} \left(\frac{1}{x}\right)^r = C_r^{10} x^{10-2r}$$

For the term to be independent of x, put $10 - 2r = 0 \Rightarrow r = 5$

Now $n = 10$ and $r = 5$

$$T_{5+1} = C_5^{10} = \frac{10.9.8.7.6.5!}{5!} = 10.9.8.7.6 = 30240 .$$

This the 6th term independent of x.

2. Find the term independent of x in the expansion of $\left(x-\dfrac{1}{x}\right)^{12}$

Solution: $\because T_{r+1} = C_r^n x^{n-r} y^r$

$$\because T_{r+1} = C_r^{12} x^{12-r} \left(\frac{1}{x}\right)^r = C_r^{12} x^{12-2r}$$

For the term to be independent of x, put $12 - 2r = 0 \Rightarrow r = 6$

$$T_{6+1} = C_6^{12} = \frac{12.11.10.9.8.7}{6.5.4.3.2.1} = 11.6.2 = 132$$

1.1.6 Greatest Binomial Coefficient

1. Find the greatest binomial coefficient in the binomial expansion of $(2x+3y)^{10}$

Solution: In the given expansion the index is 10 - an even number.

The greatest binomial coefficient will be: $\,{}^nC_{\left(\frac{n}{2}+1\right)} = {}^{12}C_{\left(\frac{12}{2}+1\right)} = {}^{12}C_7$.

2. Find the greatest binomial coefficient in the binomial expansion of $(2x-4y)^9$

Solution: In this problem the index is 9, which is odd. There will be two greatest binomial coefficients.

$${}^nC_{\frac{n+1}{2}} \text{ and } \underline{\quad}$$

$$\therefore {}^nC_{\frac{n+1}{2}} = {}^9C_{\frac{9+1}{2}} = {}^9C_5 \text{ i.e. the 5}^{\text{th}} \text{ term}$$

$${}^nC_{\frac{n+3}{2}} = {}^9C_{\frac{9+3}{2}} = {}^9C_6 \text{ i.e. the 6}^{\text{th}} \text{ term}$$

Practice Problems

1. Write the expansions of (i) $\left(3+2x^2\right)^4$, and (ii) $\left(2x+\dfrac{y}{2}\right)^5$

2. Find out the 6th term in (i) $\left(2x-\dfrac{1}{x^2}\right)^7$, and (ii) 5th tern in $(2x+3y)^8$

3. Find the middle terms of (i) $\left(2x-\dfrac{1}{y}\right)^8$, and (ii) $\left(x^4-\dfrac{1}{x^3}\right)^{11}$

4. Find the terms independent of x in (i) $\left(\dfrac{3x^2}{2}-\dfrac{1}{3x}\right)^9$, (ii) $\left(2x^2-\dfrac{1}{x}\right)^{12}$

Answers:

1. **(i)** $81 + 216x^2 + 216x^3 + 96x^6 + 16x^8$,

   (ii) $32x^5 + 40x^4 y + 20x^3 y^2 + 5x^2 y^5 + \dfrac{5}{8} xy^4 + \dfrac{1}{32} y^5$

2. **(i)** $\left(\dfrac{-84}{x^8}\right)$, **(ii)** $90720 x^4 y^4$

3. **(i)** $\left(\dfrac{1120 x^4}{y^4}\right)$, **(ii)** $-462 x^9, 462 x^2$

4. **(i)** $T_7 = \dfrac{7}{18}$, **(ii)** $T_9 = 7920$

1.1.7   Binomial Expansion For any Index

$$(1+x)^n = 1 + nx + \frac{n(n-1)}{2!} x^2 + \frac{n(n-1)(n-2)}{3!} x^3 + ... + \frac{n(n-1)...(n-r+1)}{r!} x^r + ...\infty$$

The above expansion is valid only when $-1 < x < 1$

If the first term is not 1, then make the first term 1 in the following way:

$$(a+x)^n = \left\{ a\left(1+\frac{x}{a}\right)\right\}^n = a^n \left(1+\frac{x}{a}\right)^n \ \text{if} \ \frac{x}{a} < 1 \ \text{or} \ x < a$$

or $\quad (a+x)^n = \left\{ x\left(1+\frac{a}{x}\right)\right\}^n = x^n \left(1+\frac{a}{x}\right)^n \ \text{if} \ \frac{a}{x} < 1 \ \text{or} \ a < x$

1. Replacing n by $-n$ in the following expansion:

$$(1+x)^n = 1 + nx + \frac{n(n-1)}{2!} x^2 + \frac{n(n-1)(n-2)}{3!} x^3 + ... + \frac{n(n-1)...(n-r+1)}{r!} x^r + ...\infty$$

We get : $\quad (1+x)^{-n} = 1 - nx + \dfrac{n(n+1)}{2!} x^2 - \dfrac{n(n+1)(n+2)}{3!} x^3 + ...$

$$... + (-1)^r \frac{n(n+1)...(n+r-1)}{r!} x^r + ...\infty$$

The general term in the above expansion is:

$$T_{r+1} = \frac{n(n+1)(n+2)...(n+r-1)}{r!} x^r$$

2. Replacing x by $-x$ in the following expansion

$$(1+x)^{-n} = 1 - nx + \frac{n(n+1)}{2!}x^2 - \frac{n(n+1)(n+2)}{3!}x^3 + ...$$

$$.... + (-1)^r \frac{n(n+1)...(n+r-1)}{r!}x^r + ...\infty$$

We get : $\quad (1-x)^{-n} = 1 + nx + \frac{n(n+1)}{2!}x^2 + \frac{n(n+1)(n+2)}{3!}x^3 + ...$

$$... + (-1)^r \frac{n(n+1)...(n-r+1)}{r!}x^r + ...\infty$$

The general term in the expansion is:

$$T_{r+1} = (-1)^r \frac{n(n+1)...(n+r-1)}{r!}x^r$$

3. Replacing x by $-x$ in the following expansion

$$(1-x)^n = 1 - nx + \frac{n(n-1)}{2!}x^2 - \frac{n(n-1)(n-2)}{3!}x^3 + ...$$

$$+ (-1)^r \frac{n(n-1)(n-2)...(n-r+1)}{r!}x^r + ...\infty$$

The general term in the above expansion is:

$$T_{r+1} = (-1)^r \frac{n(n-1)...(n-r+1)}{r!}x^r$$

Putting $n = 1$ .....in the expansion of $(1+x)^{-n}$

$$(1+x)^{-n} = 1 - nx + \frac{n(n+1)}{2!}x^2 - \frac{n(n+1)(n+2)}{3!}x^3 + ...$$

$$.... + (-1)^r \frac{n(n+1)...(n+r-1)}{r!}x^r + ...\infty$$

We get: $(1+x)^{-1} = 1 - x + \frac{1.2}{2!}x^2 - \frac{1.2.3}{3!}x^3 + ... + (-1)^r \frac{1.2...r}{r!}x^r + ...\infty$

$$= 1 - x + x^2 - x^3 + ... + (-1)^r x^r + ...$$

Putting $n = 2$ in the expansion of $(1+x)^{-n}$

$$(1+x)^{-n} = 1 - nx + \frac{n(n+1)}{2!}x^2 - \frac{n(n+1)(n+2)}{3!}x^3 + ...$$

$$.... + (-1)^r \frac{n(n+1)...(n+r-1)}{r!}x^r + ...\infty$$

$$(1+x)^{-2} = 1-2x+\frac{1.2.3}{2!}x^2 - \frac{1.2.3.4}{3!}x^3 +........+(-1)^r \frac{1.2...r(r+1)}{r!}x^r +...\infty.$$

$$= 1-2x+3x^2 - 4x^3 +.......+(-1)^r(r+1)x^r +...\infty$$

Putting $n = 3$ in the expansion of $(1+x)^{-n}$ i.e.

$$(1+x)^{-n} = 1-nx+\frac{n(n+1)}{2!}x^2 - \frac{n(n+1)(n+2)}{3!}x^3 +....$$

$$....+(-1)^r \frac{n(n+1)...(n+r-1)}{r!}x^r +...\infty$$

$$(1+x)^{-3} = 1-3x+\frac{3.4}{2!}x^2 - \frac{3.4.5}{3!}x^3 +........+(-1)^r \frac{3.4...(r+2)}{r!}x^r +...\infty$$

$$= 1-3x+6x^2 - 10x^3 +.......+(-1)^r \frac{(r+1)(r+2)}{2}x^r +...\infty$$

Replacing x by $-x$ in the expansions obtained above:

1. $(1+x)^{-1} = 1-x+x^2 -x^3 +........+(-1)^r x^r +...$

We get: $(1-x)^{-1} = 1+x+x^2 +x^3 +........+x^r +...$

2. $(1+x)^{-2} = 1-2x+3x^2 - 4x^3 +.......+(-1)^r(r+1)x^r +...\infty$

We get: $(1-x)^{-2} = 1+2x+3x^2 +4x^3 +........+(r+1)x^r +...$

3. $(1+x)^{-3} = 1-3x+6x^2 - 10x^3 +........+(-1)^r \frac{(r+1)(r+2)}{2}x^r +...\infty$

We get: $(1-x)^{-3} = 1+3x+6x^2 - 10x^3 +........+\frac{(r+1)(r+2)}{2}x^r +...\infty$

In all the above expansions the condition is $|x| < 1$

It may be noted that when the index n is a positive integer, number terms is finite *i.e.* (n +1)

When the index n is other than positive integer, number of terms is infinite.

Solved Examples

1. Expand the following Binomial expressions up to four terms:

(i) $(1-x)^{-5}$ **(ii)** $\left(1-\dfrac{x}{2}\right)^{-5}$ **(iii)** $\left(1+\dfrac{1}{x}\right)^{-6}$

(iv) $\dfrac{1}{(1+2x)^3}$ **(v)** $(3+x)^{-\frac{1}{2}}$

Solutions: (i) $(1-x)^{-5}$

$$\because (1+x)^n = 1 + nx + \frac{n(n-1)}{2!}x^2 + \frac{n(n-1)(n-2)}{3!}x^3 + ...$$

Replacing x by −x and n by − 5

$$\therefore (1-x)^{-5} = 1 + (-5)(-x) + \frac{(-5)(-5-1)}{2!}(-x)^2 + \frac{-5(-5-1)(-5-2)}{3!}(-x)^3 + ...$$

$$= 1 + 5x + 15x^2 + 35x^3 + ...$$

(ii) $\left(1-\dfrac{x}{2}\right)^{-5}$

$$\because (1+x)^n = 1 + nx + \frac{n(n-1)}{2!}x^2 + \frac{n(n-1)(n-2)}{3!}x^3 + ...$$

Replacing x by $-\dfrac{x}{2}$ and n by − 5

$$\therefore \left(1-\frac{x}{2}\right)^{-5} = 1 + (-5)\left(-\frac{x}{2}\right) + \frac{-5(-5-1)}{2!}\left(-\frac{x}{2}\right)^2 + \frac{-5(-5-1)(-5-2)}{3!}\left(-\frac{x}{2}\right)^3 + ...$$

$$= 1 + 5.\frac{x}{2} + \frac{5.6}{2!}.\frac{x^2}{4} + \frac{5.6.7}{3!}\frac{x^3}{8} + ...$$

$$= 1 + 5.\frac{x}{2} + 15.\frac{x^2}{4} + 35.\frac{x^3}{8} + ...$$

(iii) $\left(1+\dfrac{1}{x}\right)^{-6}$

$$\because (1+x)^n = 1 + nx + \frac{n(n-1)}{2!}x^2 + \frac{n(n-1)(n-2)}{3!}x^3 + ...$$

Replacing x by $\dfrac{1}{x}$ and n by − 6

$$\because \left(1+\frac{1}{x}\right)^{-6} = 1+(-6)\left(\frac{1}{x}\right)+\frac{-6(-6-1)}{2!}\left(\frac{1}{x}\right)^2 +\frac{-6(-6-1)(-6-2)}{3!}\left(\frac{1}{x}\right)^3 +....$$

$$= 1-\frac{6}{x}+\frac{21}{x^2}-\frac{56}{x^3}+....$$

(iv) $\dfrac{1}{(1+2x)^3}=(1+2x)^{-3}$

$$\because (1+x)^n = 1+nx+\frac{n(n-1)}{2!}x^2 +\frac{n(n-1)(n-2)}{3!}x^3 +....$$

Replacing x by $2x$ and n by -3, we get:

$$(1+2x)^{-3} = 1+(-3)(2x)+\frac{-3(-3-1)}{2!}(2x)^2 +\frac{-3(-3-1)(-3-2)}{3!}(2x)^3 +....$$

$$= 1-6x+6.4.x^2 -10.8x^3 +....$$

$$= 1-6x+24x^2 -80x^3 +....$$

(v) $(3+x)^{-\frac{1}{2}}=3^{-\frac{1}{2}}\left(1+\dfrac{x}{3}\right)^{-\frac{1}{2}}$

$$\because (1+x)^n = 1+nx+\frac{n(n-1)}{2!}x^2 +\frac{n(n-1)(n-2)}{3!}x^3 +....$$

Replacing x by $\dfrac{x}{3}$ and n by $-\dfrac{1}{2}$, we get:

$$\left(1+\frac{x}{3}\right)^{-\frac{1}{2}} = 1+\left(-\frac{1}{2}\right)\left(\frac{x}{3}\right)+\frac{-\frac{1}{2}\left(-\frac{1}{2}-1\right)}{2!}\left(\frac{x}{3}\right)^2 +\frac{-\frac{1}{2}\left(-\frac{1}{2}-1\right)\left(-\frac{1}{2}-2\right)}{3!}\left(\frac{x}{3}\right)^3 +....$$

$$= 1-\frac{1}{2}.\frac{x}{3}+\frac{1}{2}.\frac{3}{2}.\frac{1}{2}.\frac{x^2}{9}-\frac{1}{2}.\frac{3}{2}.\frac{5}{2}.\frac{x^3}{27}+....$$

$$= 1-\frac{x}{6}+\frac{x^2}{24}-5\frac{x^3}{432}+....$$

$$(3+x)^{-\frac{1}{2}} = 3^{-\frac{1}{2}}\left(1+\frac{x}{3}\right)^{-\frac{1}{2}} = 3^{-\frac{1}{2}}\left[1-\frac{x}{6}+\frac{x^2}{24}-5\frac{x^3}{432}+....\right]$$

Practice Problems

Expand the following Binomial expressions up to four terms:

(i) $(1+x)^{-3}$ (ii) $(1+3x)^{-2}$ (iii) $(1+x)^{\frac{1}{2}}$

(iv) $\left(1-x^2\right)^{\frac{1}{2}}$ (v) $(1-2x)^{\frac{1}{2}}$ (vi) $(1+x)^{-4}$

(vii) $(7-3x)^{-1}$

Answers:

(i) $1-3x+6x^2-10x^3+...$ (ii) $1-16x+27x^2-108x^3+...$

(iii) $1-\dfrac{1}{2}x+\dfrac{3}{8}x^2-\dfrac{5}{16}x^3+...$ (iv) $1+\dfrac{1}{2}x^2+\dfrac{3}{8}x^4+\dfrac{5}{16}x^6+...$

(v) $1-x-x^2-x^3-....$ (vi) $1-4x+10x^2-20x^3+...$

(vii) $\dfrac{1}{7}\left[1+\dfrac{3x}{7}+\dfrac{9x^2}{49}+\dfrac{27x^3}{343}+...\right]$

Objective Questions: Binomial Expansion

1. n! is equal to:

   (a) $n\times(n+1)$ (b) n^2

   (c) $n\times(n-1)$ (d) $n\times(n-1)!$

2. $\binom{n}{r}$ stands for:

   (a) $n!r!$ (b) $n!-r!$

   (c) $\dfrac{n!}{r!}$ (d) $\dfrac{n!}{r!(n-r)!}$

3. The coefficient of x^3 in the expansion of $(1+x)^7$ is:

   (a) 35   (b) 45

   (c) 55   (d) None of these.

4. Which of the following terms is contained in the expansion of $(x+y)^8$

   (a) x^7y (b) $56x^2y^2$

   (c) $56x^3y^5$ (d) x^7y^3

5. The coefficient of x in the expansion of $(1+x)^{n-1}$

(a) 1   (b) n

(c) $(n+1)$ (d) $(n-1)$

6. The coefficient of b^2 in the expansion of $(1+b)^n$ is:

(a) $3n$ (b) $\dfrac{n}{2}$

(c) $^nC_2$ (d) $n!$

7. The coefficient of x^3, in the expansion of $(1+2x)^6$ is:

(a) 18   (b) 80

(c) 90   (d) 54

8. Which of the terms does not exists in the expansion of $(x+2y)^5$?

(a) $10x^2 y^3$ (b) $5xy^4$

(c) $10x^3 y^2$ (d) $10x^4 y^2$

9. The third term expansion of $(x+2y)^8$ is:

(a) $112x^6 y^2$ (b) $16x^7 y$

(c) $448x^5 y^3$ (d) None of theses

10. The last term of $(a-2b)^4$ is :

(a) $16b^4$ (b) $16ab^4$

(c) $-16b^4$ (d) $32b^4$

11. The term $a^8 b^4$ occurs in the expansion of the binomial:

(a) $(a+b)^8$ (b) $(a+b)^{10}$

(c) $(a+b)^{12}$ (d) $(a+b)^{16}$

12. The coefficient of the term $a^8 b^4$ in a binomial expansion is:

(a) 495   (b) 448

(c) 554   (d) 484

13. In the expansion of $(a-b)^4$, the coefficients of the terms are:

(a) 1, 4, 5, 4, 1   (b) 1, -4, 6, -4, 1

(c) 1, 4, 6, 4, 1   (d) 1, -4, 5, -4, 1

14. sIn the expansion of $(x+b)^5$, power of b in the 3rd term is?

(a) 1   (b) 2

(c) 3   (d) 4

15. Coefficient of a^4b^6 in the expansion of $(a+b)^{10}$ is same as that of:

(a) a^5b^5 (b) a^2b^8

(c) a^6b^4 (d) a^8b^2

16. In the binomial expansion of $(x-2)^n$, 6th term is $-8064x^5$. The value of n is:

(a) 5   (b) 6

(c) 10   (d) 11

17. In the binomial expansion of $(a+b)^{15}$, what is the coefficient of the term a^2b^{13} ?

(a) $^{15}C_{13}$ (b) 105

(c) 81   (d) 90

18. There are 12 terms in the expansion $(a+b)^n$. The value of n is :

(a) 10   (b) 11

(c) 12   (d) 13

19. The greatest binomial coefficient in the binomial expansion of $(3x+2y)^{12}$ is:

(a) $^{12}C_5$ (b) $^{12}C_6$

(c) $^{12}C_7$ (d) $^{12}C_8$

Answers

1. c	2. d	3. a	4. c	5. d	6. c	7. b
8. d	9. a	10. a	11. c	12. a	13. b	14. b
15. c	16. c	17. a	18. b	19. c		

10 Trignometrical Functions

1.0 BRIEF HISTORY

The history of trigonometry dates back to early ages of Egyptian and Babylonian civilization. Greek astronomer Hipparcus compiled the trigonometric table that measured the lengths of the chord subtending various angles in circle of fixed radius. This was done in increasing degree of 71. Ptolemy took this further in 5th century. The foundation of modern day trigonometry was laid by Menelaus theorem.

Around the same period (Gupta Period) study of trigonometric functions flourished in India especially due to Aryabhatt. Instead of chords, Indian mathematician created trigonometric system based upon sine function.

The history of trigonometry also include Muslin astronomers who compiled both the studied of Greek and Indians.

In 13th century, Germans fathered trigonometry by defining trigonometric functions as ratios rather than length of lines. After the discovery of logarithms by Swedish astronomer, the history of trigonometry advanced further with Issac Newton and James Streiger. Leonhard Euler (1748) used complex numbers to explain trigonometric functions and this is seen as the modern form of trigonometry.

The word Trigonometry has its origin from three Greek words. *Tri* mean three. Gonia is for angle and metron for measure

1.1 MEASUREMENT OF AN ANGLE

An **angle** is the figure formed by two rays, called the *sides* of the angle, sharing a common endpoint, called the *vertex* of the angle. The word *angle* comes from the Latin word *angulus*, meaning "a corner

There are three system of angle measurement.

1. Sexagesimal System (**Measurement is in Degree**)

2. Centesimal System (**Measurement is in Grade**)

3. Circular System (**Measurement is in Radian**)

Degree Measure: In this English system, the unit of measurement is degree. If two perpendicular lines intersect at a point, the space is divided into four parts called Quadrants. Total angle in each quadrant is said to be of 90° i.e. one right angle.

Thus 1 right angle = 90 degree (written as 90^0)

1 degree = 60 minute (Written as 60′)

1 minute = 60 seconds (written as 60″)

Grade Measure: In this French system, the unit of measurement is grade. A right angle consists of 100 grades.

1 grade = 100 minute (100′)

1 minute = 100 seconds (100″)

Radian Measure: In this Circular system, the angle is measured in radians. It is the angle subtended the centre of the circle by an arc equal to length of the radius.

$$1 \text{ radian} = \frac{\text{Length of the arc equal to radius}}{\text{Length of radius}}$$

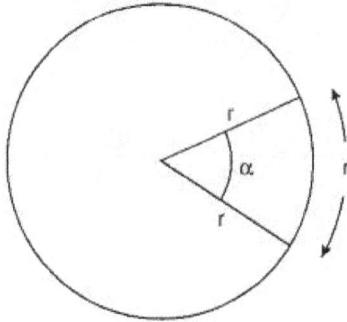

α = 1 radian

1 radian is also written as 1^c

Relation between radians and degrees of angles

$$\text{one radian} = \frac{2 \text{ } rt \text{ angles}}{\pi} = \frac{180^0}{\pi}$$

$$180^0 = \pi \text{ radian}$$

Hence one degree = $\dfrac{\pi}{180^0}$ radian

Points to Remember

1. To convert the degree into radian, multiply the degree by $\dfrac{\pi}{180^0}$

2. To convert the radian into degree, multiply the radian by $\dfrac{180^0}{\pi}$

3. (*i*) In I Quadrant range of angle is 0 to 90^0

(*ii*) **In II Quadrant range of angle is 90^0 to 180^0**

(iii) **In III Quadrant range of angle is 180^0 to 270^0**

(iv) **In IV Quadrant range of angle is $27\,0^0$ to 360^0**

Solved Examples

1. Convert the following degree measures into radians:

(*i*) 36^0 (*ii*) $50^030'$ (*iii*) 150^0 (*iv*) -520^0

Solutions: (*i*) $1^0 = \dfrac{\pi}{180^0}$ radian

$\therefore 36^0 = \dfrac{\pi}{180^0} \times 36$ radian $= \dfrac{\pi}{5}$ radian

(*ii*) $50^030' = 50^0 + 30' = 50^0 + \left(\dfrac{30}{60}\right)^0 = \dfrac{101^0}{2}$

$$\dfrac{101^0}{2} = \dfrac{101^0}{2} \times \dfrac{\pi}{180} = \dfrac{101}{360} \text{ radian}$$

(*iii*) $150^0 = 150^0 \times \dfrac{\pi}{180} = \dfrac{5\pi^c}{6}$

(*iv*) $-520^0 = -520^0 \times \dfrac{\pi}{180} = -\dfrac{26}{9}\pi^c$

2. Convert the following radian measures into degrees.

(*i*) $\dfrac{5\pi}{3}$ (*ii*) $\dfrac{\pi}{4}$ (*iii*) $\dfrac{3\pi}{5}$ (iv) **11 radian**

Solutions: (*i*) $\dfrac{5\pi}{3} = \dfrac{5\pi}{3} \times \dfrac{180^0}{\pi} = 300^0$

(*ii*) $\dfrac{\pi}{4} = \dfrac{\pi}{4} \times \dfrac{180^0}{\pi} = 45^0$

(*iii*) $\dfrac{3\pi}{5} = \dfrac{3\pi}{5} \times \dfrac{180^0}{\pi} = 108^0$

(*iv*) $11 \text{ rad} = 11 \times \dfrac{180^0}{\pi} = 11 \times \dfrac{180^0}{22} \times 7 = 630^0$

1.2 TRIGONOMETRIC RATIOS

In the right angled triangle ABC given below:

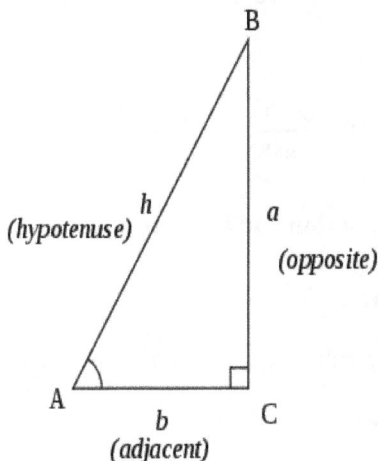

AB = Hypotenuse(h), BC= Perpendicular(a) and AC = Base(b)

$\dfrac{\text{Perpedicular}}{\text{Hypotenuse}} = \dfrac{BC}{AC}$ is called the $\sin e$ of the angle AOP.

$\dfrac{\text{Base}}{\text{Hypotenuse}} = \dfrac{AC}{BA}$ is called the cosine of the angle AOP.

$\dfrac{\text{Perpedicular}}{\text{Base}} = \dfrac{BC}{AC}$ is called the tangent of the angle AOP.

$\dfrac{\text{Hypotenuse}}{\text{Perpendicular}} = \dfrac{AB}{AC}$ is called the cosecant of the angle AOP.

$\dfrac{\text{Hypotenuse}}{\text{Base}} = \dfrac{AB}{AC}$ is called the secant of the angle AOP.

$\dfrac{\text{Base}}{\text{Perpendicular}} = \dfrac{AC}{BC}$ is called the cotangent of the angle AOP.

Remember: $\dfrac{\text{Perp.Base.Perp.}}{\text{Hypo.Hypo.Base.}} = \dfrac{PBP}{HHB}$. This gives all the trigonometric ratios.

Ratios	$\sin \theta$	$\cos \theta$	$\tan \theta$
	Perp. (P)	Base(B)	Perp.(P)
	Hyp. (H)	Hyp.(H)	Base (B)
Reciprocals	cosec θ	sec θ	cot θ

1.2.1 Reciprocal Relations/Identities

(*i*) $\operatorname{cosec} \theta = \dfrac{1}{\sin \theta}$, (*ii*) $\sec \theta = \dfrac{1}{\cos \theta}$, (*iii*) $\cot \theta = \dfrac{1}{\tan \theta}$

Quotient Relations:

(*i*) $\tan \theta = \dfrac{\sin \theta}{\cos \theta}$ (*ii*) $\cot \theta = \dfrac{\cos \theta}{\sin \theta}$

1.2.2 Trigonometric Relations or Identities

By Pythagoras theorem:

$$(\text{Perpendicular})^2 + (\text{Base})^2 = (\text{Hypotenuse})^2 \qquad \text{...(a)}$$

1. Dividing (a) by Hypotenuse:

$$\left(\frac{\text{Perpendicular}}{\text{Hypotenuse}}\right)^2 + \left(\frac{\text{Base}}{\text{Hypotenuse}}\right)^2 = \left(\frac{\text{Hypotanuse}}{\text{Hypotenuse}}\right)^2$$

$$\sin^2 \theta \quad + \quad \cos^2 \theta \quad = \quad 1 \qquad \text{...(1)}$$

2. Dividing (a) by Base:

$$\left(\frac{\text{Perpendicular}}{\text{Base}}\right)^2 + \left(\frac{\text{Base}}{\text{Base}}\right)^2 = \left(\frac{\text{Hypotanuse}}{\text{Base}}\right)^2$$

$$\tan^2 \theta \quad + \quad 1 \quad = \quad \sec^2 \theta \qquad \text{...(2)}$$

3. Dividing (a) by Perpendicular:

$$\left(\frac{\text{Perpend.}}{\text{Perpend.}}\right)^2 + \left(\frac{\text{Base}}{\text{Perpend.}}\right)^2 = \left(\frac{\text{Hypotanuse}}{\text{Perpend.}}\right)^2$$

$$1 \quad + \quad \cot^2 \theta \quad = \quad \operatorname{cosec}^2 \theta \qquad \text{...(3)}$$

Equations (1), (2) and (3) are the basic trigonometric relations. They are known as *Pythagorean Identities*

It should be noted that $\sin^2 \theta = (\sin \theta)^2$, $\cos^2 \theta = (\cos \theta)^2$ and so on.

Solved Examples

1. Prove: $\sqrt{\left(\sec^2\theta + \cos ec^2\theta\right)} = \tan\theta + \cot\theta$

Solution: Left Hand Side (LHS) = $\sqrt{\left(\sec^2\theta + \cos ec^2\theta\right)}$

$$= \sqrt{\left(\frac{1}{\cos^2\theta} + \frac{1}{\sin^2\theta}\right)} = \sqrt{\left(\frac{\sin^2\theta + \cos^2\theta}{\cos^2\theta.\sin^2\theta}\right)}$$

$$= \sqrt{\frac{1}{\cos^2\theta.\sin^2\theta}}$$

$$= \frac{1}{\cos\theta.\sin\theta}$$

Right Hand Side (RHS) = $\tan\theta + \cot\theta$

$$= \frac{\sin\theta}{\cos\theta} + \frac{\cos\theta}{\sin\theta} = \frac{\sin^2\theta + \cos^2\theta}{\cos\theta.\sin\theta} = \frac{1}{\cos\theta.\sin\theta}$$

Hence LHS = RHS

2. Prove : $\sqrt{\dfrac{1-\cos\theta}{1+\cos\theta}} = \cos ec\theta - \cot\theta$

Solution: LHS = $\sqrt{\dfrac{1-\cos\theta}{1+\cos\theta}} = \sqrt{\dfrac{(1-\cos\theta)(1-\cos\theta)}{(1+\cos\theta)(1-\cos\theta)}}$

$$= \sqrt{\frac{(1-\cos\theta)^2}{(1-\cos^2\theta)}} = \sqrt{\frac{(1-\cos\theta)^2}{\sin^2\theta}} = \frac{1-\cos\theta}{\sin\theta}$$

RHS = $\cos ec\theta - \cot\theta \quad \dfrac{1}{\sin\theta} - \dfrac{\cos\theta}{\sin\theta} = \dfrac{1-\cos\theta}{\sin\theta}$

Since LHS = RHS, hence proved

3. If $\tan\theta + \sin\theta = m$ **and** $\tan\theta - \sin\theta = n$, **show that** $m^2 - n^2 = 4\sqrt{mn}$

Solution: LHS = $m^2 - n^2 = (m+n)(m-n) = (2\tan\theta)(2\sin\theta) = 4\tan\theta.\sin\theta$ ...(i)

RHS = $4\sqrt{mn} = 4\sqrt{(\tan\theta + \sin\theta)(\tan\theta - \sin\theta)}$

$$= 4\sqrt{(\tan^2\theta - \sin^2\theta)}$$

$$= 4\sqrt{\left(\frac{\sin^2\theta}{\cos^2\theta} - \sin^2\theta\right)}$$

$$= 4\sqrt{\sin^2\theta.\left(\frac{1}{\cos^2\theta} - 1\right)}$$

$$= 4\sqrt{\sin^2\theta\left(\frac{1-\cos^2\theta}{\cos^2\theta}\right)}$$

$$= 4\sqrt{\sin^2\theta\left(\frac{\sin^2\theta}{\cos^2\theta}\right)} = 4\sin\theta.\tan\theta \quad ....(ii)$$

From (i) and (ii) we find that LHS = RHS = $4\sin\theta.\tan\theta$

4. Prove: $(\operatorname{cosec} A - \sin A)(\sec A - \cos A)(\tan A + \cot A) = 1$

Solution: LHS = $(\operatorname{cosec} A - \sin A)(\sec A - \cos A)(\tan A + \cot A)$

$$= \left(\frac{1}{\sin A} - \sin A\right)\left(\frac{1}{\cos A} - \cos A\right)\left(\frac{\sin A}{\cos A} + \frac{\cos A}{\sin A}\right)$$

$$= \left(\frac{1-\sin^2 A}{\sin A}\right)\left(\frac{1-\cos^2 A}{\cos A}\right)\left(\frac{\sin^2 A + \cos^2 A}{\sin A.\cos A}\right)$$

$$= \frac{\cos^2 A}{\sin A}.\frac{\sin^2 A}{\cos A}.\frac{1}{\sin A.\cos A} = \frac{\sin^2 A.\cos^2 A}{\sin^2 A.\cos^2 A} = 1 = \text{RHS}$$

5. Prove: $\cos^4 A - \sin^4 A + 1 = 2\cos^2 A$

Solution: LHS = $\cos^4 A - \sin^4 A + 1 = (\cos^2 A - \sin^2 A)(\cos^2 A + \sin^2 A) + 1$

$$= \cos^2 A - \sin^2 A + 1$$

$$= \cos^2 A + 1 - \sin^2 A$$

$$= \cos^2 A + \cos^2 A = 2\cos^2 A = \text{RHS}$$

6. Show that: $(\sin A + \cos A)(1 - \sin A.\cos A) = \sin^3 A + \cos^3 A$

Solution: RHS = $\sin^3 A + \cos^3 A = (\sin A + \cos A)(\sin^2 A + \cos^2 A - \sin A\cos A)$

$$= (\sin A + \cos A)(1 - \sin A\cos A) = \text{LHS}$$

Hint: $((a^3 + b^3) = (a+b)(a^2 + b^2 - ab))$

7. Show that: $\dfrac{\sin\theta}{1+\cos\theta}+\dfrac{1+\cos\theta}{\sin\theta}=2\cos ec\theta$

Solution: LHS $=\dfrac{\sin\theta}{1+\cos\theta}+\dfrac{1+\cos\theta}{\sin\theta}=\dfrac{\sin^2\theta+(1+\cos\theta)^2}{\sin\theta(1+\cos\theta)}$

$=\dfrac{\sin^2\theta+1+\cos^2\theta+2\cos\theta}{\sin\theta(1+\cos\theta)}$

$=\dfrac{2+2\cos\theta}{\sin\theta(1+\cos\theta)}$

$=\dfrac{2(1+\cos\theta)}{\sin\theta(1+\cos\theta)}$

$=\dfrac{2}{\sin\theta}=2.\text{cosec}\,\theta=\text{RHS}$

8. Prove that: $\sqrt{\left(\dfrac{1-\sin A}{1+\sin A}\right)}=\sec A-\tan A$

Solution: LHS $=\sqrt{\left(\dfrac{1-\sin A}{1+\sin A}\right)\left(\dfrac{1-\sin A}{1-\sin A}\right)}$

$=\sqrt{\dfrac{(1-\sin A)^2}{(1-\sin^2 A)}}=\dfrac{1-\sin A}{\cos A}=\dfrac{1}{\cos A}-\dfrac{\sin A}{\cos A}$

$=\sec A-\tan A$

$=\text{RHS}$

9. Prove: $\dfrac{\text{cosec}\,A}{\cot A+\tan A}=\cos A$

Solution: LHS $=\dfrac{\text{cosec}\,A}{\cot A+\tan A}=\dfrac{\text{cosec}\,A}{\dfrac{\cos A}{\sin A}+\dfrac{\sin A}{\cos A}}$

$=\dfrac{\cos ecA}{\dfrac{\cos^2 A+\sin^2 A}{\cos A\sin A}}$

$=\dfrac{\text{cosec}\,A}{\dfrac{1}{(\cos A.\sin A)}}=\dfrac{\text{cosec}\,A}{1}.(\cos A.\sin A)=\cos A\ (\because\ \text{cosec}\,A\times\sin A=1)$

10. Prove: $\left(\sec A + \cos A\right)\left(\sec A - \cos A\right) = \tan^2 A + \sin^2 A$

Solution: LHS $= \left(\sec A + \cos A\right)\left(\sec A - \cos A\right) = \sec^2 A - \cos^2 A = (1 + \tan^2 A) - \cos^2 A$

$$= \tan^2 A + (1 - \cos^2 A)$$

$$= \tan^2 \theta + \sin^2 \theta \quad \because \left(1 - \cos^2 \theta\right) = \sin^2 \theta$$

$$= \text{RHS}$$

11. Prove : $\dfrac{1}{\cot A + \tan A} = \sin A.\cos A$

Solution: LHS $= \dfrac{1}{\cot A + \tan A} = \dfrac{1}{\dfrac{\cos A}{\sin A} + \dfrac{\sin A}{\cos A}} = \dfrac{1}{\dfrac{\cos^2 + \sin^2 A}{\sin A.\cos A}} = \dfrac{1}{\dfrac{1}{\sin A.\cos A}}$

$$= \sin A.\cos A = \textbf{RHS}$$

12. Show that : $\dfrac{1 + \tan^2 A}{1 + \cot^2 A} = \tan^2 A$

Solution: LHS $= \dfrac{1 + \tan^2 A}{1 + \cot^2 A} = \dfrac{\sec^2 A}{\text{cosec}^2 A} = \dfrac{1}{\cos^2 A}.\dfrac{\sin^2 A}{1} = \tan^2 A = \text{RHS}$

(Note: (i) $1 + \tan^2 A = \sec^2 A$, (ii) $1 + \cot^2 A = \text{cosec}^2 A$)

13. Prove: $\dfrac{1 - \tan A}{1 + \tan A} = \dfrac{\cot A - 1}{\cot A + 1}$

Solution: LHS $= \dfrac{1 - \tan A}{1 + \tan A} = \dfrac{\dfrac{1}{\tan A} - 1}{\dfrac{1}{\tan A} + 1}$ (Dividing numerator and denominator by tan A)

$$= \dfrac{\cot A - 1}{\cot A + 1} = \text{RHS}$$

14. Prove: $\dfrac{1}{\sec A - \tan A} = \sec A + \tan A$

Solution: LHS $= \dfrac{1}{\sec A - \tan A} \times \dfrac{\sec A + \tan A}{\sec A + \tan A}$

$$= \dfrac{\sec A + \tan A}{\sec^2 A - \tan^2 A} = \sec A + \tan A = \text{RHS}$$

(Note: $\because \sec^2 A = 1 + \tan^2 A, \quad \therefore \sec^2 A - \tan^2 A = 1$)

15. Prove: $(\sin A + \cos A)(\cot A + \tan A) = \sec A + \operatorname{cosec} A$

Solution: LHS $= (\sin A + \cos A)(\cot A + \tan A) = (\sin A + \cos A)\left(\dfrac{\cos A}{\sin A} + \dfrac{\sin A}{\cos A}\right)$

$$= (\sin A + \cos A)\left(\dfrac{\cos^2 A + \sin^2 A}{\sin A.\cos A}\right)$$

$$= (\sin A + \cos A)\left(\dfrac{1}{\sin A.\cos A}\right)$$

$$= \left(\dfrac{\sin A}{\sin A.cos A} + \dfrac{\cos A}{\sin A\cos A}\right)$$

$$= \left(\dfrac{1}{cos A} + \dfrac{1}{\sin A}\right) = \sec A + \operatorname{cosec} A = \text{RHS}$$

16. Prove: $\sec^4 A - \sec^2 A = \tan^4 A + \tan^2 A$

Solution: LHS $= \sec^4 A - \sec^2 A = \sec^2 A(\sec^2 A - 1) = \sec^2 A.\tan^2 A$

$$= (1 + \tan^2 A).\tan^2 A$$

$$= \tan^2 A + \tan^4 A = \text{RHS}$$

17. Prove: $\cot^4 A + \cot^2 A = \operatorname{cosec}^4 A - \operatorname{cosec}^2 A$

Solution: RHS $= \operatorname{cosec}^4 A - \operatorname{cosec}^2 A = \operatorname{cosec}^2 A(\operatorname{cosec}^2 A - 1)$

$$= (1 + \cot^2 A)\left\{\left(1 + \cot^2 A\right) - 1\right\}$$

$$= \left(1 + \cot^2 A\right).\cot^2 A \; (\operatorname{cosec}^2 A = 1 + \cot^2 A)$$

$$= \cot^2 A + \cot^4 A = \textbf{LHS}$$

18. Prove that : $\sqrt{\operatorname{cosec}^2 A - 1} = \cos A.\operatorname{cosec} A$

Solution: LHS $= \sqrt{\operatorname{cosec}^2 A - 1} = \sqrt{1 + \cot^2 A - 1}$

$$= \sqrt{\cot^2 A} = \cot A$$

$$= \cos A.\dfrac{1}{\sin A} = \cos A.\cos ec A = \text{RHS}$$

19. Show that: $\sec^2 A.\text{cosec}^2 A = \tan^2 A + \cot^2 A + 2$

Solution: LHS $= \sec^2 A.\text{cosec}^2 A = (1 + \tan^2 A)(1 + \cot^2 A)$

$$= 1 + \tan^2 A + \cot^2 A + \tan^2 A.\cot^2 A$$

$$= \tan^2 A + \cot^2 A + 2 = \text{RHS}$$

20. Prove: $\tan^2 A - \sin^2 A = \sin^4 A.\sec^2 A$

Solution: LHS $= \tan^2 A - \sin^2 A = \dfrac{\sin^2 A}{\cos^2 A} - \sin^2 A = \sin^2 A\left(\dfrac{1}{\cos^2 A} - 1\right)$

$$= \sin^2 A\left(\dfrac{1 - \cos^2 A}{\cos^2 A}\right)$$

$$= \sin^4 A.\dfrac{1}{\cos^2 A}$$

$$= \sin^4 A.\sec^2 A = \text{RHS}$$

Practice Problems (Trigonometric Identities)

Prove That:

1. $\tan x + \cot x = \sec x.\text{cosec} x$

2. $\cot^2 A = \cos^2 A + (\cot A.\cos a)^2$

3. $\dfrac{1}{\sec^2 x} = \sin^2 x.\cos^2 x + \cos^4 x$

4. $\cot x.\sec x = \text{cosec} x$

5. $\sec^2 x + \text{cosec}^2 x = \dfrac{1}{\sin^2 x.\cos^2 x}$

6. $\dfrac{1 + \tan^2 x}{1 + \cot^2 x} = \tan^2 x$

7. $(\sin(\theta))4 + 2(\sin(\theta))2(\cos(\theta))2 + (\cos(\theta))4 = \tan(\theta)\cot(\theta)$

8. $\tan(\theta).\cos(\theta) = \sin(\theta)$

9. $\sin\theta + \sin\theta.\cot^2\theta = \text{cosec}\,\theta$

10. $\sin x.\cos x.\tan x = 1 - \cos^2 x$

1.2.3 Value of Trigonometric Ratio from given Trigonometric Ratio

1. If $\cos\theta = \dfrac{3}{5}$. Find other trigonometric ratios.

Solution: $\sin\theta = \sqrt{1-\cos^2\theta} = \sqrt{1-\left(\dfrac{3}{5}\right)^2} = \sqrt{1-\dfrac{9}{25}} = \sqrt{\dfrac{25-9}{25}} = \sqrt{\dfrac{16}{25}} = \dfrac{4}{5}$

$$\operatorname{cosec}\theta = \frac{1}{\sin\theta} = \frac{1}{\left(\dfrac{4}{5}\right)} = \frac{5}{4}$$

$$\sec\theta = \frac{1}{\cos\theta} = \frac{1}{3/5} = \frac{5}{3}$$

$$\tan\theta = \frac{\sin\theta}{\cos\theta} = \frac{4/5}{3/5} = \frac{4}{3}$$

$$\cot\theta = \frac{1}{\tan\theta} = \frac{3}{4}$$

2. If $\sin\theta = \dfrac{12}{13}$, Find $\cos\theta$ and $\tan\theta$

Solution: $\cos\theta = \sqrt{1-\sin^2\theta} = \sqrt{1-\left(\dfrac{12}{13}\right)^2} = \sqrt{\dfrac{169-144}{169}} = \sqrt{\dfrac{25}{169}} = \dfrac{5}{13}$

$$\tan\theta = \frac{\sin\theta}{\cos\theta} = \frac{12/13}{5/13} = \frac{12}{5}$$

3. If $\cos\theta = \dfrac{1}{4}$, Find rest of the ratios.

Solution: $\sin\theta = \sqrt{1-\cos^2\theta} = \sqrt{1-\left(\dfrac{1}{4}\right)^2} = \sqrt{\dfrac{16-1}{16}} = \sqrt{\dfrac{15}{16}} = \dfrac{\sqrt{15}}{4}$

$$\tan\theta = \frac{\sin\theta}{\cos\theta} = \frac{\sqrt{15}/4}{1/4} = \sqrt{15}$$

$$\cot\theta = \frac{1}{\tan\theta} = \frac{1}{\sqrt{15}}$$

$$\sec\theta = \frac{1}{\cos\theta} = \frac{1}{1/4} = 4$$

$$\operatorname{cosec}\theta = \frac{1}{\sin\theta} = \frac{1}{\sqrt{15}\Big/4} = \frac{4}{\sqrt{15}}$$

4. If $\tan x = 2$, Find $\sin x$ and $\sec x$

Solution: $\therefore \sec^2 x = 1 + \tan^2 x = 1 + 2^2 = 5$

$\therefore \sec x = \sqrt{5}$

$$\cos x = \frac{1}{\sec x} = \frac{1}{\sqrt{5}}$$

$$\sin x = \sqrt{1 - \cos^2 x} = \sqrt{1 - \left(\frac{1}{\sqrt{5}}\right)^2} = \sqrt{1 - \frac{1}{5}} = \sqrt{\frac{4}{5}} = \frac{2}{\sqrt{5}}$$

5. $\tan\theta = \frac{1}{\sqrt{7}}$, then find the value of $\dfrac{\operatorname{cosec}^2\theta - \sec^2\theta}{\operatorname{cosec}^2\theta + \sec^2\theta}$

Solution: $\because \tan\theta = \frac{1}{\sqrt{7}}$ $\therefore \tan^2\theta = \frac{1}{7}$. $\cot^2\theta = 7$

$$\frac{\operatorname{cosec}^2\theta - \sec^2\theta}{\operatorname{cosec}^2\theta + \sec^2\theta} = \frac{1 + \cot^2\theta - 1 - \tan^2\theta}{1 + \cot^2\theta + 1 + \tan^2\theta}$$

$$= \frac{\cot^2\theta - \tan^2\theta}{2 + \cot^2\theta + \tan^2\theta}$$

$$= \frac{7 - \dfrac{1}{7}}{2 + 7 + \dfrac{1}{7}} = \frac{48/7}{9 + \dfrac{1}{7}} = \frac{48/7}{64/7} = \frac{48}{64} = \frac{3}{4}$$

Practice Problems: (Other trigonometric ratios from a given ratio)

1. If $\tan\theta = \frac{3}{4}$, find the sine, cosine and cotangent of the angle θ .

2. If $\sec A = \frac{3}{2}$, find tan A and cosecant A

3. If $\cos\theta = \frac{3}{5}$, find $\sin\theta$ and $\cot\theta$.

Answers:

1. $\sin\theta = \frac{3}{5}, \cos\theta = \frac{4}{5}, \cot\theta = \frac{4}{3}$, **2.** $\tan\theta = \frac{5}{4}$, **3.** $\sin\theta = \frac{4}{5}, \cot\theta = \frac{4}{3}$

1.2.4 Angular Values of Trigonometric Ratios

Trig. Ratio	Value of θ				
Sin θ	0	30	45	60	90
Write column value as:	0	1	2	3	4
Divide each column value by 4 and take sq-root	$\sqrt{\dfrac{0}{4}}$	$\sqrt{\dfrac{1}{4}}$	$\sqrt{\dfrac{2}{4}}$	$\sqrt{\dfrac{3}{4}}$	$\sqrt{\dfrac{4}{4}}$
Value of Sin θ	0	$\dfrac{1}{2}$	$\dfrac{1}{\sqrt{2}}$	$\dfrac{\sqrt{3}}{2}$	1
Cos θ	0	30	45	60	90
Write column value as:	4	3	2	1	0
Divide each column value by 4 and take sq-root	$\sqrt{\dfrac{4}{4}}$	$\sqrt{\dfrac{3}{4}}$	$\sqrt{\dfrac{2}{4}}$	$\sqrt{\dfrac{1}{4}}$	$\sqrt{\dfrac{0}{4}}$
Value of Cos θ	1	$\dfrac{\sqrt{3}}{2}$	$\dfrac{1}{\sqrt{2}}$	$\dfrac{1}{2}$	0
Value of tan θ	0	$\dfrac{1}{\sqrt{3}}$	1	$\sqrt{3}$	∞

Solved Examples (on angular values)

1. If $A = 30^0$ verify

    (*i*) $\sin 3A = 3\sin A - 4\sin^3 A$,   (*ii*) $\sin 2A = 2\sin A.\cos A$

Solution: (*i*) LHS = $\sin 3A = \sin 3 \times 30^0 = \sin 90^0 = 1$

    RHS = $3\sin A - 4\sin^3 A = 3\sin 30^0 - 4\sin^3 30^0 = 3 \times \dfrac{1}{2} - 4\left(\dfrac{1}{2}\right)^3$

    $= \dfrac{3}{2} - \dfrac{4}{8} = \dfrac{12 - 4}{8} = \dfrac{8}{8} = 1 = $ LHS

(*ii*) $\sin 2A = 2\sin A.\cos A$

Solution: LHS = $\sin 2A = \sin 2 \times 30^0 = \sin 60^0 = \dfrac{\sqrt{3}}{2}$

    RHS = $2\sin A.\cos A = 2\sin 30^0.\cos 30^0 = 2 \times \dfrac{1}{2} \times \dfrac{\sqrt{3}}{2} = \dfrac{\sqrt{3}}{2}$

Hence LHS = RHS

2. If $A = 45^0$, verify (*i*) $\cos 2A = 1 - 2\sin^2 A$. (*ii*) $\tan 2A = \dfrac{2\tan A}{1 - \tan^2 A}$

(*i*) $\cos 2A = 1 - 2\sin^2 A$

Solution: LHS = $\cos 2A = \cos 2 \times 45^0 = \cos 90^0 = 0$

$$\text{RHS} = 1 - 2\sin^2 A = 1 - 2\sin^2 45^0 = 1 - 2\left(\dfrac{1}{\sqrt{2}}\right)^2 = 1 - 2 \times \dfrac{1}{2} = 1 - 1 = 0$$

Hence LHS = RHS

(*ii*) $\tan 2A = \dfrac{2\tan A}{1 - \tan^2 A}$

Solution: $\tan A = \tan 45^0 = 1$

$$\text{LHS} = \tan 2A = \tan 90^0 = \infty \text{ (infinity)}$$

$$\text{RHS} = \dfrac{2\tan A}{1 - \tan^2 A} = \dfrac{2 \times 1}{1 - 1} = \dfrac{2}{0} = \infty$$

$$\therefore \ \ \textbf{LHS = RHS}$$

3. Verify that: $\sin^2 30^0 + \sin^2 45^0 + \sin^2 60^0 = \dfrac{3}{2}$

Solution: LHS = $\sin^2 30^0 + \sin^2 45^0 + \sin^2 60^0 = \left(\dfrac{1}{2}\right)^2 + \left(\dfrac{1}{\sqrt{2}}\right)^2 + \left(\dfrac{\sqrt{3}}{2}\right)^2$

$$= \dfrac{1}{4} + \dfrac{1}{2} + \dfrac{3}{4} = \dfrac{1 + 2 + 3}{4} = \dfrac{6}{4} = \dfrac{3}{2}$$

Hence LHS= RHS $= \dfrac{3}{2}$

Practice Problems (On angular values)

1. Verify $\sin^2 \theta + \cos^2 \theta = 1$ ......if $\theta = 30^0, 60^0$

2. Verify $\cosec^2 \theta = 1 + \cot \theta$ if $\theta = \dfrac{\pi}{3}$

3. Prove: $\sin 60^0 = \dfrac{2\tan 30^0}{1 + \tan^2 30^0}$

4. Prove: $\sec 30^0 \tan 60^0 + \sin 45^0 \cos ec45^0 + \cos 30^0 \cot 60^0 = \dfrac{7}{2}$

5. If $\cos \theta = \dfrac{5}{13}$, then show that $\dfrac{13\cos \theta + 5\sec \theta}{5\tan \theta + 6\cosec \theta} = -\dfrac{2}{37}$

1.2.5   Trigonometric Ratios of Sum and Difference of two Angles

Ratios for sum of angles

$$\sin(A+B) = \sin A.\cos B + \cos A.\sin B$$

$$\cos(A+B) = \cos A.\cos B - \sin A.\sin B$$

Ratios for difference of angles

$$\sin(A-B) = \sin A.\cos B - \cos A.\sin B$$

$$\cos(A-B) = \cos A.\cos B + \sin A.\sin B$$

Solved Examples

1. Given: $\sin A = \dfrac{3}{5}$ and $\cos B = \dfrac{9}{41}$. Find the values of $\sin(A+B)$, $\sin(A-B)$, $\cos(A+B)$ and $\cos(A-B)$

Solution: $\sin A = \dfrac{3}{5}$

$$\therefore \ \cos A = \sqrt{1-\sin^2 A} = \sqrt{1-\left(\frac{3}{5}\right)^2} = \sqrt{1-\frac{9}{25}} = \sqrt{\frac{16}{25}} = \frac{4}{5}$$

$$\cos B = \frac{9}{41}$$

$$\therefore \ \sin B = \sqrt{1-\left(\frac{9}{41}\right)^2} = \sqrt{1-\frac{81}{1681}} = \sqrt{\frac{1600}{1681}} = \frac{40}{41}$$

(*i*) $\sin(A+B) = \sin A\cos B + \cos A.\sin B$

$$= \frac{3}{5}.\frac{9}{41} + \frac{4}{5}.\frac{40}{41} = \frac{27+160}{205} = \frac{187}{205}$$

(*ii*) $\sin(A-B) = \sin A\cos B - \cos A.\sin B$

$$= \frac{3}{5}.\frac{9}{41} - \frac{4}{5}.\frac{40}{41} = \frac{27-160}{205} = \frac{-133}{205}$$

(*iii*) $\cos(A+B) = \cos A\cos B - \sin A.\sin B$

$$= \frac{4}{5}.\frac{9}{41} - \frac{3}{5}.\frac{40}{41} = \frac{36-120}{205} = \frac{-84}{205}$$

(iv) $\cos(A - B) = \cos A \cos B + \sin A . \sin B$

$$= \frac{4}{5} . \frac{9}{41} + \frac{3}{5} . \frac{40}{41} = \frac{36 + 120}{205} = \frac{156}{205}$$

2. Find the values of $\sin 75^0$ and $\cos 75^0$.

Solution.: $\sin 75 = \sin(45 + 30) = \sin 45 . \cos 30 + \cos 45 . \sin 30$

Applying $\sin(A + B) = \sin A \cos B + \cos A . \sin B$)

$$= \frac{1}{\sqrt{2}} . \frac{\sqrt{3}}{2} + \frac{1}{\sqrt{2}} . \frac{1}{2} = \frac{\sqrt{3} + 1}{2\sqrt{2}}$$

$\cos 75 = \cos(45 + 30) = \cos 45 . \cos 30 - \sin 45 . \sin 30$

$$= \frac{1}{\sqrt{2}} . \frac{\sqrt{3}}{2} - \frac{1}{\sqrt{2}} . \frac{1}{2} = \frac{\sqrt{3} - 1}{2\sqrt{2}}$$

3. Prove: $\sin 105 + \cos 105 = \cos 45$

Solution: LHS $= \sin 105 + \cos 105 = \sin(60 + 45) + \cos(60 + 45)$

Apply $\sin(A + B) = \sin A \cos B + \cos A . \sin B$ and $\cos(A + B) = \cos A \cos B - \sin A . \sin B$

$$= \left[\sin 60 . \cos 45 + \cos 60 . \sin 45 \right] + \left[\cos 60 . \cos 45 - \sin 60 . \sin 45 \right]$$

$$= \left[\frac{\sqrt{3}}{2} . \frac{1}{\sqrt{2}} + \frac{1}{2} . \frac{1}{\sqrt{2}} \right] + \left[\frac{1}{2} . \frac{1}{\sqrt{2}} - \frac{\sqrt{3}}{2} . \frac{1}{2} \right] = 2 \left[\frac{1}{2} . \frac{\sqrt{3}}{2} \right] = \frac{\sqrt{3}}{2} = \cos 45 = \text{RHS}$$

$\therefore$ LHS = RHS

4. Prove: $\sin 75 - \sin 15 = \cos 105 + \cos 15$

Solution: LHS $= \sin 75 - \sin 15 = \sin(45 + 30) - \sin(45 - 30)$

$$= \left[\sin 45 \cos 30 + \cos 45 . \sin 30 \right] - \left[\sin 45 \cos 30 - \cos 45 . \sin 30 \right]$$

$$= 2 . \left(\sin 45 \cos 30 \right) = 2 . \frac{1}{\sqrt{2}} . \frac{1}{2} = \frac{1}{\sqrt{2}}$$

RHS $= \cos 105 + \cos 15$

$$= \cos(60 + 45) + \cos(60 - 45)$$

$$= 2 . \cos 60 . \cos 45 = 2 . \frac{1}{2} . \frac{1}{\sqrt{2}} = \frac{1}{\sqrt{2}}$$

$\therefore$ LHS = RHS

5. Prove: (*i*) $\sin 20^0 \cos 10^0 + \cos 20^0 \sin 10^0 = \dfrac{1}{2}$

(*ii*) $\cos 55^0 \cos 10^0 + \sin 55^0 \sin 10^0 = \dfrac{1}{\sqrt{2}}$

Solutions: (*i*) To prove: $\sin 20^0 \cos 10^0 + \cos 20^0 \sin 10^0 = \dfrac{1}{2}$

We know that $\sin A.\cos B + \cos A.\sin B = \sin(A + B)$

$\therefore \sin 20^0 \cos 10^0 + \cos 20^0 \sin 10^0 = \sin(20 + 10) = \sin 30^0 = \dfrac{1}{2}$ **Proved**

(*ii*) To prove: $\cos 55^0 \cos 10^0 + \sin 55^0 \sin 10^0 = \dfrac{1}{\sqrt{2}}$

$\because \cos A \cos B + \sin A.\sin B = \cos(A - B)$

Put A = 55^0 and B = 10^0 in the above identity

We get: $\cos 55^0 \cos 10^0 + \sin 55^0 \sin 10^0 = \cos(55^0 - 10^0) = \cos 45^0 = \dfrac{1}{\sqrt{2}}$ **Proved**

6. Evaluate (*i*) $\sin 79 \cos 19 - \cos 79 \sin 19$ **(*ii*)** $\cos 48 \cos 12 - \sin 48 \sin 12$

(*iii*) $\cos 70 \cos 40 + \sin 70 \sin 40$ (*iv*) $\sin 36 \cos 9 + \cos 36 \sin 9$

Solution: (*i*) $\sin 79 \cos 19 - \cos 79 \sin 19$

$\because \sin(A - B) = \sin A.\cos B - \cos A.\sin B$

$\because \sin 79 \cos 19 - \cos 79 \sin 19 = \sin(79 - 19) = \sin 60 = \dfrac{\sqrt{3}}{2}$

(*ii*) $\cos 48 \cos 12 - \sin 48 \sin 12$

$\because \cos(A + B) = \cos A.\cos B - \sin A.\sin B$

$\cos 48 \cos 12 - \sin 48 \sin 12 = \cos(A + B) = \cos(48 + 12) = \cos 60 = \dfrac{1}{2}$

(*iii*) $\cos 70 \cos 40 + \sin 70 \sin 40$

$\because \cos(A - B) = \cos A.\cos B + \sin A.\sin B$

$\therefore \cos 70 \cos 40 + \sin 70 \sin 40 = \cos(70 - 40) = \cos 30 = \dfrac{\sqrt{3}}{2}$

(iv) $\sin 36 \cos 9 + \cos 36 \sin 9$

$\because \sin(A+B) = \sin A.\cos B + \cos A.\sin B$

$\sin 36 \cos 9 + \cos 36 \sin 9 = \sin(36+9) = \sin 45 = \dfrac{1}{\sqrt{}}$

1.2.6 Tangents of sum and difference of two angles.

$$\tan\left(A+B\right) = \frac{\tan A + \tan B}{1 - \tan A.\tan B} \quad \text{and} \quad \tan\left(A-B\right) = \frac{\tan A - \tan B}{1 + \tan A.\tan B}$$

Solved Examples

1. Given $\tan A = \dfrac{1}{2}$ and $\tan B = \dfrac{1}{3}$. Find the values of $\tan\left(2A+B\right)$ and $\tan\left(2A-B\right)$

Solution: 1. $\tan 2A = \tan(A+A) = \dfrac{\tan A + \tan A}{1 - \tan A.\tan A} = \dfrac{2\tan A}{1 - \tan^2 A} = \dfrac{2.\dfrac{1}{2}}{1 - \left(\dfrac{1}{2}\right)^2} = \dfrac{1}{1 - \dfrac{1}{4}} = \dfrac{4}{3}$

$$\tan(2A+B) = \frac{\tan 2A + \tan B}{1 - \tan 2A.\tan B} = \frac{\dfrac{4}{3} + \dfrac{1}{3}}{1 - \dfrac{4}{3}.\dfrac{1}{3}} = \frac{\dfrac{5}{3}}{\dfrac{5}{9}} = 3$$

$$\tan(2A-B) = \frac{\tan 2A - \tan B}{1 + \tan 2A.\tan B} = \frac{\dfrac{4}{3} - \dfrac{1}{3}}{1 + \dfrac{4}{3}.\dfrac{1}{3}} = \frac{\dfrac{3}{3}}{\dfrac{13}{9}} = \frac{9}{13}$$

2. Prove $\tan(\alpha+\beta) = 1$, if $\tan \alpha = \dfrac{5}{6}$ and $\tan\beta = \dfrac{1}{11}$

Solution: $\tan(\alpha+\beta) = \dfrac{\tan\alpha + \tan\beta}{1 - \tan\alpha.\tan\beta} = \dfrac{\dfrac{5}{6} + \dfrac{1}{11}}{1 - \dfrac{5}{6}.\dfrac{1}{11}} = \dfrac{\dfrac{61}{66}}{\dfrac{61}{66}} = 1$ Proved.

3. Prove: $\dfrac{\tan 32^0 + \tan 28^0}{1 - \tan 32^0 \tan 28^0} = \sqrt{3}$

Solution: $\because \tan\left(A+B\right) = \dfrac{\tan A + \tan B}{1 - \tan A.\tan B}$

Put A $=32^0$ and B $= 28^0$ in the above identity: $\dfrac{\tan A + \tan B}{1 - \tan A.\tan B} = \tan\left(A+B\right)$

We get:

$$\frac{\tan 32^0 + \tan 28^0}{1 - \tan 32^0.\tan 28^0} = \tan\left(32^0 + 28^0\right) = \tan 60^0 = \sqrt{3} \text{ Proved}$$

4. Find the values of $\cos 15^0$

Solution: we can write $\cos 15^0 = \cos(45^0 - 30^0) = \cos 45^0.\cos 30^0 + \sin 45^0.\sin 30^0$

$$= \frac{1}{\sqrt{2}}.\frac{\sqrt{3}}{2} + \frac{1}{\sqrt{2}}.\frac{1}{2} = \frac{\sqrt{3}+1}{2\sqrt{2}}$$

5. Prove: $\cot(\frac{\pi}{4} + A).\cot(\frac{\pi}{4} - A) = 1$

Solution: $\cot(\frac{\pi}{4} + A).\cot(\frac{\pi}{4} - A) = \dfrac{1}{\tan(\frac{\pi}{4}+A)}.\dfrac{1}{\tan(\frac{\pi}{4}-A)}$

$$= \frac{1-\tan(\frac{\pi}{4})\tan A}{\tan\frac{\pi}{4}+\tan B}.\frac{1+\tan(\frac{\pi}{4})\tan A}{\tan\frac{\pi}{4}-\tan B} = \frac{1-\tan A}{1+\tan B}.\frac{1+\tan A}{1-\tan B} = 1 \qquad \tan\frac{\pi}{4} = 1$$

6. Evaluate : (*i*) $\tan 15$ (*ii*) **tan 105**

Solution : (*i*) $\tan 15 = \tan(45 - 30)$

$$\because \tan\left(A - B\right) = \frac{\tan A - \tan B}{1 + \tan A.\tan B}$$

$$\therefore \tan\left(45 - 30\right) = \frac{\tan 45 - \tan 30}{1 + \tan 45.30} = \frac{1 - \dfrac{1}{\sqrt{3}}}{1 + \dfrac{1}{\sqrt{3}}} = \frac{\sqrt{3}-1}{\sqrt{3}+1}$$

$$\times \frac{\sqrt{3}-1}{\sqrt{3}-1} = \frac{\left(\sqrt{3}-1\right)^2}{\left(\sqrt{3}\right)-1^2} = \frac{4-2\sqrt{3}}{3-2} = 2-\sqrt{3}$$

(*ii*) $\tan 105 = \tan(60 + 45)$

$$\because \tan\left(A + B\right) = \frac{\tan A + \tan B}{1 - \tan A.\tan B}$$

$$\therefore \tan\left(60 + 45\right) = \frac{\tan 60 + \tan 45}{1 - \tan 45.30} = \frac{\sqrt{3}+1}{\sqrt{3}-1} \times \frac{\sqrt{3}+1}{\sqrt{3}-1} = \frac{\left(\sqrt{3}+1\right)^2}{\left(\sqrt{3}\right)^2 - 1^2} = \frac{4+2\sqrt{3}}{2} = 2+\sqrt{3}$$

1.2.7 Trigonometric Ratios for angle 2A

1. Show: $\sin 2A = 2 \sin A . \cos A$

Proof: We know that $\sin(A + B) = \sin A.\cos B + \cos A . \sin B$

Put $B = A$ then we get $\sin(A + A) = \sin A \cos A + \cos A. \sin A$

$\therefore$ **sin 2 A = 2 sin A . cos A**

2. Show that $\cos 2A = \cos^2 A - \sin^2 A$

Proof: We know $\cos(A + B) = \cos A \cos B - \sin A \sin B$

Put $B = A$ then we get $\cos(A + A) = \cos A \cos A - \sin A \sin A$

cos 2A = cos²A – sin²A

Also we can write $\cos^2 A - \sin^2 A = (1 - \sin^2 A) - \sin^2 A = 1 - 2\sin^2 A$

Similarly $\cos^2 A - \sin^2 A = \cos^2 A - (1 - \cos^2 A) = 2\cos^2 A - 1$

$\therefore$ **cos 2A = 1 – 2sin ²A = 2cos²A – 1**

3. Prove: $\tan 2A = \dfrac{2 \tan A}{1 - \tan^2 A}$

Proof: $\tan(A + B) = \dfrac{\tan A + \tan B}{1 - \tan A \tan b}$

Put $B = A$, then we get $\tan(A + A) = \dfrac{\tan A + \tan A}{1 - \tan A \tan A} = \dfrac{2 \tan A}{1 - \tan^2 A}$

$\therefore \tan 2A = \dfrac{2 \tan A}{1 - \tan^2 A}$

Practice Problems: (Trigonometric ratios of sum and difference of angles)

1. Verify the followings:

(i) $\sin 80^0 \cos 20^0 - \cos 80^0 \sin 20^0 = \dfrac{\sqrt{3}}{2}$, (ii) $\cos(\dfrac{\pi}{2} - \theta) = \sin \theta$,

(iii) $\cos(\dfrac{\pi}{2} + \theta) = -\sin \theta$, (iv) $\sin 3x.\cos 2x + \cos 3x \sin 2x = \sin 5x$,

(v) $\cos(\pi/2 - \theta) = \sin \theta$, (vi) $\cos(\pi/2 + \theta) = -\sin \theta$

2. Find the value of $\sin 2\theta$ when (i) $\cos \theta = \dfrac{3}{5}$, (ii) $\sin \theta = \dfrac{12}{13}$

3. If $\tan A = \dfrac{5}{6}$ and $\tan B = \dfrac{1}{11}$, find the value of $(A+B)$.

4. Find the values of $\tan 15^0$ and $\tan 75^0$

Answers: 2. (i) $\dfrac{24}{25}$ **,(ii)** $\dfrac{120}{169}$ **, 3.** $(A+B)=\dfrac{\pi}{4}$ **, 4.** $(2-\sqrt{3})$ and $(2+\sqrt{3})$

1.2.8 Conversion of sum or difference of sines or cosines into their products

$$\cos(A)+\cos(B)=2\cos\left(\frac{A+B}{2}\right)\cos\left(\frac{A-B}{2}\right)$$

$$\cos(A)-\cos(B)=-2\sin\left(\frac{A+B}{2}\right)\sin\left(\frac{A-B}{2}\right)$$

$$\sin(A)+\sin(B)=2\sin\left(\frac{A+B}{2}\right)\cos\left(\frac{A-B}{2}\right)$$

$$\sin(A)-\sin(B)=-2\cos\left(\frac{A+B}{2}\right)\sin\left(\frac{A-B}{2}\right)$$

The formulae are also called as 'C-D' formulae

Solved Examples

1. Convert the sum $\cos(3A)+\cos(7A)$ into products.

Solution: $\because \cos(A)+\cos(B)=2\cos\left(\dfrac{A+B}{2}\right)\cos\left(\dfrac{A-B}{2}\right)$

$\therefore \cos(3A)+\cos(7A)=2\cos\left(\dfrac{3A+7A}{2}\right)\cos\left(\dfrac{3A-7A}{2}\right)$

$\qquad = 2\cos(5A)\cos(-2A)=2\cos(5A)\cos(2A) \quad \because \cos(-\theta)=\cos(\theta)$

2. Show that $\sin(5A)+2\sin(3A)+\sin(A)=4\cos^2(A).\sin(3A)$

Solution: LHS $= \sin(5A)+2\sin(3A)+\sin(A)$

$\qquad = \sin(5A)+\sin(A)+2\sin(3A)$

$\qquad = 2\sin\left(\dfrac{5A+A}{2}\right)\cos\left(\dfrac{5A-A}{2}\right)+2\sin(3A)$

$\qquad = 2\sin(3A)\cos(2A)+2\sin(3A)$

$\qquad = 2\sin(3A)\{\cos(2A)+1\}$

$\qquad = 2\sin(3A)\{2\cos^2(A)-1+1\} \quad \because \cos(2A)=2\cos^2(A)-1$

$\qquad = 4\sin(3A)\cos^2(A) = $ RHS

3. Show: $\dfrac{\sin 5A - \sin 3A}{\cos 3A + \cos 5A} = \tan A$

Solution: LHS $= \dfrac{\sin 5A - \sin 3A}{\cos 3A + \cos 5A} = \dfrac{2\cos\left(\dfrac{5A+3A}{2}\right)\sin\left(\dfrac{5A-3A}{2}\right)}{2\cos\left(\dfrac{3A+5A}{2}\right)\cos\left(\dfrac{3A-5A}{2}\right)}$

$$= \dfrac{2\cos(4a)\sin(A)}{2\cos(4A)\cos(A)} = \dfrac{\sin(A)}{\cos(A)} = \tan A = \text{RHS}$$

4. Show that: $\dfrac{\sin 2A + \sin 2B}{\sin 2A - \sin 2B} = \dfrac{\tan(A+B)}{\tan(A-B)}$

Solution: LHS $= \dfrac{\sin 2A + \sin 2B}{\sin 2A - \sin 2B} = \dfrac{2\sin\left(\dfrac{2A+2B}{2}\right)\cos\left(\dfrac{2A-2B}{2}\right)}{2\cos\left(\dfrac{2A+2B}{2}\right)\sin\left(\dfrac{2A-2B}{2}\right)}$

$$= \dfrac{2\sin(A+B)\cos(A-B)}{2\cos(A+B)\sin(A-B)} = \dfrac{\sin(A+B)}{\cos(A+B)} \cdot \dfrac{\cos(A-B)}{\sin(A-B)}$$

$$= \tan(A+B).\cot(A-B)$$

$$= \dfrac{\tan(A+B)}{\tan(A-B)} = \text{RHS}$$

5. Show: $\dfrac{\sin A + 2\sin 3A + \sin 5A}{\sin 3A + 2\sin 5A + \sin 7A} = \dfrac{\sin 3A}{\sin 5A}$

Solution: LHS $= \dfrac{\sin A + 2\sin 3A + \sin 5A}{\sin 3A + 2\sin 5A + \sin 7A} = \dfrac{\sin A + \sin 5A + 2\sin 3A}{\sin 3A + \sin 7A + 2\sin 5A}$

$$= \dfrac{2\sin\left(\dfrac{A+5A}{2}\right)\cos\left(\dfrac{A-5A}{2}\right) + 2\sin 3A}{2\sin\left(\dfrac{3A+7A}{2}\right)\cos\left(\dfrac{3A-7A}{2}\right) + 2\sin 5A}$$

$$= \dfrac{2\sin 3A\cos A + 2\sin 3A}{2\sin 5A\cos A + 2\sin 5A}$$

$$= \dfrac{2\sin 3A(\cos 2A + 1)}{2\sin 5A(\cos 2A + 1)}$$

$$= \dfrac{\sin 3A}{\sin 5A} = \text{RHS}$$

6. $\sin 50^0 - \sin 70^0 + \sin 10^0 = 0$

Solution : LHS $= \sin 50^0 - \sin 70^0 + \sin 10^0 = 2\cos 60\sin(-10) + \sin(10)$

$$= -2 \times \frac{1}{2} \times \sin(-10) + \sin(10)$$

$$= -\sin(-10) + \sin(10) = 0 = \text{RHS}$$

Practice Problems

1. Convert the following sums or differences into the products:

    (*i*) sin2A + sin8A   (*ii*) cosA + cos3A

    (*iii*) sin 4A – sin2A   (*iv*) cos25 – cos37

Prove the followings:

2. $\dfrac{\sin A + \sin 3A}{\cos A + \cos 3A} = \tan 2A$

3. $\dfrac{\sin 7A - \sin A}{\sin 8A - \sin 2A} = \dfrac{\cos 4A}{\cos 5A}$

4. $\dfrac{\sin 75 - \sin 15}{\cos 75 + \cos 15} = \dfrac{1}{\sqrt{3}}$

5. $\dfrac{\cos 2B + \cos 2A}{\cos 2B - \cos 2A} = \cot(A + B)$

6. $\dfrac{\cos 3A - \cos A}{\cos A + \cos 3A} = -\tan 2A \tan A$

7. $\dfrac{(\cos A - \cos 3A)(\sin 8A + \sin 2A)}{(\sin 5A - \sin A)(\cos 4A - \cos 6A)} = 1$

8. $\dfrac{\tan 5A + \tan 3A}{\tan 5A - \tan 3A} = 4\cos 4A \cos 2A$

1.3 INVERSE TRIGONOMETRIC FUNCTION

Consider the equation $\sin \theta = x$. For a given value of θ we get an unique value of x.

    For example if $\theta = \dfrac{\pi}{2}$, then $x = \sin \dfrac{\pi}{2} = 1$

    If $\theta = \dfrac{3\pi}{2}$, then $x = \sin \dfrac{3\pi}{2} = \sin\left(\pi + \dfrac{\pi}{2}\right) = -\sin \dfrac{\pi}{2} = -1$

Thus the solution of equation oscillates between -1 to +1. We can write $-1 < \sin\theta < 1$. Thus x can acquire -1 as the minimum value and +1 as the maximum value.

Also the angle θ can be expressed as $\theta = \sin^{-1} x$. $\sin^{-1} x$ is read as "sine inverse x". Its value is either in degrees or radians.

It must be remembered that $\sin^{-1} x \neq (\sin x)^{-1}$.

As said, $\sin^{-1} x$ is the measure of angle in degree or radian, where as $(\sin x)^{-1} = \dfrac{1}{\sin x}$. It is a ratio having no unit.

Similarly we can define $\cos^{-1} x$ as the angle whose cosine is equal to x.

Other trigonometric functions are $\tan^{-1} x, \cot^{-1} x, \sec^{-1} x$ and $\operatorname{cosec}^{-1} x$

These functions are known as *inverse trigonometric functions or inverse circular functions.*

$\sin^{-1} x, \cos^{-1} x, \tan^{-1} x$ etc may also be written as $\arcsin x, \arccos x, \arctan x$ ...

Solved Examples

1. Prove: $\sin^{-1}(\sin\theta) = \theta$

Proof: Let $\sin\theta = x \Rightarrow \theta = \sin^{-1} x$

LHS $= \sin^{-1}(\sin\theta) = \sin^{-1}(x) = \theta =$ **RHS** $\because \theta = \sin^{-1} x$

Hence LHS = RHS

2. Prove: $\sin\left(\sin^{-1} x\right) = x$

Proof:: *Let* $\sin^{-1} x = \theta$

$x = \sin\theta$

$\because \theta = \sin^{-1} x$

$x = \sin\left(\sin^{-1} x\right)$

LHS = RHS

3. Prove: $\cos^{-1}(\cos\theta) = \theta$

Proof: Let $\cos\theta = x$

$\therefore \theta = \cos$

$= \cos^{-1}(\cos\theta) \quad \because x = \cos\theta$

RHS = LHS

4. Prove: $\cos\left(\cos^{-1}x\right)=x$

Proof: Let $\cos^{-1}x=\theta$

$\therefore x=\cos\theta$

$=\cos\left(\cos^{-1}x\right)$

$\therefore$ RHS = LHS

5. Prove: $\sin^{-1}x=\operatorname{cosec}^{-1}\left(\dfrac{1}{x}\right)$

Proof: Let $=\sin^{-1}x=\theta$

$\therefore x=\sin\theta$

$\therefore\dfrac{1}{x}=\dfrac{1}{\sin\theta}=\operatorname{cosec}\theta$

$\therefore\operatorname{cosec}\theta=\dfrac{1}{x}$

$\therefore\theta=\operatorname{cosec}^{-1}\left(\dfrac{1}{x}\right)$

But $\theta=\sin^{-1}x$ (assumed)

$\therefore\sin^{-1}x=\operatorname{cosec}^{-1}\left(\dfrac{1}{x}\right)$

$\therefore$ LHS = RHS

6. Prove: $\operatorname{cosec}^{-1}x=\sin^{-1}\left(\dfrac{1}{x}\right)$

Proof: Let $\operatorname{cosec}^{-1}x=\theta$

$\therefore x=\operatorname{cosec}\theta$

$\dfrac{1}{x}=\dfrac{1}{\operatorname{cosec}\theta}=\sin\theta$

$\therefore\sin\theta=\dfrac{1}{x}$

$\therefore\theta=\sin^{-1}\left(\dfrac{1}{x}\right)$

But $\theta=\operatorname{cosec}^{-1}x$

Hence $\operatorname{cosec}^{-1}\quad\sin^{-1}\left(-\right)$

$\therefore$ LHS = RHS

7. Prove: $\tan^{-1} \quad \cot^{-1}\left(-\right)$

Proof: Let $\tan^{-1} x = \theta$

$\therefore x = \tan \theta$

$\therefore \cot \theta = \dfrac{1}{x}$

$\therefore \theta = \cot^{-1}\left(\dfrac{1}{x}\right)$

But $\theta = \tan$

$\therefore \tan^{-1} x = \cot^{-1}\left(\dfrac{1}{x}\right)$

LHS = RHS

8. Prove: $\sin^{-1}(-x) = -\sin^{-1}(x)$

Proof: Let $\sin^{-1}(-x) = \theta \Rightarrow -x = \sin \theta$

$-x = \sin \theta$

$\therefore x = -\sin \theta$

$\therefore x = \sin(-\theta)$

$\sin^{-1} x = -\theta = -\sin^{-1}(-x)$

$\therefore \sin^{-1}(-x) = -\sin^{-1}(x)$

LHS = RHS

9. Prove: $\cos^{-1}(-x) = \pi - \cos^{-1} x$

Proof: Let $\cos^{-1}(-x) = \theta$

$-x = \cos \theta$

$x = -\cos \theta = \cos(\pi - \theta)$

$\cos^{-1} x = \pi - \theta$

$\cos^{-1} x = \pi - \cos^{-1}(-x)$

$\cos^{-1}(-x) = \pi - \cos^{-1} x$

$\therefore$ LHS = RHS

10. Prove: $\sin^{-1} x + \cos^{-1} x = \dfrac{\pi}{2}$

Proof: Let $\sin^{-1} x = \theta$

$$\therefore x = \sin\theta = \cos\left(\frac{\pi}{2} - \theta\right)$$

$$\therefore \cos^{-1} = \frac{\pi}{2} - \theta$$

$$\therefore \cos^{-1} x = \frac{\pi}{2} - \sin^{-1} x \quad \because \theta = \sin^{-1} x$$

$$\therefore \sin^{-1} x + \cos^{-1} x = \frac{\pi}{2}$$

$\therefore$ LHS = RHS

11. Prove: $\tan^{-1} x + \cot^{-1} x = \dfrac{\pi}{2}$

Proof: Let $\tan^{-1} x = \theta$

$$\therefore x = \tan\theta = \cot\left(\frac{\pi}{2} - \theta\right) \quad \because \cot(90 - \theta) = \tan\theta$$

$$\cot^{-1} x = \frac{\pi}{2} - \theta$$

$$\theta + \cot^{-1} x = \frac{\pi}{2}$$

$$\tan^{-1} x + \cot^{-1} x = \frac{\pi}{2} \quad \because \theta = \tan^{-1} x$$

$\therefore$ LHS = RHS

12. Prove: $\tan^{-1}(-x) = -\tan^{-1}(x)$

Proof: Let $\tan^{-1}(-x) = \theta$

$$\therefore -x = \tan\theta$$

$$x = -\tan\theta = \tan(-\theta) \quad \because \tan(-\theta) = -\tan\theta$$

$$\tan^{-1} x = -\theta = -\tan^{-1}(-x)$$

or $\quad -\tan^{-1} x = \tan^{-1}(-x)$

RHS = LHS

13. Prove: $\csc^{-1}(-x) = -\csc^{-1}(x)$

Proof: Let $\csc^{-1}(-x) = \theta$

$$\therefore (-x) = \csc(\theta) = \frac{1}{\sin\theta}$$

$$x = \frac{1}{\sin(-\theta)} = \csc(-\theta) \qquad\qquad \because \frac{1}{\sin\theta} = \csc\theta$$

$$\therefore x = \csc(-\theta)$$

or $\quad \csc^{-1}x = -\csc^{-1}(-x)$

or $\quad -\csc^{-1}x = \csc^{-1}(-x)$

$\therefore$ **RHS = LHS**

14. Prove: $\tan^{-1}x + \tan^{-1}y = \tan^{-1}\left(\dfrac{x+y}{1-xy}\right)$

Proof: Let $\tan^{-1}x = \theta_1 \ \therefore x = \tan\theta_1$

Let $\tan^{-1}y = \theta_2 \ \therefore y = \tan\theta_2$

$$\therefore \tan(\theta_1 + \theta_2) = \frac{\tan\theta_1 + \tan\theta_2}{1 - \tan\theta_1\tan\theta_2}$$

$$\tan(\theta_1 + \theta_2) = \frac{x+y}{1-x.y}$$

$$(\theta_1 + \theta_2) = \tan^{-1}\left(\frac{x+y}{1-x.y}\right)$$

$$\tan^{-1}x + \tan^{-1}y = \tan^{-1}\left(\frac{x+y}{1-x.y}\right)$$

$\therefore$ **LHS = RHS**

15. Evaluate: *(i)* $\sin^{-1}\left(\sin\dfrac{5\pi}{6}\right)$, *(ii)* $\cos^{-1}\left\{\cos\left(\dfrac{-\pi}{4}\right)\right\}$, *(iii)* $\tan^{-1}\left\{\tan\left(\dfrac{3\pi}{4}\right)\right\}$

Solution: *(i)* $\sin^{-1}\left(\sin\dfrac{5\pi}{6}\right)$

We know that $\sin^{-1}(\sin\theta) = \theta$ See Solved Example 1

$$\therefore \sin^{-1}\left(\sin\frac{5\pi}{6}\right) = \sin^{-1}\left\{\sin\left(\pi - \frac{\pi}{6}\right)\right\}$$

$$= \sin^{-1}\left(\sin\frac{\pi}{6}\right) \quad \{\sin(\pi-\theta)\} = \sin\theta$$

We know that $\sin^{-1}(\sin\theta) = \theta$ See Solved Example 1

$$\therefore \sin^{-1}\left(\sin\frac{\pi}{6}\right) = \frac{\pi}{6}$$

Solution: **(ii)** $\cos^{-1}\left\{\cos\left(\frac{-\pi}{4}\right)\right\} = \cos^{-1}\left\{\cos\left(\frac{\pi}{4}\right)\right\}$ $\because \cos(-\theta) = \theta$

$$\cos^{-1}\left\{\cos\left(\frac{\pi}{4}\right)\right\} = \frac{\pi}{4} \qquad \because \sin^{-1}(\sin\theta) = \theta \text{ (solved Ex.3)}$$

(iii) $\tan^{-1}\left\{\tan\left(\frac{3\pi}{4}\right)\right\}$

Solution: $\tan^{-1}\left\{\tan\left(\frac{3\pi}{4}\right)\right\} = \tan^{-1}\left\{\tan\left(\pi-\frac{\pi}{4}\right)\right\}$

$$= \tan^{-1}\left\{\tan\left(\frac{\pi}{4}\right)\right\} \quad \because \tan(\pi-\theta) = \theta$$

$$= \frac{\pi}{4}$$

16. Evaluate: (i) $\cos\left\{\sin^{-1}\frac{3}{5}\right\}$ **,(ii)** $\sin\left\{\cos^{-1}\frac{4}{5}\right\}$, **(iii)** $\tan\left\{\cos^{-1}\frac{8}{17}\right\}$,

(iv) $\csc\left\{\cos^{-1}\left(\frac{-12}{13}\right)\right\}$

Solution: (i) $\cos\left\{\sin^{-1}\frac{3}{5}\right\}$

Let $\left\{\sin^{-1}\frac{3}{5}\right\} = \theta \Rightarrow \sin\theta = \frac{3}{5}$

$$\cos\theta = \sqrt{1-\sin^2\theta} = \sqrt{1-\left(\frac{3}{5}\right)^2} = \sqrt{1-\frac{9}{25}} = \sqrt{\frac{25-9}{25}} = \sqrt{\frac{16}{25}} = \frac{4}{5}$$

$$\cos\theta = \frac{4}{5} \Leftarrow \theta = \cos^{-1}\left(\frac{4}{5}\right)$$

$$\therefore \cos\left\{\sin^{-1}\frac{3}{5}\right\} = \cos\theta = \frac{4}{5} \quad \because \text{ we assumed } \left\{\sin^{-1}\frac{3}{5}\right\} = \theta$$

(ii) $\sin\left\{\cos^{-1}\dfrac{4}{5}\right\}$,

Solution: Let $\cos^{-1}\dfrac{4}{5}=\theta \Rightarrow \cos\theta=\dfrac{4}{5}$

$$\sin\theta = \sqrt{1-\cos^2\theta} = \sqrt{1-\left(\dfrac{4}{5}\right)^2} = \sqrt{\dfrac{9}{25}} = \dfrac{3}{5}$$

Thus $\sin\theta = \dfrac{3}{5}$

$$\sin\left\{\cos^{-1}\dfrac{4}{5}\right\} = \sin\theta = \dfrac{3}{5}$$

$$\because \theta = \sin^{-1}\left(\dfrac{3}{5}\right)$$

(iii) $\tan\left\{\cos^{-1}\dfrac{8}{17}\right\}$

Solution: Let $\cos^{-1}\dfrac{8}{17}=\theta \Rightarrow \cos\theta=\dfrac{8}{17}$

$$\sin\theta = \sqrt{1-\left(\dfrac{8}{17}\right)^2} = \sqrt{1-\dfrac{64}{289}} = \sqrt{\dfrac{225}{289}} = \dfrac{15}{17}$$

$$\tan\theta = \dfrac{\sin\theta}{\cos\theta} = \dfrac{15/17}{8/17} = \dfrac{15}{8}$$

$$\tan\left\{\cos^{-1}\dfrac{8}{17}\right\} = \tan\theta = \dfrac{15}{8} \quad \because \cos^{-1}\dfrac{8}{17}=\theta \ (\text{assumed})$$

(iv) $\csc\left\{\cos^{-1}\left(\dfrac{12}{15}\right)\right\}$

Solution: Let $\cos^{-1}\left(-\dfrac{12}{15}\right)=\theta$

$$\therefore \cos\theta = \dfrac{-12}{15}$$

$$\therefore \sin\theta = \sqrt{1-\cos^2\theta} = \sqrt{1-\dfrac{144}{225}} = \sqrt{\dfrac{81}{225}} = \dfrac{9}{15}$$

$$\therefore \cos\theta = \dfrac{15}{9} \quad \because \csc\theta = 1/\sin\theta$$

$$\therefore \csc\left\{\cos^{-1}\left(\dfrac{-12}{15}\right)\right\} = \csc\theta = \dfrac{15}{9}$$

1.3.1   Conversion into Simplest Forms

1. Write the following inverse functions in simplest forms

(i) $\tan^{-1}\sqrt{\dfrac{1-\cos\theta}{1+\cos\theta}}$ (ii) $\tan^{-1}\left(\dfrac{x}{\sqrt{a^2-x^2}}\right)$ (iii) $\tan^{-1}\left(\dfrac{\cos\theta-\sin\theta}{\cos\theta+\sin\theta}\right)$

Solution: (i) $\tan^{-1}\sqrt{\dfrac{1-\cos\theta}{1+\cos\theta}}$

$$\sqrt{\frac{1-\cos\theta}{1+\cos\theta}}=\sqrt{\frac{1-\left(1-2\sin^2\dfrac{\theta}{2}\right)}{1+\left(2\cos^2\dfrac{\theta}{2}-1\right)}}=\sqrt{\frac{2\sin^2\dfrac{\theta}{2}}{2\cos^2\dfrac{\theta}{2}}}=\frac{\sin\dfrac{\theta}{2}}{\cos\dfrac{\theta}{2}}=\tan\frac{\theta}{2}$$

Note: $\cos\theta=\left(2\cos^2\dfrac{\theta}{2}-1\right)=\left(1-2\sin^2\dfrac{\theta}{2}\right)$

$\therefore\ \tan^{-1}\sqrt{\dfrac{1-\cos\theta}{1+\cos\theta}}=\tan^{-1}\left(\tan\dfrac{\theta}{2}\right)=\dfrac{\theta}{2}$

Solution: (ii) $\tan^{-1}\left(\dfrac{x}{\sqrt{a^2-x^2}}\right)$

Let $x=a\sin\theta\Rightarrow\theta=\sin^{-1}\left(\dfrac{x}{a}\right)$

$$\left(\frac{x}{\sqrt{a^2-x^2}}\right)=\left(\frac{a\sin\theta}{\sqrt{a^2-a^2\sin^2\theta}}\right)=\left(\frac{a\sin\theta}{\sqrt{a^2\left(1-\sin^2\theta\right)}}\right)=\left(\frac{a\sin\theta}{a\cos\theta}\right)=\tan\theta$$

$\therefore\tan^{-1}\left(\dfrac{x}{\sqrt{a^2-x^2}}\right)=\tan^{-1}\left(\tan\theta\right)=\theta=\sin^{-1}\left(\dfrac{x}{a}\right)$

Solution: (iii) $\tan^{-1}\left(\dfrac{\cos\theta-\sin\theta}{\cos\theta+\sin\theta}\right)$

$$\left(\frac{\cos\theta-\sin\theta}{\cos\theta+\sin\theta}\right)=\left(\frac{(\cos\theta-\sin\theta)/\cos\theta}{(\cos\theta+\sin\theta)/\cos\theta}\right)\quad\text{(Dividing Nr \& Denominator by }\cos\theta)$$

$$=\left(\frac{(1-\tan\theta)}{(1+\tan\theta)}\right)=\left(\frac{\tan\dfrac{\pi}{4}-\tan\theta}{1-\tan\dfrac{\pi}{4}.\tan\theta}\right)=\tan\left(\frac{\pi}{4}-\theta\right)$$

$$\left[\text{Note}: \ \tan(x-y) = \frac{\tan x - \tan y}{1 + \tan x. \tan y} \right]$$

$$\therefore \ \tan^{-1}\left(\frac{\cos\theta - \sin\theta}{\cos\theta + \sin\theta}\right) = \tan^{-1}\left\{\tan\left(\frac{\pi}{4} - \theta\right)\right\} = \left(\frac{\pi}{4} - \theta\right)$$

Practice Problems

1. Prove : $2\sin^{-1} x = \sin^{-1}\left\{2x\sqrt{1-x^2}\right\}$, $-1 < x < 1$

2. Prove: $2\cos^{-1} x = \cos^{-1}\left(2x^2 - 1\right)$

3. Prove: $2\tan^{-1} x = \tan^{-1}\left(\dfrac{2x}{1-x^2}\right)$

4. Prove: $\sin^{-1} = \cos^{-1}\sqrt{1-x^2}$

5. Prove: $\tan^{-1} x = \sin^{-1}\left(\dfrac{x}{\sqrt{1+x^2}}\right)$

6. Prove : $\tan^{-1} x = \cos^{-1}\left(\dfrac{1}{\sqrt{1+x^2}}\right)$

7. Prove: $\tan^{-1}\dfrac{1}{7} + \tan^{-1}\dfrac{1}{13} = \tan^{-1}\dfrac{2}{9}$

8. Prove: $\sin^{-1}\left(\dfrac{-4}{5}\right) = \tan^{-1}\left(\dfrac{4}{5}\right)$

9. Find the value of x if

(*i*) $\tan^{-1}(1-x) + \tan^{-1}(1+x) = \dfrac{\pi}{2}$

(*ii*) $\tan^{-1}\left(\dfrac{a}{x}\right) + \tan^{-1}\left(\dfrac{b}{x}\right) = \dfrac{\pi}{2}$

10 Evaluate the following functions:

(*i*) $\sin^{-1}\left(\dfrac{1}{2}\right)$, (ii) $\cos\left(\tan^{-1}\dfrac{3}{4}\right)$, (iii) $\tan^{-1}\left(\tan\dfrac{\pi}{4}\right)$, (iv) $\sin\left(\cot^{-1} x\right)$

Answers: 9 (*i*) 0 (ii) $\sqrt{ab}$ **10. (i)** $\dfrac{\pi}{6}$, (*ii*) $\dfrac{3}{4}$, (*iii*) $\dfrac{\pi}{4}$, (iv) $\dfrac{1}{1+x^2}$

1.4 HEIGHT AND DISTANCE

One of the application of trigonometry is to find the distance of a point from a given point and height of an object viz. pole, pillar, building, cliff etc.

Angle of Elevation: Let Q be position of Object above the eye level O.

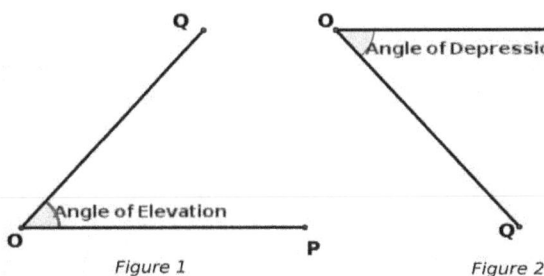

Figure 1   Figure 2

Then perpendicular AO is the height of the object and AB is the distance of the Eye(viewer).

The angle O (θ) subtended by the object B at the eye is called the angle of depression.

∴ Angle of elevation = Angle ABO = $\angle \theta$

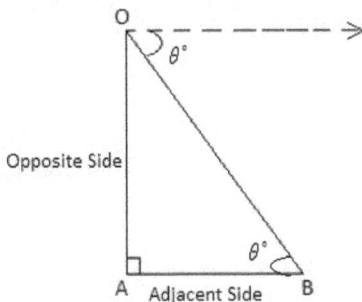

Note: The angle of elevation or depression is measured from horizontal line

Example 1. The shadow of the tower against the sun is $\sqrt{3}$ times of its height. Find the angle of elevation of the top of the tower

Solution: Let height of the tower AO = x. Hence length of the shadow AB = $\sqrt{3}\, x$

ABO is a right angled triangle. The angle of elevation is Angle ABO= θ

$$\tan \theta = \frac{\text{Perpendicular}}{\text{Base}} = \frac{AO}{AB} = \frac{x}{\sqrt{3}x} = \frac{1}{\sqrt{3}}$$

From the table of angular value, $\tan 30^0 = \dfrac{1}{\sqrt{3}}$

Hence angle of elevation $\theta = 30^0$

Example 2: The ladder of length 9.2m is leaning against a wall. The foot of the ladder is 4.6m away from the wall. Find the angle of elevation of the ladder.

Solution: Given that the length of the ladder AB = 9.6. BC represents the wall. The ladder touches the wall at B and the ground at A. Given AC = 4.6m

ABC is a right-angled triangle. Let $\angle BAC = \alpha$ is the angle of elevation.

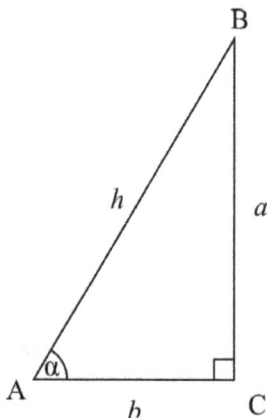

$$\cos \alpha = \frac{\text{Base}}{\text{Hypotenuous}} = \frac{AC}{AB} = \frac{4.6}{9.2} = \frac{1}{2}$$

From the table of angular value, $\cos 60^0 = \dfrac{1}{2}$

Hence angle of elevation $\alpha = 60^0$

Example 3: Find angle of elevation of a tree top from a point on the ground.. The 4m tree is 7.2m away the point.

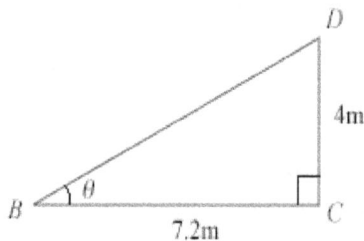

θ = Angle of Elevation of D as seen from B

Solution: Let DC is the tree and B is the point of view.

Join B to D. Thus ABC is a right angled triangle.

$\angle CBD = \theta$ is the angle of elevation and $\angle C = 90^0$.

Given CB = 7.2m and DC = 4m

$$\tan \theta = \frac{\text{Perpendicular}}{\text{Base}} = \frac{DC}{BC} = \frac{4}{7.2} = 0.56$$

Example 4: If the shadow of the tower is 10-m when the sun's angle of elevation is 60^0.

Find the height of the tower.

Solution: In the right-angled triangle ABC as shown above-

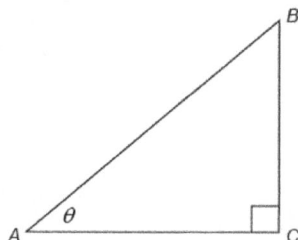

BC is the tower and AC is the length of the shadow =10m. $\angle BAC = \theta = 60^0$

Hence from rt-angled $\triangle$ ABC -

$$\tan 60^0 = \frac{Perpendicular}{Base} = \frac{BC}{10}$$

$$\therefore \frac{BC}{10} = \tan 60 = \sqrt{3}$$

$$\therefore BC = 10\sqrt{3}$$

Example 5: From a cliff 30m high, the angle of depres\sion of two ships in the same direction are found to be 30^0 and 60^0 .Find the distance between two ships.

Solution: Let AB =30m is the cliff and D and C are the ships.

Let DB = b and CB = a .

The distance between the ships = b - a

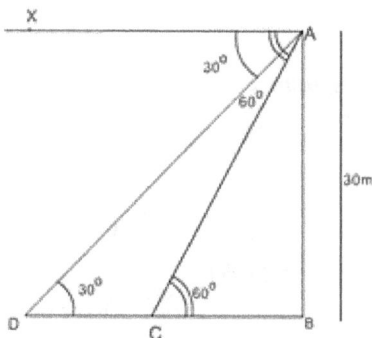

From the figure, $\angle ACB = 60^0$ and $\angle ADB = 30^0$

$\triangle s$ ADB and ACB are rt-angled at B..

From $\triangle ADC$ $\tan 60^0 = \dfrac{AB}{CB} = \dfrac{30}{a} \Rightarrow a = \dfrac{30}{\sqrt{3}}$

From $\triangle ADB$ $\tan 30^0 = \dfrac{AB}{DB} = \dfrac{30}{b} \Rightarrow b = \dfrac{30}{\tan 30^0} = \dfrac{30}{1/\sqrt{3}} = 30\sqrt{3}$

$AB = b - a = 30\sqrt{3} - \dfrac{30}{\sqrt{3}} = \dfrac{30 \times 3 - 30}{\sqrt{3}}$

$$= \dfrac{60}{\sqrt{3}} = \dfrac{60}{\sqrt{3}} \times \dfrac{\sqrt{3}}{\sqrt{3}} = \dfrac{60\sqrt{3}}{3} = 20\sqrt{3}m$$

Practice Problems : (height & distance)

1. The shadow of a pole is 20m. The angle of elevation of its top is 60^0. Find the height of the pole.

2. A boy is flying a kite. The string of the kite subtends an angle of 30^0 with the ground. The kite is flying at the height of 18 m. Find the length of the string.

3. Find the angle of elevation of the moon, if the length of the shadow of a pole is equal to its shadow.

4. The angle of elevation of the peak of the poll is 30^0 from a point 30 m away from the foot of the pole. Find the height of the pole

Answers:

1. $20\sqrt{3}$ m,
       2. 36 m,

3. 45^0,
       4. 17.2m

Objective Questions : Trigonometry

1. If sin A = 3/5, then the value of tan A is
   - (a) 4/5
   - (b) 3/4
   - (c) 4/3
   - (d) 5/3

2. If sin A =1/2, then value of angle A is
   - (a) 30°
   - (b) 45°
   - (c) 60°
   - (d) 90°

3. $12\tan\theta = 5$, then value of cos is:
   - (a) 12/13
   - (b) 5/12
   - (c) 13/12
   - (d) 15/13

4. The value of $\left(\sin\theta + \cos\theta\right)^2 + \left(\sin\theta - \cos\theta\right)^2$ is:

(a) 1

(b) 2

(c) 0

(d) $4\sin\theta\cos\theta$

5. $\tan\theta = 3\cot\theta$, then value of 0 is:

(a) $45°$

(b) $30°$

(c) $60°$

(d) $90°$

6. If $\sec^2\theta + \tan^2\theta = 7$, then value of 0 is:

(a) $30°$

(b) $45°$

(c) $60°$

(d) $90°$

7. $\sin\alpha = \dfrac{3}{5}$, then value of $\sin 2\alpha$ is:

(a) $\dfrac{16}{25}$

(b) $\dfrac{24}{25}$

(c) $\dfrac{4}{5}$

(d) $\dfrac{8}{25}$

8. $\sin 2A - 2\sin A$, is equivalent to :

(a) $\sin A\,(\sin A - 2)$

(b) $2\sin A(\,\sin A - 1)$

(c) $\sin A(2\cos A - 1)$

(d) $2\sin A(\cos A - 1)$

9. If $\sin(x + 20) = \cos x$, then value of x is:

(a) $45°$

(b) $30°$

(c) $60°$

(d) $70°$

10. $\dfrac{\tan\theta}{\csc\theta}$ is equal to:

(a) $\dfrac{\cos^2\theta}{\sin\theta}$

(b) $\dfrac{\sin^2\theta}{\cos\theta}$

(c) $\dfrac{\sin\theta}{\cos^2\theta}$

(d) $\dfrac{\cos\theta}{\sin^2\theta}$

11. $\dfrac{1 - \tan^2\theta}{1 + \tan^2\theta}$ is equal to:

(a) $\sin^2\theta - \cos^2\theta$

(b) $\cos^2\theta - \sin^2\theta$

(c) $\cot^2\theta - \tan^2\theta$

(d) $\cot^2\theta + \tan^2\theta$

12. $\sin\theta - \cos\theta = 0$, then the value of $\left(\sin^4\theta + \cos^4\theta\right)$ is:

(a) 1

(b) 3/4

(c) 1/2

(d) 1/4

13. If $4\tan A = 3$ m then $\dfrac{\left[4\sin A - \cos A\right]}{\left[4\sin A + \cos A\right]}$ is equal to:

(a) 1

(b) 3/4

(c) 1/2

(d) 1/4

14. If $\sin a = \dfrac{1}{2}$ and $\cos b = \dfrac{1}{2}$ then value of $\left(a\ b\right)$ is:

(a) 30

(b) 45

(c) 60

(d) 90

15. $\left(\sin\theta + \cos\theta\right)^2 + \left(\sin\theta - \cos\theta\right)^2$ is equal to:

(a) 1

(b) 2

(c) 0

(d) $4\sin\theta\cos\theta$

16. The value of $\dfrac{1 - \tan 15}{1 + \tan 15}$ is:

(a) 30°

(b) 45°

(c) 60°

(d) 90°

17. $7\sec^2\theta - 7\tan^2\theta$ is equal to:

(a) 1

(b) 2

(c) 0

(d) 7

18. If $\left(\sec\theta - \tan\theta\right) = \dfrac{1}{3}$, then $\left(\sec\theta + \tan\theta\right)$ is equal to:

(a) 1

(b) 2

(c) 3

(d) 4

19. If $\sin\theta = \cos\theta$, then value of $\operatorname{cosec}\theta$ is:

(a) 1

(b) 2

(c) $\sqrt{2}$

(d) 4

20. $\tan\alpha = \dfrac{1}{2}$ and $\tan\beta = \dfrac{1}{3}$, then $\left(\alpha + \beta\right)$ is:

(a) 30°

(b) 45°

(c) 60°

(d) 90°

21. $\left(1-\cos^2\theta\right)\cot^2\theta$ is equivalent to:

(a) $\sin^2\theta$

(b) $\cos^2\theta$

(c) $\tan\theta$

(d) $\cot\theta$

22. $\sin 10x\cos 5x - \cos 10x\sin 5x$ is equal to:

(a) $\sin 5x$

(b) $\cos 5x$

(c) $5\sin x$

(d) $5\cot 5x$

23. $\cos 35\cos 25 - \sin 35\sin 25$ is equal to

(a) 1

(b) $\dfrac{\sqrt{3}}{2}$

(c) $\dfrac{1}{\sqrt{2}}$

(d) $\dfrac{1}{2}$

24. The value of $\cos 75^0$ is equal to:

(a) $\dfrac{\sqrt{6}-\sqrt{2}}{4}$

(b) $\dfrac{\sqrt{6}+\sqrt{2}}{4}$

(c) $\dfrac{\sqrt{6}-\sqrt{2}}{2}$

(d) $\dfrac{\sqrt{6}+\sqrt{2}}{2}$

25. $\sin 52x\cos 22x - \cos 52x\sin 22x$ is equal to:

(a) 1

(b) $\dfrac{1}{2}$

(c) $\dfrac{1}{\sqrt{2}}$

(d) $\dfrac{\sqrt{3}}{2}$

26. $\dfrac{\cos 2B - \cos 2A}{\sin 2B + \sin 2A}$ is equal to:

(a) $\tan\left(A-B\right)$

(b) $\cot\left(A-B\right)$

(c) $\tan(A+B)$

(d) $\cot\left(A+B\right)$

27. $\dfrac{\sin x + \sin y}{\cos x + \cos y}$ is equal to :

(a) $\tan\left(A-B\right)$

(b) $\tan(A+B)$

(c) $\cot\left(A-B\right)$

(d) $\cot\left(A+B\right)$

28. $\dfrac{\sin 5x - \sin 3x}{\cos 3x + \cos 5x}$ is equal to :

    (a) $\tan x$ (b) $\cot x$

    (c) $\cos x$ (d) $\sin x$

29. $\sin 50 - \sin 70 + \sin 10$ is equal to:

    (a) 0   (b) 1

    (c) 2   (d) 3

Answers

1. b	2. a	3. a	4. b	5. c	6. c	7. b	8. d
9. d	10. b	11. b	12. a	13. a	14. d	15. b	16. a
17. d	18. c	19. a	20. a	21. b	22. b	23. a	24. a
25. b	26. b	27. c	28. a	29. b			

11 Cartesian Coordinates – 3D

1.1 COORDINATE SYSTEM IN 3-DIMENTIONS

Let O be a point. Through O draw three mutually perpendicular lines $X'OX$, $Y'OY$ and $Z'OZ$. These lines are called $x-axis(X'OX)$, $y-axis(Y'OY)$ and $z-axis(Z'OZ)$.

The planes bound by $x-axis$ and $y-axis$ is known as the $xy-plane$ or $XOY-plane$

The planes bound by $y-axis$ and $z-axis$ is known as the $yz-plane$ or $YOZ-plane$

The planes bound by $z-axis$ and $x-axis$ is known as the $zx-plane$ or $ZOX-plane$

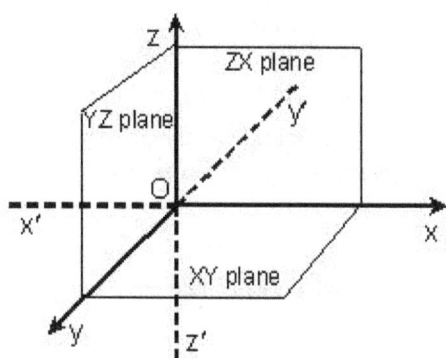

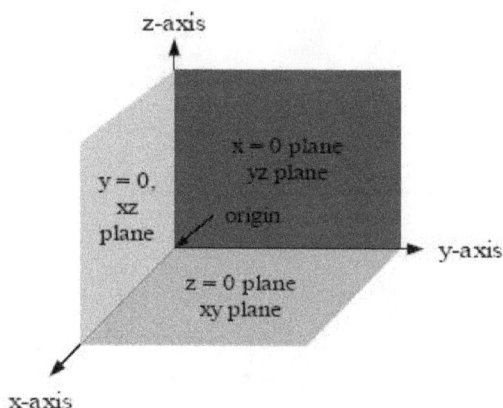

Co-ordinates of a point

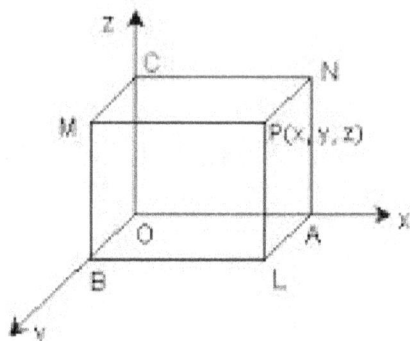

Let P be a point in 3d coordinate system. Drop perpendiculars PL, PM and PN from P on $xy-plane$, $yz-plane$ and $zx-plane$ respectively.

From the diagram above we see:

$$PM = BL = OA = x-\text{coordinate}$$

$$PN = MC = OB = y-\text{coordinate}$$

$$PL = AN = OC = z-\text{coordinate}$$

Hence the coordinates of point P are (x, y, z) or we can write as $P(x, y, z)$

Thus the coordinates x, y, z of a point P are the perpendicular distances of P from the three rectangular coordinate planes – xy, yz and zx respectively.

Octants

The co-ordinate planes divide the whole space into eight compartments called **octants.**

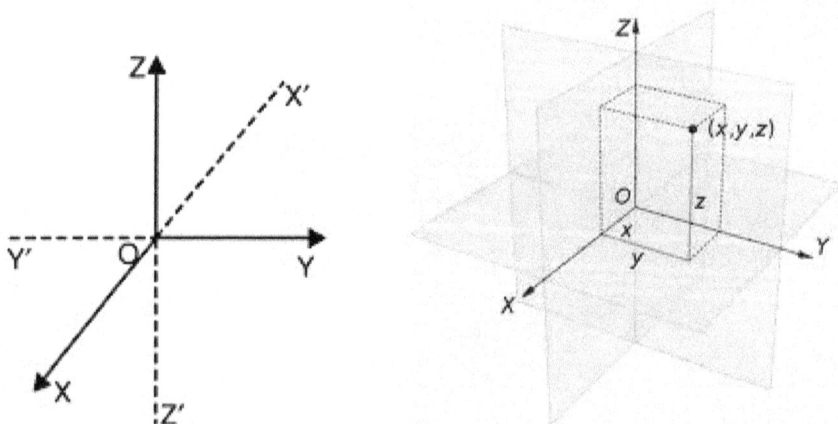

The eight octants formed are :

$OXYZ,\ OXY'Z, OXY'Z',\ OXYZ',\ OX'YZ',\ OX'Y'Z,\ OX'YZ',\ OX'Y'Z'$

The signs of co-ordinates of a point situated in any of the octants are given below:

Octants Co- ordinates	$OXYZ$	$OX'YZ$	$OXY'Z$	$OXYZ'$	$OX'Y'Z$	$OX'YZ'$	$OX'Y'Z'$	$OX'Y'Z'$
x	+	–	+	+	–	+	–	–
y	+	+	–	+	–	–	+	–
z	+	+	+	–	+	–	–	–

In the above table

X, Y, Z _ represents the positive sides of the respective axes.

X', Y', Z' _ represents the negative sides of the respective axes.

The octant $OX'YZ$ shows that point is on negative side of x – axis, but on positive sides of y and z – axes . For example the point (–2, 3, 5) is in $OX'YZ$

Similarly other octants can be explained.

Image of a Point

The image of a point is formed in the opposite direction of the axis.

If the point lies in the octant $OXYZ$ then its image will lie in the octant $OX'Y'Z'$

Similarly image of a point lying in the octant $OX'YZ$, will be in the octant $OX'Y'Z'$

In the position of the image the direction of the axis is changed

Points to Remember

1. (*i*) If a point lies on x – axis , then its y – coordinate and z – coordinate are zero.

    (*ii*) If a point lies on y – axis , then its z – coordinate and x – coordinate are zero.

    (*iii*) If a point lies on z – axis , then its x – coordinate and y – coordinate are zero.

2. (*i*) If a point lies on xy – plane , then its z – coordinate is zero.

    (*ii*) If a point lies on yz – plane , then its x – coordinate is zero.

    (*iii*) If a point lies on zx – plane , then its y – coordinate is zero.

Solved Examples

1. Name the Octant in which the following points lie:

   (*i*) (3, 2, 1) (*ii*) (0, 2, –3) (*iii*) (0, 0, –5) (*iv*) (– 3, 4, –5)

Solutions: (*i*) the point (3, 2, 1) lies in the octant $OXYZ$

(*ii*) the point (0, 2, – 3) lies in the octant $OXYZ'$

(*iii*) the point (0, 0, –5) lies in the octant $OXYZ'$

(*iv*) the point (– 3, 4, –5) lies in the octant $OX'YZ'$

2. Find the location of the images of the following points:

(*i*) (3, 4, 5) (*ii*) (1, 2, –3) (*iii*) (2, –3, –5) (*iv*) (– 3, 5, –6)

Solutions: (*i*) The point (3, 4, 5) lies in the octant: $OXYZ$, where all the axes are positive.

Hence its image will lie in the octant: $OX'Y'Z'$, where all axes are in negative direction. The location of image is (–3, –4, –5)

(*ii*) The point (1, 2, –3) lies in the octant $OXYZ'$ ie. X-axis and Y-axis are positive direction and Z is negative direction.

Hence its image will lie in the octant: $OX'Y'Z$, where X and Y axes are negative and Z is in positive direction. The location of image is (–1, –2, 3).

(*iii*) The point (2, –3, –5) lies in the octant $OXYZ'$, hence its image will lie in the octant $OX'YZ$. The location of the image is (–2, 3, 5)

(*iv*) The point (– 3, 5, –6) lies in the octant $OX'YZ'$, its image (3, – 5, 6) will lie in $OXY'Z$

1.2 DISTANCE BETWEEN TWO POINTS

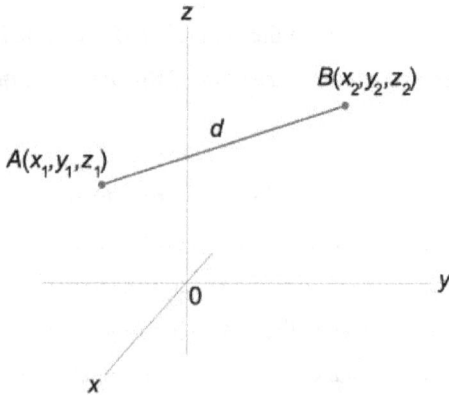

Let the points are $\left(x_1, y_1, z_1\right)$ and $\left(x_2, y_2, z_2\right)$, then distance between them is given by:

$$d = \sqrt{\left(x_2 - x_1\right)^2 + \left(y_2 - y_1\right)^2 + \left(z_2 - z_1\right)^2}$$

Solved Examples

1. Find the distance between the points A(-4, 3, 6) and B(1, 2, 3)

Solution: $AB = \sqrt{(x_2 - x_1)^2 + (y_2 - y_1)^2 + (z_2 - z_1)^2}$

$$= \sqrt{(-4-1)^2 + (3-2)^2 + (6-5)^2} = \sqrt{(-5)^2 + (1)^2 + (1)^2} = \sqrt{27} = 3\sqrt{3}$$

2. Find the distance between the points A(0, 0, 0) and B(-1, 1, 1)

Solution: $AB = \sqrt{(x_2 - x_1)^2 + (y_2 - y_1)^2 + (z_2 - z_1)^2}$

$$= \sqrt{(-1-0)^2 + (1-0)^2 + (1-0)^2} = \sqrt{(-1)^2 + (1)^2 + (1)^2} = \sqrt{3}$$

3. Find the value of **a** if the distance between the points A (5, -1, 7) and B (a, 5, 1) is 9.

Solution: $AB = \sqrt{(a-5)^2 + (5+1)^2 + (1-7)^2} = 9$

$$\sqrt{(a-5)^2 + (5+1)^2 + (1-7)^2} = 9$$

$$\left(a^2 - 10a + 25\right) + 36 + 36 = 81$$

$$a^2 - 10a + 97 = 81$$

$$a^2 - 10a + 16 = 0$$

$$(a-2)(a-8) = 0 \Rightarrow a = 2, 8$$

The value of a = 2 or 8

4. Show that the points A (3, 2, 4), B(4, 5, 2) and C(5, 8, 0) are collinear.

Solution: If the determinant formed by the points is zero, then the points are collinear.

$$D = \begin{vmatrix} 3 & 2 & 4 \\ 4 & 5 & 2 \\ 5 & 8 & 0 \end{vmatrix} = 3\begin{vmatrix} 5 & 2 \\ 8 & 0 \end{vmatrix} - 4\begin{vmatrix} 2 & 4 \\ 8 & 0 \end{vmatrix} + 5\begin{vmatrix} 2 & 4 \\ 5 & 2 \end{vmatrix} = 3(0-16) - 4(0-32) + 5(4-20)$$

$$= -48 + 128 - 80 = 0$$

5. Hence points A, B, and C are collinear. Show that the points A (1, −3, 4), B(3, 11, −6) and C (2, 4, −1) are collinear.

Solution:
$$AB = \sqrt{(1-3)^2 + (-3-11)^2 + (4+6)^2}$$
$$= \sqrt{(-2)^2 + (-14)^2 + (10)^2} = \sqrt{4 + 196 + 100} = \sqrt{300} = 10\sqrt{3}$$

$$BC = \sqrt{(3-2)^2 + (11-4)^2 + (-6+1)^2}$$
$$= \sqrt{(1)^2 + (7)^2 + (-5)^2} = \sqrt{1 + 49 + 25} = \sqrt{75} = 5\sqrt{3}$$

$$CA = \sqrt{(1-2)^2 + (-3-4)^2 + (4+1)^2}$$
$$= \sqrt{(-1)^2 + (-7)^2 + (5)^2} = \sqrt{1+49+25} = \sqrt{75} = 5\sqrt{3}$$

From the above we see AC + CB = AB

Hence the points A , B, C are collinear.

6. Show that points A (0 , 1, 2) , B (2, -2, 3) and C (1, -3, 1) are the vertices of an isosceles right angled triangle.

Solution:
$$AB = \sqrt{(2-0)^2 + (-1-1)^2 + (3-2)^2}$$
$$= \sqrt{(2)^2 + (-2)^2 + (1)^2} = \sqrt{4+4+1} = \sqrt{9} = 3$$

$$BC = \sqrt{(1-2)^2 + (-3+1)^2 + (1-3)^2}$$
$$= \sqrt{(-1)^2 + (-2)^2 + (-2)^2} = \sqrt{1+4+4} = \sqrt{9} = 3$$

$$CA = \sqrt{(0-1) + (1+3)^2 + (2-1)^2}$$
$$= \sqrt{(-1)^2 + (4)^2 + (1)^2} = \sqrt{1+16+1} = \sqrt{18} = 3\sqrt{2}$$

We see that: (i) $AB^2 + BC^2 = CA^2$ i.e. ABC is a right angled triangle.

(ii) AB = BC i.e. ABC is isosceles triangle.

7. Find the coordinates of a point on z – axis and which is equidistant from the points A (3, 1, 4) and B (5, 5, 2)

Solution: Let point P (0, 0, z) lies on z – axis. (On z-axis, x = 0 and y = 0)

$$PA = \sqrt{(3-0)^2 + (1-0)^2 + (4-z)^2}$$

$$PB = \sqrt{(5-0)^2 + (5-0)^2 + (2-z)^2}$$

Given PA = AB

$$\therefore (PA)^2 = (PB)^2$$

$$(3-0)^2 + (1-0)^2 + (4-z)^2 = (5-0)^2 + (5-0)^2 + (2-z)^2$$
$$9+1+z^2 +16-8z = 25+25+z^2 +4-4z$$
$$26-8z = 54-4z$$

$$\therefore z = -7$$

Hence the coordinates of P are (0, 0, –7).

8. Find the value of x, if the distance between the points $(x, 2, 1)$ and $(3, 4, 5)$ is 6.

Solution: The distance between two points = 6

$$\sqrt{(x-3)^2 +(2-4)^2 +(1-5)^2} = 6$$

$$(x-3)^2 +(2-4)^2 +(1-5)^2 = 36$$

$$(-3) +4+16 = 36$$

$$(x-3)^2 = 16$$

$$(x-3) = \pm 4$$

$$x = \pm 4 + 3$$

$$x = 7 \ or -1$$

9. Show that the points A $(4, 2, 4)$, B $(10, 2, -2)$ and C $(2, 0, -4)$ form an equilateral triangle.

Solution: $AB = \sqrt{(10-4)^2 +(2-2)^2 +(-2-4)^2} = \sqrt{36+0+36} = 6\sqrt{2}$

$$BC = \sqrt{(2-10)^2 +(0-2)^2 +(-4+2)^2} = \sqrt{64+4+4} = \sqrt{72} = 6\sqrt{2}$$

$$CA = \sqrt{(4-2)^2 +(2-0)^2 +(4+4)^2} = \sqrt{4+4+64} = \sqrt{72} = 6\sqrt{2}$$

We see that AB =BC = CA. Hence points A, B, C form an equilateral triangle.

Point to Remember

1. The perpendicular distance of appoint P (a, b, c) from:

(*i*) x – axis is $\sqrt{b^2 +c^2}$

(*ii*) y – axis is $\sqrt{a^2 +c^2}$

(*iii*) z – axis is $\sqrt{a^2 +b^2}$

1.3   DIVISION OR SECTION FORMULA

To find the co – ordinates of the point C (x, y, z) dividing the line joining two points

A (x_1, y_1, z_1) and B (x_2, y_2, z_2) in the ratio $m_1 : m_2$

In case of internal division: C lies between A and B, and m_1 *and* m_2 **are both positive**

The co-ordinates of C are given by:

$$x = \frac{m_1 x_2 + m_2 x_1}{m_1 + m_2}$$

$$y = \frac{m_1 y_2 + m_2 y_1}{m_1 + m_2}$$

$$z = \frac{m_1 z_2 + m_2 z_1}{m_1 + m_2}$$

In case of external division: C lies outside the line segment AB , and m_1 *and* m_2 **are of opposite signs.**

The co-ordinates of C are given by:

$$x = \frac{m_1 x_2 - m_2 x_1}{m_1 - m_2}$$

$$y = \frac{m_1 y_2 - m_2 y_1}{m_1 - m_2}$$

$$z = \frac{m_1 z_2 - m_2 z_1}{m_1 - m_2}$$

Solved Examples

1. Find the coordinates of the point which divides the line joining the points $(2, -1, 3)$ and $(4, 3, 1)$ in the ratio 3 : 4 internally.

Solution: Let C (x, y, z) be the point that divides the line. $m_1 = 3$ and $m_2 = 4$

$$x = \frac{m_1 x_2 + m_2 x_1}{m_1 + m_2} = \frac{3 \times 4 + 4 \times 2}{3 + 4} = \frac{20}{7}$$

$$y = \frac{m_1 y_2 + m_2 y_1}{m_1 + m_2} = \frac{3 \times 3 + 4 \times -1}{3 + 4} = \frac{5}{7}$$

$$z = \frac{m_1 z_2 + m_2 z_1}{m_1 + m_2} = \frac{3 \times 1 + 4 \times 3}{3 + 4} = \frac{15}{7}$$

2. Find the coordinates of the point which divides the line joining the points $(-1, -3, 2)$ and $(1, -1, 2)$ in the ratio 2 : 3 externally.

Solution: Let C (x, y, z) be the point that divides the line. $m_1 = 2$ and $m_2 = 3$

$$x = \frac{m_1 x_2 - m_2 x_1}{m_1 - m_2} = \frac{2 \times 1 - 3 \times -1}{2 - 3} = \frac{5}{-1} = -5$$

$$y = \frac{m_1 y_2 - m_2 y_1}{m_1 - m_2} = \frac{2 \times -1 + 3 \times -3}{2 - 3} = \frac{-11}{-1} = 11$$

$$z = \frac{m_1 z_2 - m_2 z_1}{m_1 - m_2} = \frac{2 \times 2 - 3 \times 2}{2 - 3} = \frac{-2}{-1} = 2$$

Hence coordinates of C (-5, 11, 2).

3. Given that P (3, 2, -4), Q (5, 4, -6) and R (9, 8, -10) are collinear. Find the ratio in which Q divides PR?

Solution: Let the ratio be $m_1 : m_2$. x-coordinate of Q $= 5$

By division formula: $x = \dfrac{m_1 x_2 + m_2 x_1}{m_1 + m_2}$

$$5 = \frac{m_1 9 + m_2 3}{m_1 + m_2}$$

$$5(m_1 + m_2) = 9 m_1 + 3 m_2$$

$$5 m_2 - 3 m_2 = 9 m_1 - 5 m_1$$

$$2 m_2 = 4 m_1 \Rightarrow \frac{m_1}{m_2} = \frac{2}{4} \Rightarrow m_1 : m_2 :: 1 : 2$$

4. Find the ratio in which the line joining the points (2, 4, 5) and (3, 5, -4) is divided by the yz–plane.

Solution: Let the ratio be $m_1 : m_2$. The coordinates of the point are

$$x = \frac{m_1 x_2 + m_2 x_1}{m_1 + m_2} \ , \ y = \frac{m_1 y_2 + m_2 y_1}{m_1 + m_2} \ \text{and} \ z = \frac{m_1 z_2 + m_2 z_1}{m_1 + m_2}$$

By on yz–plane , x–coordinate is zero, hence $x = \dfrac{m_1 x_2 + m_2 x_1}{m_1 + m_2} = 0$

$$\frac{m_1 x_2 + m_2 x_1}{m_1 + m_2} = 0$$

$$\Rightarrow m_1 x_2 + m_2 x_1 = 0$$

$$\Rightarrow m_1 x_2 = -m_2 x_1$$

$$\Rightarrow \frac{m_1}{m_2} = -\frac{x_1}{x_2} = \frac{-2}{3} \qquad \because x_1 = 2 \ \text{and} \ x_2 = 3$$

5. Find the coordinates of the points which trisect the line joining the points A (2, 1, -3) and B (5, -8, 3)

Solution: Let the points are C (x_0, y_0, z_0) and D (x', y', z') such that AC : CB :: 1:3

And AD : DB : 3 :1

For coordinates of C (x_0, y_0, z_0), we have $m_1 : m_2 = 1:3$ and $x_1 = 2$, $y_1 = 1$, $z_1 = -3$ and $x_2 = 5$, $y_2 = -8$, $z_2 = 3$

$$x_0 = \frac{m_1 x_2 + m_2 x_1}{m_1 + m_2} = \frac{1 \times 5 + 3 \times 2}{1 + 3} = \frac{11}{4}$$

$$y_0 = \frac{m_1 y_2 + m_2 y_1}{m_1 + m_2} = \frac{1 \times -8 + 3 \times 1}{1 + 3} = \frac{-5}{4}$$

$$z_0 = \frac{m_1 z_2 + m_2 z_1}{m_1 + m_2} = \frac{1 \times 3 + 3 \times -3}{1 + 3} = \frac{-6}{4} = \frac{-3}{2}$$

For coordinates of D (x', y', z'), we have $m_1 : m_2 = 3:1$ and $x_1 = 2$, $y_1 = 1$, $z_1 = -3$ and $x_2 = 5$, $y_2 = -8$, $z_2 = 3$

$$x' = \frac{m_1 x_2 + m_2 x_1}{m_1 + m_2} = \frac{3 \times 5 + 1 \times 2}{3 + 1} = \frac{17}{4}$$

$$y' = \frac{m_1 y_2 + m_2 y_1}{m_1 + m_2} = \frac{3 \times -8 + 1 \times 1}{3 + 1} = \frac{-23}{4}$$

$$z' = \frac{m_1 z_2 + m_2 z_1}{m_1 + m_2} = \frac{3 \times 3 + 1 \times -3}{3 + 1} = \frac{6}{4} = \frac{3}{2}$$

6. Find the ratio in which the join of the points A(2, 1, 5) and B (3, 4, 3) is divided by the plane $2x + 2y - 2z = 1$

Solution: Let the given plane intersect the line at C (x, y, z) in the ratio k : 1

$$x_1 = 2, \ y_1 = 1, \ z_1 = 5 \ \text{and} \ x_2 = 3, \ y_2 = 4, \ z_2 = 3$$

$$x = \frac{k.3 + 1.2}{k + 1} = \frac{3k + 2}{k + 1}$$

$$y = \frac{k.4 + 1.2}{k + 1} = \frac{4k + 2}{k + 1}$$

$$z = \frac{k.3 + 1.5}{k + 1} = \frac{3k + 5}{k + 1}$$

Since point C lies in the plane $2x + 2y - 2z = 1$, hence substituting the coordinates of C in the equation of the plane, we have–

$$2\left(\frac{3k+2}{k+1}\right)+2\left(\frac{4k+1}{k+1}\right)-2\left(\frac{3k+5}{k+1}\right)=1$$

$$2(3k+2)+2(4k+1)-2(3k+5)=k+1$$

$$6k+4+8k+2-6k-10=k+1$$

$$8k-4=k+1$$

$$7k=5\Rightarrow k=\frac{5}{7}$$

Hence the ratio is k : 1 = 5:7

7. A point P lies on the line whose end points are A (2, 3, 2) and (5, 3, 5). If z-coordinate of P is 4, then find its other coordinates .

Solution: Let the coordinates of P are (x, y, z) and let it divides AB in the ratio k:1.

Its z –coordinate is given by: $z = \dfrac{kz_2 + 1.z_1}{k+1}$

Putting z = 4 , $z_1 = 2$, $z_2 = 5$, we get –

$$4 = \frac{k.5 + 1.2}{k+1} \Rightarrow 4k + 4 = 5k + 2 \Rightarrow k = 2$$

The ratio in which P divides AB is 2 : 1

Now x –coordinate of P is:

$$x = \frac{kx_2 + 1.x_1}{k+1} = \frac{2.5 + 1.2}{2+1} = \frac{12}{3} = 4$$

Now y –coordinate of P is:

$$y = \frac{ky_2 + 1.y_1}{k+1} = \frac{2.3 + 1.3}{2+1} = \frac{9}{3} = 3$$

Hence Coordinates of P are (4, 3, 4).

Practice Problems

1. Find the distance between the points: (i) (4, 3, –6) and (–2, 1, –3),

2. Show that the following points are collinear: (3, 2, 4), (4, 5, 2) and (5, 8, 0)

3. Show that the points (–1, 0, –4) , (0, 1,–6) and (1, 2, 5) form a right – angled triangle.

4. Find the coordinates of the point dividing the line joining the points (3, 2, 1) and (6, –5, 3) in the ratio 3 : 5 (i) internally and (ii) externally.

5. Find the ratio in which the co-ordinate plane- zx divides the line joining the points (-2, 4, 7) and (3, -5, 8).

6. Find the distances of the point (3, 4, 5) from coordinate axes.

Answers:

    1. 7 **4.** (*i*) (33/8,–5/8, 7/4) , (*ii*) (–3/2, 25/2, –2)   **5.** 4 : 5   **6.** $\sqrt{41}$, $\sqrt{31}$, 5

1.4   DIRECTION COSINES AND DIRECTION RATIOS

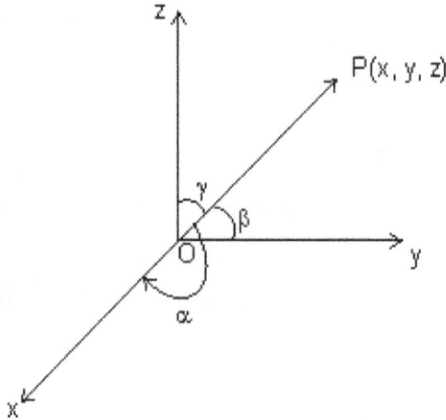

Let the line OP from the origin drawn parallel to the given line AB , makes angles α, β and γ in anticlockwise direction with the positive direction of axes - OX, OY and OZ. Then, $\cos\alpha$, $\cos\beta$ and $\cos\gamma$ are known as the *direction cosines* (d.c.s) of OP and hence that of AB

Direction Cosines are generally denoted by letters l , m, n.

Thus $l = \cos\alpha$, $m = \cos\beta$ and $n = \cos\gamma$ and $l^2 + m^2 + n^2 = 1$

The sum of squares of direction cosines of a line is 1

Let a , b , c are numbers proportional to direction cosines l , m, n. of a line. Then numbers a, b, c are known as *direction ratios* (d.r.s)

Thus $\dfrac{l}{a} = \dfrac{m}{b} = \dfrac{n}{c} = \pm\dfrac{\sqrt{l^2 + m^2 + n^2}}{\sqrt{a^2 + b^2 + c^2}} = \pm\dfrac{1}{\sqrt{a^2 + b^2 + c^2}}$ (by ratio and proportion)

Hence $l = \pm\dfrac{a}{\sqrt{a^2 + b^2 + c^2}}$, $m = \pm\dfrac{b}{\sqrt{a^2 + b^2 + c^2}}$ and $n = \pm\dfrac{c}{\sqrt{a^2 + b^2 + c^2}}$

Let A (x_1, y_1, z_1) and B (x_2, y_2, z_2) be two points. Then the direction ratios of the line AB may be given as: $a = (x_2 - x_1)$, $b = (y_2 - y_1)$ and $c = (z_2 - z_1)$

1.5   ANGLE BETWEEN TWO LINES

Let (l_1, m_1, n_1) and (l_2, m_2, n_2) be the direction cosines two points of a line and θ be angle between them. Then:

$$\cos\theta = l_1 l_2 + m_1 m_2 + n_1 n_2$$

Conditions for perpendicularity and parallelism

(*i*) If the lines are perpendicular the $\theta = 90$ and $\cos\theta = \cos 90^0 = 0$

$$l_1 l_2 + m_1 m_2 + n_1 n_2 = 0$$

(*ii*) If the lines are parallel then $\theta = 0$ and $\cos\theta = \cos 0^0 = 1$

$$l_1 l_2 + m_1 m_2 + n_1 n_2 = 1$$

If (a_1, b_1, c_1) and (a_2, b_2, c_2) are the direction ratios two points of a line, then
(*i*) the lines are perpendicular if :

$$a_1 a_2 + b_1 b_2 + c_1 c_2 = 0$$

(*ii*) the lines are parallel if : $\dfrac{a_1}{a_2} = \dfrac{b_1}{b_2} = \dfrac{c_1}{c_2}$

Solved Examples

1. The direction ratios of a line are: $1, -2, -2$. Find its direction cosines.

Solution: We are given direction ratios: $a = 1$, $b = -2$, $c = -2$

$$\sqrt{a^2 + b^2 + c^2} = \sqrt{(1)^2 + (-2)^2 + (-2)^2} = \sqrt{9} = \pm 3$$

The direction cosines are :

$$l = \pm \frac{a}{\sqrt{a^2 + b^2 + c^2}} = \pm\frac{1}{3}$$

$$m = \pm \frac{b}{\sqrt{a^2 + b^2 + c^2}} = \pm\frac{-2}{3} = \mp\frac{2}{3}$$

$$n = \pm \frac{c}{\sqrt{a^2 + b^2 + c^2}} = \pm\frac{-2}{3} = \mp\frac{2}{3}$$

Hence direction cosines are: $\dfrac{1}{3}, \dfrac{-2}{3}, \dfrac{-2}{3}$ or : $\dfrac{-1}{3}, \dfrac{2}{3}, \dfrac{2}{3}$

2. Can a line have direction angles of 30^0, 45^0 and 60^0.

Solution: $l = \cos\alpha = \cos 30^0 = \pm\dfrac{\sqrt{3}}{2}$

$$m = \cos\beta = \cos 45^0 = \pm\dfrac{1}{\sqrt{2}}$$

$$n = \cos\gamma = \cos 60^0 = \dfrac{1}{2}$$

$$l^2 + m^2 + n^2 = \left(\pm\dfrac{\sqrt{3}}{2}\right)^2 + \left(\pm\dfrac{1}{\sqrt{2}}\right)^2 + \left(\dfrac{1}{2}\right)^2 = \dfrac{3}{4} + \dfrac{1}{2} + \dfrac{1}{4} = \dfrac{3+2+1}{4} = \dfrac{7}{4} > 1$$

Hence a line can not have direction angles of 30^0, 45^0 and 60^0.

3. Prove that, 1, 1, 1 can not be direction cosines of a line.

Solution: $l^2 + m^2 + n^2 = 1+1+1 = 3$, which is not possible. $(l^2 + m^2 + n^2 = 1)$

Hence 1, 1, 1 can not be the direction cosines of a line.

4. Find the direction cosines and direction ratios of the line joining the points A (0, 0, 0) and B (4, 8, -9).

Solution: Direction ratios are: $a = (x_2 - x_1) = 4 - 0 = 4$

$$b = (y_2 - y_1) = 8 - 0 = 8$$

$$c = (z_2 - z_1) = \text{-8} - 0 = \text{- 8}$$

$$\sqrt{a^2 + b^2 + c^2} = \sqrt{(4)^2 + (8)^2 + (-8)^2} = \sqrt{16 + 64 + 64} = \sqrt{144} = 12$$

Direction cosines are:

$$l = \pm\dfrac{a}{\sqrt{a^2 + b^2 + c^2}} = \pm\dfrac{4}{12} = \pm\dfrac{1}{3}$$

$$m = \pm\dfrac{b}{\sqrt{a^2 + b^2 + c^2}} = \pm\dfrac{8}{12} = \pm\dfrac{2}{3}$$

$$n = \pm\dfrac{c}{\sqrt{a^2 + b^2 + c^2}} = \pm\dfrac{-8}{12} = \mp\dfrac{2}{3}$$

5. If the coordinates of A and B are (2, 3, 4) and (1, -2, 1) respectively, then prove that line OA is perpendicular to line OB, where O is the origin.

Solution: Direction ratios of OA are: $a_1 = (0-2) = \text{-2}$, $b_1 = (0-3) = -3$, $c_1 = (0-4) = -4$

Direction ratios of OA are: $\mathbf{a_2} = (0-1) = \text{-}1$, $\mathbf{b_2} = (0+2) = 2$, $\mathbf{c_2} = (0-1) = -1$

The condition of perpendicularity is $a_1 a_2 + b_1 b_2 + c_1 c_2 = 0$

$$a_1 a_2 + b_1 b_2 + c_1 c_2 = (-2)(-1) + (-3)(2) + (-4)(-1) = 2 - 6 + 4 = 0$$

Hence OA is perpendicular to OB at O.

6. Show that the join of the points A (1, 2, 3) and B (4, 5, 7) is parallel to the join of the points C (-4, 3, -6) and D (2, 9, 2).

Solution: Direction ratios of AB are: $\mathbf{a_1} = (4-1) = 3$, $\mathbf{b_1} = (5-2) = 3$, $\mathbf{c_1} = (7-3) = 4$

Direction ratios of CD are: $\mathbf{a_2} = (2+4) = 6$, $\mathbf{b_2} = (9-3) = 6$, $\mathbf{c_2} = (2+6) = 8$

$$\frac{a_1}{a_2} = \frac{3}{6} = \frac{1}{2}$$

$$\frac{b_1}{b_2} = \frac{3}{6} = \frac{1}{2}$$

$$\frac{c_1}{c_2} = \frac{4}{8} = \frac{1}{2}$$

$\therefore \dfrac{a_1}{a_2} = \dfrac{b_1}{b_2} = \dfrac{c_1}{c_2} = \dfrac{1}{2}$, hence the lines are parallel.

8. Find the angles between the lines whose d.r.s are: (5, -12, 13) and (-3, 4, 5)

Solution: Let θ be the angle between the lines.

$$\cos\theta = a_1 a_2 + b_1 b_2 + c_1 c_2 = (5)(-3) + (-12)(4) + (13)(5)$$

$$= -15 - 48 + 65 = 2$$

$$\theta = \cos^{-1}(2)$$

9. Find the value of x such that line joining the points A (4, 1, 2) and B (5, x, 0) is parallel to the line joining the points C(2, 1, 3) and D (3, 3, 1).

Solution: Direction ratios of AB are: $\mathbf{a_1} = (5-4) = 1$, $\mathbf{b_1} = (x-1)$, $\mathbf{c_1} = (0-2) = -2$

Direction ratios of CD are: $\mathbf{a_2} = (3-2) = 1$, $\mathbf{b_2} = (3-1) = 2$, $\mathbf{c_2} = (1-3) = -2$

$$\frac{a_1}{a_2} = \frac{1}{1}$$

$$\frac{b_1}{b_2} = \frac{(x-1)}{2}$$

$$\frac{c_1}{c_2} = \frac{-2}{-2} = \frac{1}{1}$$

Since lines are parallel, hence $\dfrac{a_1}{a_2} = \dfrac{b_1}{b_2} = \dfrac{c_1}{c_2}$

$$\therefore \frac{1}{1} = \frac{x-1}{2} = \frac{1}{1}$$

$$\therefore x - 1 = 2$$

$$\therefore x = 2 + 1 = 3$$

10. Show that the lines whose direction ratios are $(1,\ 1,\ 2)$ and $\left(\sqrt{3}-1, -\sqrt{3}-1, 4\right)$, are inclined to each other at an angle of $\frac{\pi}{3}$.

Solution: Given $a_1 = 1$, $b_1 = 1$, $c_1 = 2$

$$\therefore \sqrt{a_1^2 + b_1^2 + c_1^2} = \sqrt{1+1+4} = \sqrt{6}$$

$$l_1 = \frac{a_1}{\sqrt{a_1^2 + b_1^2 + c_1^2}} = \frac{1}{\sqrt{6}} , \quad m_1 = \frac{b_1}{\sqrt{a_1^2 + b_1^2 + c_1^2}} = \frac{1}{\sqrt{6}} , \quad n_1 = \frac{c_1}{\sqrt{a_1^2 + b_1^2 + c_1^2}} = \frac{2}{\sqrt{6}}$$

$$a_2 = \sqrt{3}-1 , \quad b_2 = -\sqrt{3}-1 \quad c_2 = 4$$

$$\therefore \sqrt{a_2^2 + b_2^2 + c_2^2} = \sqrt{\left(\sqrt{3}-1\right)^2 + \left(-\sqrt{3}-1\right)^2 + \left(4\right)^2}$$

$$= \sqrt{3+1-2\sqrt{3}+3+1+2\sqrt{3}+16} = \sqrt{24} = 2\sqrt{6}$$

$$l_2 = \frac{a_2}{\sqrt{a_2^2 + b_2^2 + c^2}} = \frac{\sqrt{3}-1}{2\sqrt{6}}$$

$$m_2 = \frac{b_2}{\sqrt{a_2^2 + b_2^2 + c^2}} = \frac{-\sqrt{3}-1}{2\sqrt{6}}$$

$$n_2 = \frac{c_2}{\sqrt{a_2^2 + b_2^2 + c^2}} = \frac{4}{2\sqrt{6}}$$

$$l_1 l_2 = \left(\frac{1}{\sqrt{6}}\right)\left(\frac{\sqrt{3}-1}{2\sqrt{6}}\right) = \frac{\sqrt{3}-1}{12}$$

$$m_1 m_2 = \left(\frac{1}{\sqrt{6}}\right)\left(\frac{-\sqrt{3}-1}{2\sqrt{6}}\right) = \frac{-\sqrt{3}-1}{12}$$

$$n_1 n_2 = \left(\frac{2}{\sqrt{6}}\right)\left(\frac{4}{2\sqrt{6}}\right) = \frac{8}{12}$$

$$\cos\theta = l_1 l_2 + m_1 m_2 + n_1 n_2 = \frac{\sqrt{3}-1-\sqrt{3}-1+8}{12} = \frac{6}{12} = \frac{1}{2} = \cos 60^0$$

$$\therefore \theta = 60^0 = \frac{\pi}{3}$$

11. Show that three lines with direction cosines $\left(\dfrac{12}{13}, \dfrac{-3}{13}, \dfrac{-4}{13}\right), \left(\dfrac{4}{13}, \dfrac{12}{13}, \dfrac{3}{13}\right)$ and $\left(\dfrac{3}{13}, \dfrac{-4}{13}, \dfrac{12}{13}\right)$ are at right angle to each other.

Solution: We have: $l_1 = \dfrac{12}{13}, \quad m_1 = \dfrac{-3}{13}, \quad n_1 = \dfrac{-4}{13}$

$$l_2 = \frac{4}{13}, \quad m_2 = 12, \quad n_1 = \frac{3}{13}$$

$$l_2 = \frac{3}{13}, \quad m_2 = \frac{-4}{13}, \quad n_1 = \frac{12}{13}$$

$$l_1 l_2 + m_1 m_2 + n_1 n_2 = \left(\frac{12}{13}\right)\left(\frac{4}{13}\right) + \left(\frac{-3}{13}\right)\left(\frac{12}{13}\right) + \left(\frac{-4}{13}\right)\left(\frac{12}{13}\right) = \frac{48 - 36 - 12}{169} = 0 = \cos 90^0$$

Similarly:

$$l_2 l_3 + m_2 m_3 + n_2 n_3 = \left(\frac{4}{13}\right)\left(\frac{3}{13}\right) + \left(\frac{12}{13}\right)\left(\frac{-4}{13}\right) + \left(\frac{3}{13}\right)\left(\frac{12}{13}\right) = \frac{36 - 48 + 12}{169} = 0 = \cos 90^0$$

$$l_3 l_1 + m_3 m_1 + n_3 n_1 = \left(\frac{3}{13}\right)\left(\frac{12}{13}\right) + \left(\frac{-4}{13}\right)\left(\frac{-3}{13}\right) + \left(\frac{12}{13}\right)\left(\frac{-4}{13}\right) = \frac{36 + 12 - 48}{169} = 0 = \cos 90^0$$

Hence the three lines are mutually perpendicular to each other.

12. If α, β, γ are the angles which a line makes with the axes, then show that:

$$\sin^2 \alpha + \sin^2 \beta + \sin^2 \gamma = 2$$

Solution: We know that $l = \cos\alpha, \ m = \cos\beta, \ n = \cos\gamma$ and $l^2 + m^2 + n^2 = 1$

$$\therefore \cos^2 \alpha + \cos^2 \beta + \cos^2 \gamma = 1$$

$$1 - \sin^2 \alpha + 1 - \sin^2 \beta + 1 - \sin^2 \gamma = 1$$

$$-\sin^2 \alpha - \sin^2 \beta - \sin^2 \gamma = 1 - 3 = -2$$

$$\therefore \sin^2 \alpha + \sin^2 \beta + \sin^2 \gamma = 2$$

Practice Problems

1. Find the direction cosines of the line whose direction ratios are 2, 3, -6 **2**.

2. Find the direction cosines of the line which is equally inclined to the axes.

3. If the point P(2, 3, –6) and Q (3, –4, 5) , then find the angle between OP and OQ,

where O is the origin. $\cos\theta = \dfrac{-18\sqrt{2}}{35}$

4. Show that the line joining points (0, 1, 2) and (3, 4, 5) is parallel to the line joining the points (–4, 3, -6) and (5, 12, 6)

5. If points P, Q are (2, 3, 4) and (1, –2, 1), then prove that OP is perpendicular to OQ, where O is the origin (0, 0, 0)

6. For what value of **k** ,the line joining the points A (k, 1, –1) and B (2k, 0, 2) is perpendicular to line B to C(3, –5, 4)

Answers:

1. (2/7, 3/7, -6/7) **2.** $\pm\dfrac{1}{\sqrt{3}}, \pm\dfrac{1}{\sqrt{3}}, \pm\dfrac{1}{\sqrt{3}}$ **3.** $\cos\theta = \dfrac{-18\sqrt{2}}{35}$ **6.** $k = 3$

1.6 EQUATION OF A STRAIGHT LINE PASSING THROUGH A GIVEN POINT

(*i*) Let a line passes through the point P (x_1, y_1, z_1) and has *direction cosines* (l, m, n) .

Then the equation of the straight line would be:

$$\frac{x-x_1}{l} = \frac{y-y_1}{m} = \frac{z-z_1}{n}$$

(*ii*) Let a line passes through the point P (x_1, y_1, z_1) and has *direction ratios* (a, b, c) .

Then the equation of the straight line would be:

$$\frac{x-x_1}{a} = \frac{y-y_1}{b} = \frac{z-z_1}{c}$$

Solved Examples

1. Find the equation of a straight line passing through the point $(1, 2, 3)$ and having *direction cosines* proportional to $1, -3, 7$

Solution: Applying the formula: $\dfrac{x-x_1}{l} = \dfrac{y-y_1}{m} = \dfrac{z-z_1}{n}$

Given: $x_1 = 1,\ y_1 = 2,\ z_1 = 31$

$$l = 1, m = -3, n = 7$$

Substituting the given values in the formula, the required equation is:

$$\frac{x-1}{1} = \frac{y-2}{-3} = \frac{z-3}{7}$$

2. Find the equation of a straight line passing through the point $(1, 0, 2)$ and whose direction ratio are $3, -1, 5$.

Solution: The equation of the straight line passing through a point and having direction ratios is:

$$\frac{x-x_1}{a}=\frac{y-y_1}{b}=\frac{z-z_1}{c}$$

Given: $x_1 = 1, y_1 = 0, z_1 = 2$

$$l = 3, m = -1, n = 5$$

Substituting the given values in the formula, the required equation is:

$$\frac{x-1}{3}=\frac{y-0}{-1}=\frac{z-2}{5}$$

3. Find the point where the line: $\dfrac{x-1}{3}=\dfrac{y+2}{4}=\dfrac{z-3}{5}$ meets the plane $x+y+z=14$.

Solution: Let $\dfrac{x-1}{3}=\dfrac{y+2}{4}=\dfrac{z-3}{5}=r$

$$x = 3r+1$$
$$y = 4r-1$$
$$z = 5r+3$$

Since the line meets the pane: $x+y+z=14$, the above coordinate will satisfy the equation of the plane.

Substituting the values of coordinates in the equation of the plane, we get:

$$(3r+1)+(4r-2)+(5r+3)=14$$
$$12r=14-2=12 \Rightarrow r=1$$

The Coordinates are:

$$x = 3\times1+1 = 4$$
$$y = 4\times1-2 = 2$$
$$z = 5\times1+3 = 8$$

Hence the point is (4, 2, 8)

4. Find the distance of the point (–1, –5, –10) from the point of intersection of the line

$$\frac{x-2}{3}=\frac{y+1}{4}=\frac{z-2}{12}$$ and the plane $x-y+z=5$

Solution: Let $\dfrac{x-2}{3}=\dfrac{y+1}{4}=\dfrac{z-2}{12}=r$

$$x = 3r+2$$
$$y = 4r-1$$
$$z = 12r+2$$

Since these coordinates lie on the plane also,

$$(3r+2)-(4r-1)+(12r+2)=5$$

$$11r=0 \Rightarrow r=0$$

$$x=3r+2=2$$

$$y=4r-1=-1$$

$$z=12r+2=2$$

The point of intersection is (2, –1, 2)

Hence distance between (–1, –5 , –10) and (2, –1, 2) is:

$$d=\sqrt{(-1-2)^2+(-5+1)^2+(-10-2)^2}=\sqrt{9+16+144}=\sqrt{169}=13$$

Hence distance between the given point and point of intersection is =13

5. Find the distance of the point (2, –3, 4) from the plane $x+y-z=5$ measured parallel to the line $\dfrac{x}{3}=\dfrac{y}{4}=\dfrac{z}{5}$.

Solution: Any line passing through (2, -3, 4) and parallel to a given line can be expresses as:

$$\frac{x-2}{3}=\frac{y+3}{4}=\frac{z-4}{5}$$

Now let $\dfrac{x-2}{3}=\dfrac{y+3}{4}=\dfrac{z-4}{5}=r$

We get:

$$x=3r+2$$

$$y=4r-3$$

$$z=5r+4$$

The above coordinates will satisfy the equation of the given line: $x+y-z=5$

$$(3r+2)+(4r-3)-(5r+4)=5$$

$$2r-5=5 \Rightarrow r=\frac{10}{5}=2$$

Hence the values of coordinates are:

$$x=3r+2=3\times5+2=17$$

$$y=4r-3=4\times5-3=17$$

$$z=5r+4=5\times5+4=29$$

Now we have to find the distance between the points (17, 17, 29) and 2, –3, 4).

$$\text{Distance}=\sqrt{(17-2)^2+(17+3)^2+(29-4)^2}=\sqrt{225+225+625}=\sqrt{1250}=25\sqrt{2}$$

1.7 EQUATION OF A STRAIGHT LINE, PASSING THROUGH TWO POINTS.

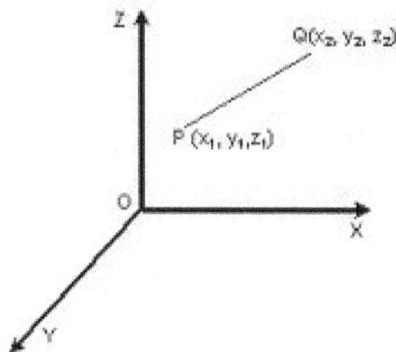

Let a line passes through the points P (x_1, y_1, z_1) and Q (x_2, y_2, z_2).

Then the equation of the straight line would be:

$$\frac{x - x_1}{x_2 - x_1} = \frac{y - y_1}{y_2 - y_1} = \frac{z - z_1}{z_2 - z_1}$$

$x_2 - x_1, y_2 - y_1$ *and* $z_2 - z_1$ are the direction ratios.

Solved Examples

1. Find the equation of the straight line passing through the points (4, -5, -2) and (-1, 5, 3).

Solution: The direction ratios of the line are.

$$a = x_2 - x_1 = -1 - 4 = -5$$
$$b = y_2 - y_1 = 5 + 5 = 10$$
$$c = z_2 - z_1 = 3 + 2 = 5$$

The equation of the lines is given by:

$$\frac{x - x_1}{a} = \frac{y - y_1}{b} = \frac{z - z_1}{c}$$

$$\frac{x - 4}{-5} = \frac{y + 5}{10} = \frac{z + 2}{5}$$

$$\frac{x - 4}{-1} = \frac{y + 5}{2} = \frac{z + 2}{1}$$

2. Find the value of k so that the line $\dfrac{1 - x}{3} = \dfrac{7y - 14}{2k} = \dfrac{z - 3}{2}$ and $\dfrac{7 - 7x}{3k} = \dfrac{y - 5}{1} = \dfrac{6 - z}{5}$ are perpendicular to each other.

Solution: The given lines can be written in standard form:

$$\frac{x-1}{-3} = \frac{y-2}{2k/7} = \frac{z-3}{2} \qquad \text{...(1)}$$

and
$$\frac{x-1}{-3k/7} = \frac{y-5}{1} = \frac{z-6}{-5} \qquad \text{...(2)}$$

The direction ratios of line (1) are:

$$a_1 = -3, \quad b_1 = 2k/7, \quad c_1 = 2$$

The direction ratios of line (2) are :

$$a_2 = -3k/7, \quad b_2 = 1, \quad c_2 = -5$$

The sum of products of direction ratios are

$$a_1 a_2 + b_1 b_2 + c_1 c_2 = (-3)\left(\frac{-3k}{7}\right) + \left(\frac{2k}{7}\right)(1) + (2)(-5) = \frac{9k}{7} + \frac{2k}{7} - 10$$

For the lines to be perpendicular:

$$a_1 a_2 + b_1 b_2 + c_1 c_2 = 0$$

$$\therefore \quad \frac{9k}{7} + \frac{2k}{7} - 10 = 0$$

$$\frac{11k}{7} = 10$$

or

$$\text{or} \quad k = \frac{70}{11}$$

3. Show that the lines $\dfrac{x-5}{7} = \dfrac{y+2}{-5} = \dfrac{z}{1}$ and $\dfrac{x}{1} = \dfrac{y}{2} = \dfrac{z}{3}$ are perpendicular to each other.

Solution: Comparing the equations with standard for: $\dfrac{x-x_1}{a} = \dfrac{y-y_1}{b} = \dfrac{z-z_1}{c}$

We find: The direction ratios of first line are : $a_1 = 7, \quad b_1 = -5, \quad c_1 = 1$

The direction ratios of second line are: $a_2 = 1, \quad b_2 = 2, \quad c_2 = 3$

$$a_1 a_2 + b_1 b_2 + c_1 c_2 = (7)(1) + (-5)(2) + (1)(3)$$

$$= 7 - 10 + 3 = 0$$

The lines are perpendicular since $a_1 a_2 + b_1 b_2 + c_1 c_2 = 0$

Practice Problems

1. Write the symmetric equation of a line passing through the points: (*i*) (1, 2, 3) and (1, 5, 8) (*ii*) (9, 3, 2) and (4, 5, −1)

2. Find the points where the line joining the points (2, 1, 3) and (4, –2, 5) cuts the plane $2x + y - z = 3$

Answers:

1. (*i*) $\dfrac{y-2}{3} = \dfrac{z-3}{5}$ (*ii*) $\dfrac{x-9}{-5} = \dfrac{y-3}{2} = \dfrac{z+2}{1}$ **(2)** (0, 4, 1)

Objective Questions

1. The number of co-ordinate planes is:
   (a) 3   (b) 4
   (c) 6   (d) 8

2. The co-ordinate planes divide the whole space into compartments. The number of compartments is:
   (a) 4   (b) 6
   (c) 8   (d) 2

3. The ratios in which yz – plane cuts the line joining the points (– 2 , 4, 7) and (3, –5, 8) is:
   (a) 2 : 3   (b) 3 : 2
   (c) – 2 : 3   (d) 2 : –3

4. The distance of the point (1, 2, 3) from y- axis is:
   (a) $\sqrt{14}$ (b) $\sqrt{10}$
   (c) $\sqrt{5}$ (d) $\sqrt{13}$

5. The distance of appoint (4, 5, 3) from the origin is:
   (a) $2\sqrt{5}$ (b) $2\sqrt{3}$
   (c) $5\sqrt{2}$ (d) 12

6. If α, β, γ are the angles made by a line then $\sin^2 \alpha + \sin^2 \beta + \sin^2 \gamma = ...$
   (a) 1   (b) 2
   (c) 3   (d) 4

7. D.Cs of al line equally inclined with the coordinate axes are:
   (a) 1, 1, 1   (b) $\dfrac{1}{\sqrt{3}}, \dfrac{1}{\sqrt{3}}, \dfrac{1}{\sqrt{3}}$

   (c) $\dfrac{-1}{\sqrt{3}}, \dfrac{-1}{\sqrt{3}}, \dfrac{-1}{\sqrt{3}}$ (d) –1,–1,–1

8. Which of the followings may be the DCs of a line:

   (a) $1, 1, 1$

   (b) $1, -1, 1$

   (c) $1, 1, -1$

   (d) $\dfrac{1}{\sqrt{3}}, \dfrac{1}{\sqrt{3}}, \dfrac{1}{\sqrt{3}}$

9. The DCs of the line joining the points $(-2, 1, -8)$ and $(4, 3, -5)$

   (a) $2, 4, -3$

   (b) $6, 2, 3$

   (c) $\dfrac{6}{7}, \dfrac{2}{7}, \dfrac{3}{7}$

   (d) $-6, -2, 3$

10. The direction cosines of a line whose direction ratios are $(2, 3, -6)$ are:

    (a) $\dfrac{2}{7}, \dfrac{3}{7}, \dfrac{-6}{7}$

    (b) $\dfrac{3}{7}, \dfrac{2}{7}, \dfrac{-6}{7}$

    (c) $\dfrac{3}{7}, \dfrac{-6}{7}, \dfrac{2}{7}$

    (d) $\dfrac{-6}{7}, \dfrac{2}{7}, \dfrac{3}{7}$

11. The angle between two lines whose D.rs are $(1, 1, 2)$ are $(-1, 0, 1)$ is:

    (a) 0^0

    (b) 30^0

    (c) 45^0

    (d) 60^0

12. The line joining the points $(1, 2, -1)$ and $(-1, 0, 1)$ is given by $\dfrac{x-1}{l} = \dfrac{y-2}{m} = \dfrac{z+1}{n}$.
    The value of l, m, n are:

    (a) $(-1, 0, 1)$

    (b) $(1, 1, -1)$

    (c) $(1, 2, -1)$

    (d) $(0, 1, 0)$

13. The point of intersection of the lines $\dfrac{x-4}{5} = \dfrac{y-1}{2} = \dfrac{z}{1}$ and $\dfrac{x-1}{2} = \dfrac{y-2}{3} = \dfrac{z-3}{4}$ is:

    (a) $(1, -1, -1)$

    (b) $(-1, -1, 0)$

    (c) $1, -1, -1)$

    (d) $(-1, 1, -1)$

14. If a_1, b_1, c_1 and a_2, b_2, c_2 are the d.r.s of two mutually perpendicular lines, then:

    (a) $a_1 a_2 + b_1 b_2 + c_1 c_2 = 0$

    (b) $a_1 a_2 + b_1 b_2 + c_1 c_2 = -1$

    (d) $a_1 a_2 + b_1 b_2 + c_1 c_2 = -1$

    (d) $\dfrac{a_1}{a_2} = \dfrac{b_1}{b_2} = \dfrac{c_1}{c_2}$

15. If the line $\dfrac{x-2}{2} = \dfrac{y+1}{3} = \dfrac{z}{4}$ and the origin lie on the plane $4x + 4y - kz = 0$. Then k = ...

    (a) 1

    (b) 3

    (c) 5

    (d) 7

16. The point P divides the line joining the points $(1, 2, -1)$ and $(-1, 0, 1)$ externally. The coordinates of P are:

(a) $(3, 4, 3)$

(b) $\left(\dfrac{1}{3}, \dfrac{4}{3}, \dfrac{-1}{3}\right)$

(c) $\left(1, \dfrac{2}{3}, \dfrac{-1}{3}\right)$

(d) none of these

17. The lines $3x - 6y - 2z = 7$ and $2x + y - kz = 5$ are perpendicular. The value of k is:

(a) 0

(b) 1

(c) 2

(d) 3

18. The shortest distance of point (a, b, c) from x–axis is:

(a) $\sqrt{a^2 + b^2}$

(b) $\sqrt{b^2 + c^2}$

(c) $\sqrt{c^2 + a^2}$

(d) $\sqrt{a^2 + b^2 + c^2}$

19. The angle between two lines $3x - 4y + 5z = 0$ and $2x - y - 2z = 5$ is:

(a) $\dfrac{\pi}{3}$

(b) $\dfrac{\pi}{2}$

(c) $\dfrac{\pi}{6}$

(d) $\dfrac{\pi}{4}$

Answers

1. (a)	2. (c)	3. (a)	4. (b)	5. (a)	6. (b)	7. (b), (c)
8. (d)	9. (c)	10. (a)	11. (d)	12. (b)	13. (a)	14. (c)
15. (c)	160 (b)	17. (a)	18. (b)	19. (b)		

Literature Consulted

1. Plane Trigonometry: S.L.Loney, S. Chand & Co. New Delhi

2. Inter Trigonometry: B.C.Das & B.N.Mukherjee, U.N.Dhur & Sons Pvt.Ltd, Kolkata-73

3. Co-ordinate Geometry : Gorakh Prasad & H.C.Gupta, Pothisala Prakashan ,Allhabad

4. Differential Calculus: B.C.Das & B.N.Mukherjee, U.N.Dhur & Sons Pvt.Ltd, Kolkata-73

5. Integral Calculus : Gorakh Prasad

6. Mathematics For Class XI &XII: R.D.Sharma

8. Secondary School Mathematics: R.S.Agarwal, Dhanpat Rai Publications, New Delhi-2

9. I.S.C Mathematics : O.P.Malhotra & S.K.Gupta S.Chand & Co. New Delhi

Literature Consulted

1. Plane Trigonometry: S.L. Loney, S. Chand & Co. New Delhi

2. Inter Trigonometry: B.C. Das & B.N. Mukherjee,
   U.N. Dhur & Sons Pvt. Ltd, Kolkata-73

3. Co-ordinate Geometry : Gorakh Prasad & H.C. Gupta, Pothisala Prakashan, Allahabad

4. Differential Calculus: B.C. Das & B.N. Mukherjee, U.N.Dhur &
   Sons Pvt. Ltd., Kolkata-73

5. Integral Calculus : Gorakh Prasad

6. Mathematics For Class XI & XII: R.D. Sharma

8. Secondary School Mathematics: R.S. Agarwal, Dhanpat Rai Publications,
   New Delhi-2

9. I.S.C Mathematics O.P. Malhotra & S.K. Gupta, S. Chand & Co. New Delhi